INDUSTRIAL LOCOMOTIVES

including preserved and
minor railway locomotives

HANDBOOK
19EL

INDUSTRIAL RAILWAY SOCIETY
A Charitable Incorporated Organisation
Registered Charity (England & Wales) No. 1177413
www.irsociety.co.uk

Published by the INDUSTRIAL RAILWAY SOCIETY
at 24 Dulverton Road, Melton Mowbray, Leicestershire, LE13 0SF

© INDUSTRIAL RAILWAY SOCIETY 2024

ISBN   978 1 912995 19 6   (hardbound)

ISBN   978 1 912995 20 2   (softbound)

British Library Cataloguing-in-Publication Data
A catalogue record for this book is available from the British Library

Visit the Society at   www.irsociety.co.uk

The content of this book is based on records held by the Industrial Railway Society, originally created by the late Eric Hackett. These records have now been continuously updated over the years by Alex Betteney, Ian Bendall, John Beechey, Colin Billinghurst, Adrian Booth, Bob Darvill, Ted Knotwell, Mick Morgan, George Morton and Andrew Waldron.

The Society obtains update information from a wide variety of sources, but primarily from reports of visits and other observations submitted by our readership, both members and non-members, plus some information kindly supplied by locomotive manufacturers, owners and operators. The publishers are extremely grateful for all this essential co-operation, without which production of our Handbooks would be impossible. All readers are encouraged to submit data, correctional, additional, or confirmatory, to the most seemingly appropriate Assistant Records Officer (see list on page 5) or by email to : hist.records@irsociety.co.uk

Production co-ordinated by Alex Betteney

Distributed by IRS Publications, 24 Dulverton Road, Melton Mowbray, Leicestershire, LE13 0SF

Produced for the IRS by Print Rite, Witney, Oxon. 01993 881662.

CONTENTS

Cover Photographs :

Front Cover – Hunslet Engine Company locomotive TH V325 of 1987 is seen shunting bogie limestone hopers at Tunstead Quarry, Buxton, Derbyshire. [Photo : Hunslet Engine Company]

Rear Cover – Hunslet Engine Company locomotive 'FORTH' AB 649 of 1980 is seen hauling tank wagons at INEOS Grangemouth Refinery, Falkirk. [Photo : Hunslet Engine Company]

The text of this book incorporates all amendments notified before 1st April 2024

FOREWORD TO THE 19TH EDITION

19EL – 55 YEARS OF PUBLISHING "INDUSTRIAL LOCOMOTIVES"

Founded 75 years ago in 1949 as the Birmingham Locomotive Club – Industrial Locomotive Information Section we later demerged from the BLC under the title Industrial Railway Society. From the early days we published Pocket Books covering a specific county or region that listed all known industrial locomotives that had worked in that particular area. These continue to be published as Handbooks, still covering all known locomotives in a specific county or region at home and countries overseas.

With the end of main line steam operations in 1968, attention and cameras were turned towards the industrial steam locomotive that was still at work in UK industry. At the time, very little information was available to the general public apart from the Pocket Books and there was nothing contained in one volume. It was decided to publish details of the existing industrial locomotives in the UK and make the information available to a wider audience.

The very first "British Industrial Locomotives" was published in January 1969 and now 55 years later we are proud to publish this 19th edition – "Industrial Locomotives" 19EL. Over the years the editions have developed to take advantage of the improvements in printing and publishing technology. The first edition manuscript was produced on a typewriter, now, one of the biggest advantages is the word-processor with its variety of fonts, but also the ease of producing colour photographs and, with digital technology, the ease of capturing and transmitting an image.

Fifty-five years has seen many changes in industry and the rise of the heritage railway. All these changes have been diligently recorded by our members and readers. This information is gathered by the society to enable this and our other publications to be as accurate as possible. Changes and news items that are regularly reported to us help to form the contents of our regular Bulletins that keep members informed and up-to-date.

The lists contained in this publication are in accordance with our records to 1st April 2024. We wish to stress the importance of this continuous supply of information and researches of both members and other readers so please continue to report your visits even if there are no changes to the railway operations and locomotives you observe. Please send your reports and observations (and any questions) to whoever you judge to be the most appropriate Hon. Records Officer selected from the list below or email our sorting office at: hist.records@irsociety.co.uk

Current Observations
I.R. Bendall (Assistant Hon. Records Officer)
25 Byron Way
Melton Mowbray
Leicestershire
LE13 1NY
chairman@irsociety.co.uk

Preservation & Heritage Railways
M. Morgan (Assistant Hon. Records Officer)
29 Groby Road
Ratby
Leicestershire
LE6 0JL
mick.d.morgan@btinternet.com

Northern Ireland & Eire
A.J. Waldron (Assistant Hon. Records Officer)
Riverside, 1 Hazel pear Close
Fairlawns
Horwich
Bolton
Greater Manchester
BL6 5GS
Ireland@irsociety.co.uk

All Historical Matters
R. Waywell (Hon. Records Officer)
29 Caldbeck Close
Gunthorpe
Peterborough
PE4 7NE
records@irsociety.co.uk

The Society also has Assistant Hon. Records Officers who collate information which is appropriate to various industrial railway subjects such as the Ministry of Defence, rolling stock, etc. Information supplied to any of the above officers will be circulated to the appropriate officer. All information received is circulated to members of the Industrial Railway Society by means of regular Bulletins thus enabling this book to be kept up to date. The original reports are archived for the benefit of researchers in the future.

INTRODUCTION

Within the main Country Sections of this book, information is presented in a similar format to that used in previous editions, with locomotives listed under owners' titles arranged alphabetically, within sub-groups of Industrial and Preservation Sites. Within these sub-groups, each location is sequenced alphabetically, using its main title, or surname if a person's name is included.

LOCATION HEADINGS include both National Grid References and postcodes where known; details of missing grid references and/or postcodes will be most welcome.

Postcodes and Grid References are particularly useful for readers with Internet access, as either code can be requested at one of the free mapping sites (such as www.streetmap.co.uk or http://www.multimap.com) to obtain a printable map of the precise location of the railway site.

In general, we try to fix the Grid Reference/s to indicate the locomotive shed or stabling point, particularly at the larger sites.

DIVISION OF COUNTRIES. In recent years there have been a number of reorganisations of the Administrative Areas in England, Scotland and Wales, in some cases the traditional "geographic counties" have been re-mapped and divided to create small independent areas now administered by Unitary Authorities of one sort or another. In order to avoid fragmenting our listing data, in this edition we have amalgamated some listings under headings of their "most recent" larger-landmass titles (many of which are still in current use for certain administrative purposes).

In the case of Wales and Northern Ireland, which have in fact relatively few locomotive locations, we have re-listed the entries on a Land Area basis, supported by maps indicating the geographical position of the locations and by extending the location titles to include the name of the actual Authority in which the site is now located. The boundaries of the Land Areas have been selected to coincide with the coverage of books in our companion "Historic Handbook" series. We hope that readers will find this arrangement more convenient to use and, if so, it may be extended to some other areas in the next edition.

Locations are indicated on such maps by numbers, which correspond to the serial numbers that appear immediately to the right of the location titles.

LOCOMOTIVES AT SCRAPYARDS purely for scrapping are not generally listed, unless they have been 'resident' for a year or more. Those at dealers' yards are, however, listed in order that readers can keep a full picture of their movements.

CONTRACTORS' LOCOMOTIVES can be found on sites in all parts of the country, working on sewer and tunnel schemes, etc. The locomotives of such firms are shown, in the usual way, in the list of the County in which the firm's plant depot (or main plant depot if more than one is in use) is situated. A list of such Contractors and Plant Hire specialists and their relevant base counties appears later in this Introduction, following the table listing Locomotive Builders.

LOCOMOTIVE OWNERSHIP. The information within this publication does not confirm ownership of any of the locomotives listed, only their physical presence at a location. When a locomotive is sold, the transaction may have taken place sometime before the date that we record the locomotive moving to another location.

   Industrial/Commercial location: Some locomotives may be hired or leased from third parties.
   Where the hirer or leasing company is known, this information is shown in a footnote.
   Preservation Sites: Locomotives may be owned by individuals, groups, trusts, etc and may not
   be the property of the host site under which they are listed.

HIRE FLEET OPERATORS. A small number of firms maintain a fleet of locomotives which are hired out to both Industrial Sites and Main-line Train Operating (TOC) Companies. The principal firms involved are listed on page 24. Within the listings of these fleet-owners, we detail their full fleet (in so far as we know it) under an entry for their operational base. Such locomotives are duplicated in other Industrial or Preservation Site lists, to record their known current locations.

LOCATIONS THAT ARE CLOSED, but where locomotives are still present, are indicated either by "(Closed)" appearing beside their title or subtitle, or else in a line of text immediately beneath that title. When a location is still in operation but no longer uses rail traffic, although the locomotives remain on site, this is indicated by "R.T.C." after the gauge concerned. The letters RTC are an abbreviation of the phrase "Rail Traffic Ceased".

TRACK GAUGE. The gauge(s) of the railway system(s) are given at the head of the locomotive list(s). At preservation sites and museums, etc, where several gauges differ by only a small fraction they are usually all listed under the nominal gauge of the majority.

WEBSITE. The official website relating to a location is listed. Unofficial websites or affiliated websites are not normally listed. Websites are subject to change from time to time. Details of websites were correct at the time of going to press.

POSTCODE. These are shown in bold type above the Grid Reference. Postcodes shown in brackets, are the nearest known, but not necessarily officially used locations. Some postcodes relate to a street or village and the respective location may be some distance away. All postcodes were checked and correct at the time of going to press.

SEQUENCE OF LOCOMOTIVES WITHIN THE LISTS is generally in accordance with our basic formula of "list steam first, followed by non-steam" and, in each of these groups, "list ex-mainline locomotives first in main-line running number order, followed by non-mainline locomotives in builder/works number order". Permanent-way motorised trolleys and other miscellaneous units are usually relegated to the end of the lists. However, in some cases, the nature or details of the actual fleet dictates a different arrangement of listing.

LOCOMOTIVE NUMBER and NAME. The title of the locomotive - number, name or both - is given in the first two columns. A name unofficially bestowed and used by staff but not carried on the locomotive is indicated by quotes (inverted commas "   "). Locomotives under renovation at preservation sites, etc, may not currently carry their intended title; nevertheless these are shown unless it is definitely known that the name will not be retained.

Ex British Railways / Train Operating Company / Leasing Company (ROSCO) locomotives are further identified by the inclusion of formerly carried numbers in brackets even if not carried now.

TYPE   The type of locomotive is given in column three.

The Whyte system of wheel classification is used in the main, but when driving wheels are connected not by outside rods but by chains or other means (as in various 'Sentinel' steam locomotives and diesel locomotives) they are shown as 4w (four-wheeled), 6w (six-wheeled), or if only one axle is motorised it is shown as 2w-2.

Trapped Rail System locomotives are shown using wheel type "ad" (meaning "axles driven") – thus "3ad" indicates a locomotive which may have 3 or more axles, but only 3 are driven.

For ex British Railways / Train Operating Company / Leasing Company (ROSCO) diesel and electric locomotives, the usual development of the Continental notation is used.

The following abbreviations are used :-

CT	Crane Tank - a T type locomotive fitted with load lifting apparatus
F	Fireless steam locomotive
IST	Inverted Saddle Tank
PT	Pannier Tank - side tanks not fastened to the frame
ST	Saddle Tank
STT	Saddle Tank with Tender
T	Side Tank or similar - a tank positioned externally and fastened to the frame
VB	Vertical Boilered locomotive
WT	Well Tank - a tank located between the frames under the boiler
BE	Battery powered Electric locomotive
BH	Battery powered Electric locomotive - Hydraulic transmission
CA	Compressed Air powered locomotive
CE	Conduit powered Electric locomotive
D	Diesel locomotive - unknown transmission
DC	Diesel locomotive - Compressed air transmission
DE	Diesel locomotive - Electrical transmission
DH	Diesel locomotive - Hydraulic transmission

DM	Diesel locomotive - Mechanical transmission
F	(as a suffix, for example BEF, DMF) – Flameproof (see following paragraph)
FE	Flywheel Electric locomotive
GTE	Gas Turbine Electric locomotive
P	Petrol or Paraffin locomotive - unknown transmission
PE	Petrol or Paraffin locomotive - Electrical transmission
PH	Petrol or Paraffin locomotive - Hydraulic transmission
PM	Petrol or Paraffin locomotive - Mechanical transmission
R	Railcar - a vehicle primarily designed to carry passengers
RE	Third Rail powered Electric locomotive
WE	Overhead Wire powered Electric locomotive

FLAMEPROOF locomotives, usually battery but sometimes diesel powered, are denoted by the addition of the letter **F** to the wheel arrangement in column three.

CYLINDER POSITION is shown in column four for steam locomotives.
In each case, a prefix numeral (**3, 4, etc**) denotes more than the usual two cylinders.

IC	Inside cylinders
OC	Outside cylinders
VC	Vertical cylinders
G	Geared transmission - suffixed to IC, OC or VC

RACK DRIVE. Certain locomotives are fitted with rack-drive equipment to enable them to climb steep inclines and were mainly used in the UK underground in coal mines. The pattern of rack used was the "Abt" design. Locomotives that were built with (and are thought to still retain) this equipment are indicated by the word **RACK** in column four. However, due to contraction of the coal industry, such locomotives have often been transferred to sites where the rack equipment is no longer in use.

ROAD/RAIL. Certain locomotives are capable of working on either rail or road, many being of either Unilok or Unimog manufacture. These are indicated by **R/R** in column four.

STEAM OUTLINE. Non-steam locomotives, with a steam locomotive appearance added, are shown by **s/o** in column four.

MAKERS. The builder of the locomotive is shown in column five; the abbreviations used are listed on pages 10 to 22.

MAKERS NUMBER and DATE. The sixth column shows the works number, the next column the date which appears on the plate, or the date the locomotive was built if none appears on the plate.

DOUBTFUL INFORMATION. Information known to be doubtful is denoted as such by the wording, or printed in brackets with a question mark. Thus, Wkm (7573 ?) 1956 denotes that the locomotive is a Wickham built in 1956 and possibly of the works number 7573.

MISCELLANEOUS NOTES.

c	denotes circa, i.e. about the time of the date quoted
Dsm	indicates a "long-term" dismantled locomotive (as an aid to photographers, etc)
	Note: battery locomotives not carrying a battery are not regarded as Dsm
DsmT	used for motorised trolleys which have been converted to engineless trailers
Pvd	denotes Preserved on site
reb	denotes the locomotive was rebuilt (by and when, as per list)
	Note: "rebuilt" infers significant alteration to transmission or appearance; not a routine overhaul no matter how thorough that may have been
rep	indicates a major repair, and is ONLY detailed in the listings when a "repair plate" is (or was) affixed to the locomotive (and thus an aid to identification).
OOU	"Out Of Use" - indicates a locomotive which has, apparently permanently or at least for a significant period of time, ceased work
u/c	under construction (locomotive in advanced state of construction)
M	indicates that the locomotive is considered to be 'Mainline' (Shunting locomotives)

There are cases where all locomotives at a site are OOU but the railway, siding etc, is still in use, utilising a road tractor, cable-haulage or man-power for example. Electric locomotives at steelworks are usually to be found working at Coke Ovens, the grid reference for the Ovens being shown in the usual place, when known.

HINTS ON RECORDING OBSERVATIONS

As already explained, the Society is largely reliant on the observations of enthusiasts to keep the records up to date and we are always pleased to receive reports of visits, which should be sent to the Hon. Records Officer or his assistants. The following will, we hope, be of assistance to those sending in reports.

Report anything you see. With regard to locomotives this means reporting locomotives present, even if there is no change to the published list. Someone else may visit in the next six months and find a change, the date of which can then be narrowed down to this six months. In addition to locomotives, details of rolling stock, track layouts and items of historical information gleaned are all welcome and will be added to our "Historic Handbook" files. It is surprising how often these items are required later by other enthusiasts. There is nothing to beat a note made on the spot; a note now is far better than trying to get the information from the recollection of employees in five years time.

Care should always be taken to distinguish actual observation from personal inference. If you see a 4wDH with worksplate say S 10019 and numbered 19, all well and good; but if it does not carry a worksplate, that fact should be stated. You may infer the loco is S 10019 from the running number, or by reference to this Handbook. If you say 'number on bonnet', 'number on cabside' or 'locomotive assumed to be S 10019' as the case may be, we will know the position precisely, which is important when bonnets are exchanged or cabs rebuilt, for example. If a locomotive carries no identification, you will probably be able to guess its make from design features and a note of livery may help. Also with diesel locomotives, a note should be made of the make of its engine, its type or model and its serial number, if these can be obtained. "Anonymous" locomotives can often be positively identified from our records given such details.

A thorough search of all premises is worthwhile, as locomotives (particularly if OOU) are frequently hidden away - and how often have surprises turned up in this way. Further, if you search diligently and a locomotive is missing, it may be presumed to have gone and enquiries can be made, as to where it has gone or if it has been scrapped. Please try to ascertain such information from the staff. Similarly, in the case of new arrivals, exact dates of arrival should always be obtained if possible.

Firms' titles change from time to time, so please check from the office or board at the entrance. Subsidiary companies frequently display the title of a parent company, but we always use the name under which the company trades, i.e. the subsidiary where such exists.

Preservation locations often take locomotives on short term hire, or hire locomotives out to other railways. Such movements are generally not noted within the EL series of books, unless it is known that the locomotive is intended to become a resident of a railway for at least a year or more, however, any movements should be recorded with dates, as often a temporary movement can become a permanent one.

It goes without saying that it is essential to obtain permission to enter premises to see locomotives and this applies with equal force to 'preserved' locations other than public parks and museums. Established systems which operate trains or have 'Open Days' will usually permit inspection of locomotives by arrangement.

Locomotives shown as at private homes or farms are very often in storage or in the course of restoration and, as a rule, our members and readers are strongly advised to NOT, in any way possible, embarrass the owners by writing for permission to view, but rather to wait until the owners announce that their machines are available for inspection.

You may note that whilst we list details of all known existing locomotives, some appear under incomplete or imprecise heading titles. This is regrettable, but is entirely a direct consequence of a few individuals having ignored the advice given in the paragraph above.

Finally, please try to establish friendly relations with the firms visited, as we are only allowed access by their courtesy. Do not be a nuisance or hold up production, nor expose yourself to danger in any way, but show a healthy interest in the processes being carried out. In this way, not only you, but other enthusiasts, will also be welcome there at a later date.

BUILDERS and REBUILDERS of LOCOMOTIVES

A

A&O	Alldays & Onions Ltd, Birmingham
AB	Andrew Barclay, Sons & Co Ltd, Caledonia Works, Kilmarnock, Ayrshire
ACAB	Adam Charles Albert Barber, Tan-Y-Dool Works, Llangollen
ACCars	A.C.Cars Ltd, Thames Ditton, Surrey
ACI	ACI Élévation, Isles sur Suippe, France
Acton	Acton Works, Acton, Greater London
	London Transport Executive / London Underground Ltd (from /1987)
ADC	Associated Daimler Co Ltd, Coventry
AE	Avonside Engine Co Ltd, Fishponds, Bristol
AEC	Associated Equipment Co Ltd, Southall, Middlesex
AEG	Allgemeine Elektrizitäts Gesellschaft, Berlin-Hennigsdorf, Germany
AEI	Associated Electrical Industries Ltd, Trafford Park, Greater Manchester
AFB	Société Anglo Franco-Belge des Ateliers de la Croyère, Seneffe et Godarville,
	Belgium
Afd	Ashford Works, Kent
	South Eastern & Chatham Railway / Southern Railway / British Railways
AH	A. Horlock & Co, North Fleet Iron Works, Kent
AK	Alan Keef Ltd, Cote Farm, Cote, near Bampton, Oxfordshire,
	later at Lea Line, Ross-on-Wye, Herefordshire
	(successor to SMH)
Albion	Albion Motors Ltd, Scotstoun, Glasgow W4
Alco	American Locomotive Co, USA; and/or Montreal, Canada
Alco(C)	American Locomotive Co, Cooke Works, Paterson, New Jersey, USA
Alex	W. Alexander & Co (Coachbuilders) Ltd, Falkirk, Scotland
AllenJ	Jason Allen, Grimoldby, Lincolnshire
ALR	Abbey Light Railway, P.N.Lowe & Sons, Bridge Road, Kirkstall, Leeds
Alstom	Alstom Rail Ltd, Washwood Heath, Birmingham
Alstom(L)	Alstom, La Rochelle, France
AP	Aveling & Porter Ltd, Invicta Works, Rochester, Kent
APEL	Adrian Phillips Engineering Ltd, Pontypool, Torfaen
Aquarius	Aquarius Railroad Technologies, Mickley, North Yorkshire
ARC	Amalgamated Roadstone Corporation Ltd, Stanton Harcourt Depot,
	near Witney, Oxfordshire
Artemis	Artemis Intelligent Power Ltd (a Danfos Scotland and Mitsubishi Heavy Industries Ltd JV),
	Bo'ness, West Lothian, Scotland
Artisair	Artisair Ltd, Moorwell Road, Yaddlethorpe, Scunthorpe, Lincolnshire.
AS&W	Allied Steel & Wire Ltd, Castle Works, Cardiff
Ashbury	Ashbury Railway Carriage & Iron Co Ltd, Openshaw, Manchester, Lancashire
Ashbyl	I. Ashby, Valley Nurseries, Evesham, Worcestershire
Ashford	Ashford Plant Depot, Newtown Road, Ashford, Kent (B.R./Balfour Beatty/Network Rail)
Atlas	Atlas Loco & Mfg Co Ltd, Cleveland, Ohio, U.S.A.
AtW	Atkinson Walker Wagons Ltd, Frenchwood Works, Preston, Lancashire
AVB	A.V. Access Ltd, Bridgnorth, Shropshire
AW	Sir W.G.Armstrong, Whitworth & Co (Engineers) Ltd, Newcastle-upon-Tyne
Ayle	Ayle Colliery Co Ltd, Alston, Cumbria

B

B&S	Bellis & Seekings Ltd, Birmingham
BabyDeltic	Baby Deltic Project, Barrow Hill, Derbyshire
Bala	Bala Lake Railway, Llanuwchllyn, Gwynedd
Balfour Beatty	Balfour Beatty Rail Services, Raynesway, Derby
Bance	R. Bance & Co.Ltd, Cockrow Hill House, St. Mary's Road, Surbiton, Surrey
Barlow	H.N. Barlow, Southport, Lancashire
Barnard	Barnard & Co, Norwich, Norfolk
Barnes	A. Barnes & Co, Albion Works, Rhyl, North Wales (a subsidiary of Rhyl Amusements Ltd)

Barry	Barry Tourist Railway, Barry Island, Vale of Glamorgan
BateJ	John L. H. Bate, Reigate, West Sussex
Battison	Samuel Battison, c/o Tathams Ltd, Nottingham Road, Ilkeston, Derbyshire
BBC(S)	AG Brown Boveri & Cie, Baden, Aargau, Switzerland
BBM	Officine Mecciniche BBM S.p.A. Rossano Veneto, Via Mottinello, Italy
BBT	Brush Bagnall Traction Ltd, Loughborough, Leicestershire and Stafford
B(C)	Peter Brotherhood, Engineers, Chippenham, Wiltshire
BCM	Big Country Motioneering Ltd, Kingswinford, West Midlands
BD	Baguley-Drewry Ltd, Burton-on-Trent, Staffordshire
BE	Brush Electrical Engineering Co Ltd, Loughborough, Leicestershire
Beamish	Beamish Museum, Stanley, Co. Durham
BEAZ	Beaz-Solutions BV, Spanbroek, Netherlands
BEMO	BemoRail, Debbemeerweg 59, 1749 DK Alkmaar-Warmenhuizen, Netherlands
Benford	Benford (Terex) Ltd, Warwick, Warwickshire
Berwyn	Berwyn Engineering, Thickwood, Chippenham, Wiltshire
BES	Beech Engineering Services, Unit 4, Wetmore Industrial Estate, Wharf Road, Burton-on-Trent, Staffordshire
BEV	British Electric Vehicles Ltd, Churchtown, Southport, Lancashire ("BEV" branded locomotives were later built by Wingrove & Rogers – see "WR")
Bg	E.E. Baguley Ltd, Burton-on-Trent, Staffordshire
BgC	Baguley Cars Ltd, Shobnall Road Works, Burton-on-Trent, Staffordshire
Bg/DC	built by Bg for DC; makers numbers identical
BGB	Becorit (Mining) Ltd, Grove Street, Mansfield Woodhouse, Nottingham, then Hallam Fields Road, Ilkeston, Derbyshire from /1984
BH	Black, Hawthorn & Co Ltd, Gateshead
BHSC	Barrow Haematite Steel Co Ltd, Barrow-in-Furness, Lancashire
Bickton	Bickton Engineering, Coverdale, Bickton, Hampshire
BIS	War Department, Arncott Workshops, Bicester Depot, Oxfordshire and including Ministry of Defence, Bicester Depot Workshops
BL	W.J. Bassett Lowke Ltd, Northampton
Bluebell	Bluebell Railway Co Ltd, Sheffield Park, East Sussex
BLW	Baldwin Locomotive Works, Philadelphia, Pennsylvania, USA
BM	Brown, Marshalls & Co Ltd, Adderley Park, Birmingham (in 1902 became part of what later became Metropolitan-Cammell Ltd)
BMR	Brecon Mountain Railway Co Ltd, Pant, Mid Glamorgan
BnM	Bord na Móna (Irish Turf Board) : Various of the larger sites (e.g. Bl, Bo, Dg & M) have built their own locomotives & railcars – see Section 4 (Ireland)
Bombardier(BN)	Bombardier (BN Construction), Bruges, Belgium
Bonnymount	Mr Taylor, Bonnymount Farm, Siston Common, Bristol
BoothR	R. Booth, Isle of Man Railways, Douglas, Isle of Man
BoothWKelly	Booth W Kelly, Boatbuilders, Ramsey, Isle of Man
Borsig	A. Borsig GmbH, Berlin-Tegel, Germany
Bow	Bow Locomotive Works, North London Railway
Bowman	N.Bowman, Launceston Steam Railway, Cornwall
BP	Beyer, Peacock & Co Ltd, Gorton, Manchester
BPH	Beyer, Peacock (Hymek) Ltd, Gorton, Manchester
BradleyS	Scott Bradley, Trafford Park, Greater Manchester
Braithwaite	Bevan Braithwaite / Bressingham Steam Preservation Company, Bressingham, near Diss
BRCW	Birmingham Railway Carriage & Wagon Co Ltd, Smethwick, West Midlands
BRE	British Rail Engineering Ltd
BRE(C)	British Rail Engineering Ltd, Crewe Works, Cheshire
BRE(D)	British Rail Engineering Ltd, Litchurch Lane Works, Derby
BRE(S)	British Rail Engineering Ltd, Shildon Works, Co.Durham
Bredbury	Bredbury & Romiley Urban District Council, Cheshire [Greater Manchester]
Brennan	Louis Brennan, Gillingham, Kent
BriddonA	Andrew Briddon, AFRPS, Scunthorpe, Lincolnshire
BrightonTC	Brighton Corporation Tramways Dept, Brighton, Sussex
Brookside	Brookside Engine Co Ltd, Whaley Bridge, Derbyshire
BrownGM	G.M.Brown Ltd, Stanhopeburn, Stanhope, Co.Durham
BrownJ	Joe Brown, Farington Moss, Preston, Lancashire

Bruff	Bruff Rail Ltd, Suckley, Worcestershire
Brunning	H. Brunning (? of Crewe) – see Rhiw Valley Railway, Mid Wales
Bryce	B. Bryce, Corrofin, Co. Clare, Ireland
BS(S)	British Steel Ltd, Scunthorpe Works, Lincolnshire
BT	Brush Electrical Machines Ltd, Traction Division, Falcon Works, Loughborough, Leicestershire
BTH	British Thomson-Houston Co Ltd, Rugby, Warwickshire
Bton	Brighton Works, Sussex
	London, Brighton and South Coast Railway / Southern Railway / British Railways
Bulrush(B)	Bulrush Peat Co Ltd, Bellaghy Works, Magherafelt, Co. Derry, Northern Ireland
BuryC&K	Bury, Curtis & Kennedy, Clarence Foundry, Love Lane, Liverpool
Bush Mill Rly	Bush Mill Railway, Port Arthur, Tasmania, Australia
BV	Brook Victor Electric Vehicles Ltd, Burscough Bridge, Ormskirk, Lancashire
BVR	Bure Valley Railway Ltd, Aylsham, Norfolk
Byers	R.S. Byers Ltd, Houghton, Carlisle, Cumbria
Byworth	Byworth Engineering Ltd, Keighley, West Yorkshire

C

CAF	Construcciones y Auxiliar de Ferrocarriles, Beasain, Spain
Campagne	E. Campagne & Cie, Paris, France
Cannon	Cannon, Dudley, West Midlands
Carland	Carland Engineering Ltd, Harold Wood, Essex
Carter	D.Carter, Tucking Mill Tramway, Midford, Somerset
Case	Case IH Agriculture, Steyrer Strasse 32, St. Valentin, USA
CastleGKN	Castle Works (GKN Sankey Ltd), Hadley, Telford, Shropshire
Castleline	Castleline, Nottingham
CCLR	Cleethorpes Coast Light Railway, Cleethorpes, Lincolnshire
Cdf	Cardiff West Yard Locomotive Works (Taff Vale Railway), Glamorgan
CE	Clayton Equipment Ltd, Burton-on-Trent, Staffordshire
	previously at Hatton, Derbyshire, where also traded as NEI Mining Equipment Ltd and originally Clarke Chapman Ltd
Chance	Chance Manufacturing Co Inc, Wichita, Kansas, USA
Chaplin	Alexander Chaplin & Co Ltd, Cranstonhill Works, Glasgow
CheshamSG	1ST Chesham Scout Group, Chesham, Buckinghamshire
Chieftain	Chieftain Trailers Ltd, Dungannon, Co. Tyrone, Northern Ireland
Chrz	Fabryka Lokomotywim "Feliksa Dzierzynskiego" Chrzanów, Poland
Civil	T.D.A. Civil, near Uttoxeter, Staffordshire
ClarkeE	E. Clarke, London
Clarkson	H. Clarkson & Son, York
ClayCross	Clay Cross Co Ltd, Spun Pipe Plant, Clay Cross Iron Works, Derbyshire
CM	Century Millwrights, Platts Eyot's Works, Sunbury-on-Thames, Surrey
CMAR	Construction Mechanique Automatisme Rivard SAS, Durtal, France
CNES	Corus Northern Engineering Services, Scunthorpe, Lincolnshire
Coalbrookdale	Coalbrookdale Co, Coalbrookdale Ironworks, Shropshire
Cockerill	Société pour L'Exploitation des Etablissements John Cockerill, Seraing, Belgium
Coferna	Coferna, Constructions Ferroviaires et Navales de l'Ouest, Sables d'Olonne, Vendée, France
ColdChon	Cold Chon Ltd, Oranmore, Co Galway
ColebySim	Coleby-Simkins Engineering, Melton Mowbray, Leicestershire
CollinsD	Dennis Collins, Kanturk, Co. Cork, Ireland
Cometi	Cometi S.r.l., Italy
Consett	Consett Iron Co Ltd, Consett Works, Co.Durham
Consillia	Consillia Ltd, Shardlow, Derbyshire
Corpet	L. Corpet, Avenue Philippe-Auguste, Paris, France
Couillet	Société Anonyme des Usines Métallurgiques du Hainaut, Couillet, Marcinelle, near Charleroi, Belgium
Cowlairs	Cowlairs Works, Glasgow
	North British Railway / London & North Eastern Railway / British Railways
CPM	CPM S.r.l., Marmalada, Milano, Italy
Cranmore	East Somerset Railway Co Ltd, Cranmore, Somerset

CravenEA	E.A. Craven, North Shields, Northumberland
CravenJ	J. Craven, Walesby, Nottinghamshire
Cravens	Cravens Ltd, Darnall, Sheffield
Crewe	Crewe Works, Cheshire
	London and North Western Railway / London, Midland & Scottish Railway / British Railways
Crome/Loxley	R. Crome & R. Loxley, Doncaster
CRS	Curry Rail Services, Hollidaysburg, Philidelphia, USA
CTL	CTL Seals Ltd, Ecclesfield, South Yorkshire
CurleA	A. Curle, Skipsea, East Yorkshire
CurwenD	A. Curwen, All Cannings, near Devizes, Wiltshire
Cushen	Paul Cushen Engineers Ltd, Stradbally, Co. Laois, Ireland
CW	Cowlishaw Walker Engineering Co Ltd, Biddulph, Staffordshire

D

D	Dübs & Co, Glasgow Locomotive Works, Glasgow
D&BST	Dublin & Blessington Steam Tramway, Tenure Works, Dublin, Ireland
D&S Services	D & S Services
Dar	Darlington Works, Co.Durham
	North Eastern Railway / London & North Eastern Railway / British Railways
Darlington	A1 Locomotive Trust, Hopetown Carriage Works, Darlington, Co. Durham
Dav	Davenport Locomotive Works, Davenport, Iowa, USA
DB	Sir Arthur P. Heywood, Duffield Bank Works, Derbyshire
DC	Drewry Car Co Ltd, London (supply agents only)
ĐĐak	Đuro Đakovic, Industrije Lokomotive, Strojeva I Mostova, Slavonski Brod, Yugoslavia
Dec	Société Nouvelle des Etablissements Decauville Aîne, Petit Bourg, Corbeil, Essonne, France
	(later Société Anonyme Decauville, Corbeil, Essonne, France)
Derby	Derby Locomotive Works
	Midland Railway / London, Midland & Scottish Railway / British Railways
DerbyC&W	Derby Carriage & Wagon Works, Litchurch Lane, Derby
	Midland Railway / London, Midland & Scottish Railway / British Railways
DeW	DeWinton & Co, Union Works, Caernarfon, North Wales
Dieci	Deici S.r.L., Majorana, Montecchio Emilia, Italy
Diema	Diepholzer Maschinenfabrik Fritz Schöttler GmbH., Diepholz, Niedersachsen, Germany
DK	Dick Kerr & Co Ltd, Preston, Lancashire
	(Britannia Works, Kilmarnock, Ayrshire until 1919)
DL	Dorman Long (Steel) Ltd, Middlesbrough, North Yorkshire
DLR	Difflin Lake Railway, Raphoe, Co. Donegal, Ireland
DM	Davies & Metcalfe Ltd, Romiley, Stockport, Greater Manchester
Dodman	Alfred Dodman & Co, Highgate Works, Kings Lynn, Norfolk
Don	Doncaster Works, South Yorkshire
	Great Northern Railway / London & North Eastern Railway / British Railways
Donelli	F.L. Donelli SpA, Reggio Emilia, Italy
Donelon	J.F. Donelon Ltd, Horwich, Greater Manchester
DonM	British Rail, Marshgate Permanent Way Depot, Doncaster, South Yorkshire
Dorothea	Dorothea Restorations, Mevril Spring Works, New Road, Whaley Bridge, Derbyshire
Dotto	Dotto Trains, Castelfranco, Italy
DP	Davey, Paxman & Co Ltd, Colchester, Essex
Dtz	Motorenfabrik Deutz AG / Humboldt-Deutz-Motoren AG / Klöckner-Humboldt-Deutz AG, Köln (Cologne), Nordrhein-Westfalen, Germany
DugginCG	C. G. Duggin, Littlehampton, West Sussex
Dundalk	Dundalk Works, Great Northern Railway of Ireland, Co. Louth, Republic of Ireland
DunEW	Dundalk Engineering Works, Dundalk, Co. Louth, Republic of Ireland
	Locomotives built from parts supplied by Gleismac

E

EAGIT	East Anglian Group for Industrial Training, Norwich, Norfolk
EARM	East Anglian Railway Museum, Chappel & Wakes Colne Station, near Colchester, Essex
EB	E. Borrows & Sons, St.Helens, Lancashire

EBW	E.B. Wilson & Co, Railway Foundry, Leeds
Eclipse	Eclipse Peat Co Ltd, Ashcott, Somerset
Eddy	M. Eddy
EdwardsE&J	E & J Edwards
EE	English Electric Co Ltd, London
EEDK	English Electric Co Ltd, Dick Kerr Works, Preston, Lancashire
EES	English Electric Co Ltd, Stephenson Works, Darlington (successors to RSHD)
EEV	English Electric Co Ltd, Vulcan Works, Newton-le-Willows, Lancashire (successors to VF)
EEV-AEI	English Electric Co Ltd & Associated Electrical Industries Ltd, Vulcan Foundry, Newton-le-Willows, Lancashire (predecessors to GECT)
EIMCO	Eastern Iron & Metal Corporation, Salt Lake City, Utah, USA
Eiv de Brive	L'Etablissement Industriel Équipement SNCF de Brive, Brive-de-Gaillarde, Corrèze, France
EK	Märstaverken, Eksjö, Sweden
ELC	East Lancashire Coachbuilders
Electro	Uzinele Electroputere, Craiova, Romania
Elh	Eastleigh Works, Hampshire London and South Western Railway / Southern Railway / British Railways
ENG/GEM	ENG/GEM, Wisbech, Cambridgeshire
ESCA	ESCA Engineering Ltd, 6 Wetheral Close, Hindley Industrial Estate, off Swan Lane, Hindley Green, Wigan, Greater Manchester
ESR	Exmoor Steam Railway, Devon
EvansC	Colin Evans, Littlehampton, West Sussex
EvansJ	J. Evans, Northampton
EV	Ebbw Vale Steel, Iron & Coal Co Ltd, Ebbw Vale Works, Gwent
EVM	Elham Valley Museum, Peene Yard, Peene, Folkestone, Kent
EVRA	Ecclesbourne Valley Railway Association, Wirksworth, Derbyshire
Express	Express Service OOD, Ruse, Bulgaria

F

Fairbourne	Fairbourne Railway Co, Fairbourne, Gwynedd, North Wales
Fairmont	Fairmont Railway Motors Ltd, Toronto, Ontario, Canada
FaulknerB	Brian Faulkner, Churcham, Gloucestershire
FE	Falcon Engine Works, Loughborough, Leicestershire
Ferndale	K. Watson & K. Tingle, Ferndale Engineering, Canning Vale, Western Australia
FFP	Fire Fly Project, Great Western Preservations Ltd, Bristol (1987) and Didcot, Oxfordshire (from 1989)
FH	F.C. Hibberd & Co Ltd, Park Royal, London ("Planet" locomotives) later at Butterley Works, Ripley, Derbyshire
Fiat	Fabrica Italiana Automobili Torino, Italy
Fife	Fife Heritage Railway, Leven, Fife, Scotland
Fisons	Fisons Ltd, British Moss Works, Swinefleet, near Goole, East Yorkshire constructed from parts supplied by Diema
Fitz	Fitzgerald Rail & Construction Services Ltd, Cwmbran, South Wales
FJ	Fletcher Jennings & Co, Lowca Engine Works, Whitehaven, Cumbria
FlooksG	George Flooks, Watford, Hertfordshire
FlourMill	The Flour Mill, Bream, Forest of Dean, Gloucestershire
FMB	F.M.B. Engineering Co Ltd, Unit 10, Southlands, Latchford Lane, Oakhanger, near Bordon, Hampshire
Ford	Ford Motor Co Ltd, Dagenham, Essex
FordTTC	Ford Motor Co Ltd, Thameside Technical Training Centre, Dagenham, Essex
ForshawJJ	J.J. Forshaw, Clifton, Bedfordshire
FosterRastrick	Foster, Rastrick & Co, Stourbridge, Worcestershire
Foulds/Collins	Andrew Foulds & Robert Collins, c/o E.A. Foulds Ltd, Albert Works, Clifton Street, Colne, Lancashire
FoxA	A. Fox
FRCo	Festiniog Railway Company, Boston Lodge Works, Porthmadog, North Wales
Frenze	Frenze Engineering, (Diss area), Norfolk
Freud	Stahlbahnwerke Freudenstein & Co, Berlin-Tempelhof, Germany
FRgroup4	Festiniog Railway, Group 4, Birmingham

Frichs	A/S Frichs Maskinfabrik & Kedelsmedie, Århus, Denmark
FRSociety	Festiniog Railway Society, North Staffs Group
FRT	Furness Railway Trust, Marconi Marine / Vickers Shipbuilding & Engineering Ltd, Barrow-in-Furness
Funkey	C H Funkey & Co. (Pty) Ltd., Alrode, Transvaal, Republic of South Africa [now in Gauteng]; formerly at Alberton, Transvaal; later a part of Dorbyl Transport Products, Rolling Stock Division, Boksburg, Transvaal, Republic of South Africa
FW	Fox, Walker & Co, Atlas Engine Works, Bristol

G

G&S	G.& S. Light Engineering Co Ltd, Stourbridge, Worcestershire
Gartell	Alan Gartell, Common Lane, Yenston, near Templecombe, Somerset
GB	Greenwood & Batley Ltd, Albion Ironworks, Armley, Leeds
GE	George England & Co Ltd, Hatcham Ironworks, London
GEU	General Electric Co, Erie, Pennsylvania, U.S.A.
GEC	General Electric Co Ltd, Witton, Birmingham
GEC-Alsthom	GEC-Alsthom, France
GECT	G.E.C.Traction Ltd, Vulcan Works, Newton-le-Willows, Lancashire
Geevor	Geevor Tin Mines Ltd, Pendeen, near St. Just, Cornwall
Geismar	Geismar (UK) Ltd, Salthouse Road, Brackmills Industrial Estate, Northampton and 68006 Colmar, Alsace, France
GGR	Groudle Glen Railway Co Ltd, Onchan, Isle of Man
GH	Gibb & Hogg, Victoria Engine Works, Airdrie, North Lanarkshire
Ghd	Gateshead Works, Co.Durham North Eastern Railway / London & North Eastern Railway / British Railways
GIA	GIA Industria AB, Grängesberg, Sweden
GibbonsCL	C.L. Gibbons, c/o Bury Transport Museum, Bury Depot, Greater Manchester
Gleismac	Gleismac Italiana SpA, Viale Delia Stazione 3, 46030 Bigarello, Mantova, Italy
GluyasC	Craig Gluyas, c/o Windmill Farm Railway, Burscough, Lancashire
GM	General Motors Ltd, Electro-Motive Division, La Grange, Illinois, U.S.A.
GMC	General Motors Ltd, Electro-Motive Division, London, Ontario, Canada
Gmd	Gmeinder & Co GmbH, Mosbach, Germany
GMT	Gyro Mining Transport Ltd, Victoria Road, Barnsley, South Yorkshire, then Bramley Way, Hellaby Industrial Estate, Hellaby, near Rotherham, South Yorkshire from c/1987
GNS	Great Northern Steam Co Ltd, Unit 3, Forge Way, Cleveland Industrial Estate, Darlington, Co.Durham
Goold	J.R. Goold Engineering Ltd, Camerton, near Radstock, Somerset
Gorton	Gorton Works, Manchester Great Central Railway / London & North Eastern Railway / British Railways
GOS	GOS Tool & Engineering Services Ltd, Blaenavon, Torfaen, South Wales
Govan	Govan Workshops, Broomloan Road, Glasgow Glasgow Corporation Transport Department, Underground Railway
GR	Grant, Ritchie & Co, Townholme Engine Works, Kilmarnock, Ayrshire
GRC&W	Gloucester Railway Carriage & Wagon Co Ltd, Gloucester
Greaves	J.W. Greaves & Sons Ltd, Llechwedd Quarry, Blaenau Ffestiniog, North Wales
Greensburg	Greensburg Machine Co, Greensburg, Pennsylvania, USA
Group Eng	Group Engineering
GS	George Stephenson
GS(H)	George Stephenson, Hetton, Co.Durham
GS(K)	George Stephenson, West Moor Workshops, Killingworth, Northumberland
Guest	Guest Engineering & Maintenance Co Ltd, Stourbridge, Worcestershire
GuinnessNL	Nigel L. Guinness, Cobham, Surrey
Gullivers	Gullivers Land, Milton Keynes, Bedfordshire
GWS	Great Western Society, Didcot, Oxfordshire

H

H	James & Fredk. Howard Ltd, Britannia Ironworks, Bedford
HAB	Hunslet-Barclay Ltd, Caledonia Works, Kilmarnock, Ayrshire
Hackworth	Timothy Hackworth, Soho Works, Shildon, Co.Durham

Hako	Hako GmbH, Bad Oldesloe, Germany
HallamP	Peter Hallam, St. Austell, Cornwall
HallT	T.Hall, North Ings Farm Museum, Dorrington, near Ruskington, Lincolnshire
Hano	Hannoversche Maschinenbau-AG, vormals Georg Egestorff, Hannover-Linden, Niedersachsen, Germany
Harbin	Harbin Forest Machinery Factory, Harbin, Heilongjiang Province, China
HardyK	K. Hardy, Brookhouse, Badgeworth, near Cheltenham, Gloucestershire
Harmill	Harmill Systems Ltd, Leighton Buzzard, Bedfordshire
Harsco	Harsco Track Technologies, Unit 1, Chewton Street, Eastwood, Nottinghamshire. (successors to Permaquip).
Hart	Sächsische Maschinenfabrik, vormals Richard Hartmann AG, Chemnitz, Germany
Haydock	Haydock Foundry Co Ltd, Haydock, Lancashire
Hayling	East Hayling Light Railway, Hayling Island, Hampshire
HaylockJ	J. Haylock, Moors Valley Railway, Ringwood, Dorset
Hayne	N. Hayne, Sheppards Tea Rooms & Boat House, near Saltford, Somerset, later Blaise Castle, Henbury, Bristol [Gloucestershire]
Haytor	M.P. Haytor & Son, Frensham, Surrey
HB	Hudswell Badger Ltd, Railway Foundry, Hunslet, Leeds
HC	Hudswell, Clarke & Co Ltd, Railway Foundry, Hunslet, Leeds
HE	Hunslet Engine Co Ltd, Hunslet, Leeds
	Graycar Industrial Estate, Barton-under-Needwood, Staffordshire. (A Division of LH Group Services Ltd)
	Statfold, Tamworth, Staffordshire
Heath	Robert Heath & Sons Ltd, Norton Ironworks, Stoke-on-Trent
Heatherslaw	P. Smith, Heatherslaw Light Railway Co, Cornhill-on-Tweed, Northumberland
Hedley	William Hedley, Wylam Colliery, Northumberland
Hegenscheidt	Hegenscheidt-MFD GmbH & Co. KG, Hegenscheidt Platz, Erkelenz, Germany (supply agents only)
Helical	Helical Technology Ltd, Lytham St. Annes, Lancashire
Hen	Henschel & Sohn GmbH, Kassel, Germany later traded as Thyssen Henschel and Adtranz
Herschell	Allan Herschell, North Tonawanda, New York, USA
H(L)	Hawthorns & Co, Leith Engine Works, Edinburgh
HL	R.& W.Hawthorn, Leslie & Co Ltd, Forth Bank Works, Newcastle-upon-Tyne
HLH	Hunslet Locomotive Hire Ltd, Station Road, Killamarsh, Derbyshire
HLT	Hughes Locomotive & Tramway Engine Works Ltd, Loughborough, Leicestershire
HN(R)	Harry Needle Railroad Company, Barrow Hill, Derbyshire
HopleyCP	C. P. Hopley
Hor	Horwich Works, Lancashire — Lancashire and Yorkshire Railway / London, Midland & Scottish Railway / British Railways
House	B. House
HPET	H.P.E.Tredegar Ltd, Tafarnaubach Industrial Estate, near Tredegar, Gwent
HSE	Harry Steer Engineering, Breaston, near Derby
HT	Hunslet Taylor Consolidated (Pty.) Ltd, Germiston, Transvaal, South Africa
HU	Robert Hudson Ltd, Leeds
HuntTG	T.G. Hunt, Oldbury, West Midlands
Hutchings	R. Hutchings
HW	Head, Wrightson & Co, Teesdale Ironworks, Thornaby-on-Tees, North Yorkshire
Hy-rail	This is a trademark of Harsco Track Technologies / Permaquip (see- Harsco / Perm)

I

IFA	Industrieverband Farzeugbau Association, VEB IFA-Automobilwerke, Ludwigsfeldt, Germany
Inchicore	Inchicore Works, Dublin — Great Southern & Western Railway / Great Southern Railways / Córas Iompair Éireann / Iarnród Éireann
IronHorse	Iron Horse Engineering, Southbourne, Dorset
Iso	Iso Speedic Co Ltd, Fabrications & Electric Vehicles, Charles Street, Warwick

Jaco	Jaco Engineering Co Ltd, Edwards Road, Birmingham
Jaywick	Jaywick Light Railway, near Clacton, Essex
Jesty	Bedford & Jesty Ltd, Doddings Farm, Bere Regis, Dorset
Jenbach	Jenbachwerke A G, Jenbach, Austria.
JF	John Fowler & Co (Leeds) Ltd, Hunslet, Leeds
JF(B)	John Fowler & Co (Leeds) Ltd, Old Hall Farm, Bouth, Cumbria
JMR	J.M.R.(Sales) Ltd, 173 Liverpool Road South, Birkdale, Southport, Lancashire
Jung	Arnold Jung Lokomotivfabrik GmbH, Jungenthal bei Kirchen-an-der-Sieg, Germany

K	Kitson & Co, Airedale Foundry, Leeds
Kambarka	Kambarka Machine, Kambarka, Republic of Udmurtia, Russia
Kawasaki	Kawasaki Heavy Industries Ltd, Motorcycle & Engine Company, Japan
KC	Kent Construction & Engineering Co Ltd, Ashford, Kent
Kearsley	Kearsley Power Station (Central Electricity Generating Board), Radcliffe, Gtr. Manchester
Keltec	Keltec Engineering Ltd, Co. Limerick, Ireland
Kennan	Thos Kennan & Son, Dublin, Ireland
Kershaw	Kershaw Manufacturing Co. Inc, Montgomery, Alabama, USA
Kew	Kew Bridge Steam Museum, Green Dragon Lane, Brentford, London
Keyte-Smith	Keyte Smith Ltd, Sutton-in-Ashfield, Nottinghamshire
Kierstead	Kierstead Systems & Controls Ltd, Ketley Bank Hall, Telford, Shropshire
Kilmarnock	Kilmarnock Regional Civil Engineers Workshops (British Rail, Scottish Region) Kilmarnock, Ayrshire (conversions only)
King	King Rail, Market Harborough, Leicestershire (UK agent for Zagro)
Kitching	A.Kitching, Hope Town Foundry, Darlington, Co.Durham
KLR	Kirklees Light Railway, Clayton West, West Yorkshire
KMB	Knutsford Motor Bodies, Knutsford, Cheshire
Knowell	M. Knowell
Krauss	Lokomotivfabrik Krauss & Co
KraussL	Lokomotivfabrik Krauss & Co, Linz, Austria
KraussM	Lokomotivfabrik Krauss & Co, Marsfeld Works, München (Munich), Germany
KraussS	Lokomotivfabrik Krauss & Co, Sendeling Works, München (Munich), Germany
Krupp	Friedrich Krupp, Maschinenfabriken Essen, Abt. Lokomotivbau, Essen, Nordrhein-Westfalen, Germany
KS	Kerr, Stuart & Co Ltd, California Works, Stoke-on-Trent, Staffordshire

L	R.A.Lister & Co Ltd, Dursley, Gloucestershire
LaMeuse	Société Anonyme des Ateliers de Construction de la Meuse, Sclessin, Liège, Belgium
Lake&Elliot	Lake & Elliot Ltd, Braintree, Essex
Lancing	Lancing Carriage Works, Sussex Southern Railway / British Railways
LancTan	Lancashire Tanning Co Ltd, Littleborough, Lancashire
Landore	British Rail, Landore Depot, Swansea
Landrover	Landrover Ltd, Solihull, West Midlands
Lawson	C. Lawson, 11 Okeley Lane, Highfield Estate, Tring, Hertfordshire
LB	Lister Blackstone Traction Ltd, Dursley, Gloucestershire (successor to L)
LBNGRS	Leighton Buzzard Narrow Gauge Railway Society, Stonehenge Workshops, Leighton Buzzard, Bedfordshire
Leake	J. Leake, Lytchett Matravers, Dorset
Leatham&Co	Leatham & Co, Heath, near Wakefield, West Yorkshire
LemonB	J.Lemon-Burton, Paynesfield, Albourne Green, West Sussex, and Shelmerdine & Mulley Ltd, Edgeware Road, Cricklewood, London NW2
Lesmac	Lesmac (Fasteners) Ltd, Dykehead Street, Glasgow
LewAJ	A.J. Lew, Fordingbridge, Hampshire
Lewin	Stephen Lewin, Dorset Foundry, Poole, Dorset
Leyland	Leyland Vehicles Ltd, Workington, Cumbria

LHGroup	L.H.Group Services Ltd, Barton-under-Needwood, Staffordshire
Lima	Lima Locomotive Works Inc., Lima, Ohio, U.S.A.
Lind	James Lind & Sons
LJ	Lester Jones, Hobart, Tasmania.
Llangollen	Llangollen Railway Engineering Services, Llangollen, Denbighshire, Wales
LlanwernBSC	Llanwern Works (British Steel Corporation), Newport, Gwent
LMM	Logan Mining & Machinery Co Ltd, Dundee
LO	Lokomo Oy, Tampere, Finland
LocoEnt	Locomotion Enterprises (1975) Ltd, Bowes Railway, Springwell, Gateshead, Co.Durham
Locospoor	NV Locospoor International, Den Haag, Netherlands; formerly CV Locospoor
Longcross	Longcross Film Studios, Chertsey, Surrey
Longhedge	Longhedge Works, London — South Eastern & Chatham Railway
Longleat	Longleat Light Railway, Longleat, Warminster, Wiltshire
Lowther	A.J. Lowther & Son Ltd, Ross-on-Wye, Herefordshire
LSLL	Large Scale Locomotives Ltd, Neath, South Wales
Lumb	James Lumb & Son, Elland, Leeds, West Yorkshire

M

M&P	Mather & Platt Ltd, Park Works, Manchester
M-TEK	M-TEK Engineering Ltd, Rotherham, South Yorkshire
Mace	C.Mace, The Woodland Railway, Kent
Maffei	J.A. Maffei AG, Locomotiv & Maschinenfabrik, München (Munich), Bavaria
MaK	Maschinenbau Kiel GmbH, Kiel-Friedrichsort, Germany
MaLoWa	MaLoWa Bahnwerkstatt GmbH, Klostermansfeld, Germany
MalyanG	Garry Malyan, Lappa Valley Railway, St. Newlyn East, near Newquay, Cornwall
MarshallJ	John Marshall, Spring Lane, Hockley Heath, Warwickshire
MarshJP	J.P. Marsh, Frosterley, Co. Durham
Massey	G.D. Massey, 57 Silver Street, Thorverton, Exeter, Devon
Matisa	Matisa Material Industriel SA, Arc-En-Ciel 2, Crissier, Lausanne, Switzerland
MatisaSPA	Matisa SpA, S.Palomba, Rome, Italy
MaxEng	Max Engineering, Station Road, Epworth, Doncaster, South Yorkshire
Maxfield	Arthur Maxfield
Maxitrack	Maxitrack, "Rothiemay", Offham Road, West Malling, Kent
McCulloch	W & D McCulloch, Ballantrae, Ayrshire
McDowall	Wallace McDowall Ltd, Monkton, Ayrshire
McGarigle	P.McGarigle, Niagara Falls, near Buffalo, New York, USA ("Cagney" locomotives)
Mechan	Mechan Ltd, Davy Industrial Park, Sheffield, South Yorkshire
MER	Manx Electric Railway, Derby Castle Works, Douglas, Isle of Man
Mercedes	Mercedes-Benz AG, Stuttgart, Germany
Mercia	Mercia Fabrications Ltd, Steel Fabrications, Units K1 & K3, Dudley Central Trading Estate, Shaw Road, Dudley, West Midlands
Mercury	Mercury Truck & Tractor Co Ltd, Gloucester
Meridian	Meridian (Motioneering) Ltd, Bradley Way, Hellaby Industrial Estate, Hellaby, near Rotherham, South Yorkshire
Metalair	Metalair Ltd, Wokingham, Berkshire
MetallbauE	Metallbau Emmeln GmbH, Haren (Ems), Germany
MetAmal	Metropolitan Amalgamated Railway Carriage & Wagon Co Ltd (until 6/1912)
MetC&W	Metropolitan Carriage, Wagon & Finance Co Ltd (6/1912 to 12/1928)
MetCam	Metropolitan-Cammell Carriage, Wagon & Finance Co Ltd (1/1929 to 10/1934) Metropolitan-Cammell Carriage & Wagon Co Ltd (1/1935 to 12/1964) Metropolitan-Cammell Ltd (from 1/1965 to 5/1989) — all located in Birmingham
MH	Muir-Hill (Engineers) Ltd, Trafford Park, Manchester
Middleton	Middleton Railway Trust, Hunslet, Leeds, West Yorkshire
Milner	Milner Engineering, Higher Kinnerton, Flintshire
Minilok	allrad-Rangiertecknik GmbH, D-5628 Heiligenhaus Bez, Dusseldorf, Nordrhein-Westfalen, Germany *[Note that the firm spells their name with a small 'a']*
Minirail	Minirail Ltd, Frampton Cotterell, Bristol
Mitsubishi	Mitsubishi Group, Japan
Mkm	Markham & Co Ltd, Chesterfield, Derbyshire

Moës	S.A. Moteurs Moës, Waremme, Belgium
Montania	Gerlach and König, Nordhausen, Germany
	(later became OK, but continued using the Montania name until 1945)
MoorsValley	Moors Valley Railway, Moors Valley Country Park, Horton Road, Ashley Heath,
	Ringwood, Hampshire
MorrisRP	R.P. Morris, 193 Main Road, Longfield, Kent
Morse	R.H. Morse, Potter Heigham, Norfolk
Mortimer	Mortimer Manufacturing Ltd, Cottesmore, Rutland
MossAJ	A.J. Moss, 97 Martin Lane Ends, Scarisbrick, Lancashire
MossDW	Derek W Moss, Burscough, Lancashire
Motala	AB Motala Verkstad, Motala, Sweden
Moyse	Locotracteurs Gaston Moyse, La Courneuve, Seine St.Denis, France
	Société Anonyme Moyse, La Courneuve, Seine St.Denis (after 1965)
MPES	Motive Power & Equipment Solutions, Greenville, South Carolina, USA
MR	Motor Rail & Tramcar Co Ltd / Motor Rail Ltd, Simplex Works, Bedford
MRWRS	MRW Railways Ltd, Sheffield, South Yorkshire
MSI	The Museum of Science & Industry in Manchester, Liverpool Road,
	Castlefield, Manchester
Multicar	VEB Fahrzeugwerk Waltershausen, Germany (until 1991)
	Multicar - Hako-Werke GmbH, Waltershausen, Germany
	(see also Hako - from 1996)
MV	Metropolitan-Vickers Electrical Co Ltd, Trafford Park, Manchester
MW	Manning, Wardle & Co Ltd, Boyne Engine Works, Hunslet, Leeds

N	Neilson & Co, Hyde Park Works, Springburn, Glasgow
NB	North British Locomotive Co Ltd, Glasgow
NBH	North British Locomotive Co Ltd, Hyde Park Works, Glasgow
NBQ	North British Locomotive Co Ltd, Queens Park Works, Glasgow
NBRES	NBR Engineering Services Ltd, Scarborough, North Yorkshire
	then Darlington, Co. Durham (from 2017)
NC	Northern Counties Coach Builders, Wigan, Lancashire
NCC	Northern Counties Committee (of London, Midland & Scottish Rly etc)
	York Road Works, Belfast, Northern Ireland
NDLW	North Dorset Locomotive Works, Motcombe, Dorset
Neasden	Neasden Works, London
	Metropolitan Railway
NemethJ	Joe Nemeth Engineering Ltd, Washingpool Farm, Main Road, Easter Compton, Bristol
Newag	Newag, Ripshorster Strasse, Oberhausen, Nordrhein-Westfalen, Germany
New Holland	New Holland Agricultural
Nissan	Nissan Motor Co Ltd, Japan
Niteq	Niteq BV, Overspoor 21, 1688 J.G. Nibbexwould, Netherlands
NMW	National Museum of Wales, Industrial & Maritime Museum, Butetown, Cardiff,
	South Glamorgan
NNM	Noord Nederlandsche Machinefabriek BV, Winschoten, Netherlands
Nohab	Nydquist & Holm AB, Trollhättan, Sweden
NR	Neilson Reid & Co, Glasgow
NW	Nasmyth, Wilson & Co Ltd, Bridgewater Foundry, Patricroft, Manchester

OAE	Olympic Aquatic Engineers, Norwich, Norfolk
Oerlikon	Maschinenfabrik Oerlikon, Zurich, Switzerland
OIM	OIM Ltd, Munster Railway Works, Kealkill, Co. Cork, Ireland
OK	Orenstein & Koppel AG, Berlin-Drewitz and Abt.Montania, Nordhausen, Germany
	then, from 1945 : –
	Orenstein-Koppel und Lübecker Maschinenbau AG, Dortmund-Dorstfeld, Germany
OldburyC&W	Oldbury Carriage & Wagon Co Ltd, Birmingham

P	Peckett & Sons Ltd, Atlas Locomotive Works, St.George, Bristol
P&T	Plasser & Theurer GmbH, Austria
Parkside	Parkside Electronics, Valley Mills, Nelson, Lancashire
ParmenterC	C. Parmenter, Launceston, Cornwall
PBR	Pleasure Beach Railway, Blackpool, Lancashire
Pendre	Pendre Works (Talyllyn Railway Co), Tywyn, Gwynedd, North Wales
Perm	The Permanent Way Equipment Co Ltd, Pweco Works, Lillington Road North, Bulwell, Nottingham — later at 1 Giltway, Giltbrook, Nottingham
Perriton	David Perriton, Hughley Railway, Much Wenlock, Shropshire
PL Services	Professional Lifting Services Ltd, Sheffield, South Yorkshire
Plasser	Plasser Railway Machinery (GB) Ltd, Drayton Green Road, West Ealing, London
Plymouth	Plymouth Locomotive Works, Plymouth, Ohio, USA
Porter	H. K. Porter Co, Inc, Pittsburgh, Pennsylvania, USA
PortTalbotBSC	Port Talbot Works (British Steel), Port Talbot, West Glamorgan
Potter	D.C. Potter, Yaxham Park, Yaxham, near Dereham, Norfolk
Powell	Alan Powell, Norwich, Norfolk
PPM	Parry People Movers Ltd, Corngreaves Trading Estate, Overend Road, Cradley Heath, West Midlands
Prestige	Prestige Engineering, Abbotskerwell, Newton Abbot, Devon
PRoyal	Park Royal Vehicles, Park Royal, London
Pritchard	William Pritchard, c/o Manchester, Bury, Rochdale & Oldham Steam Tramway, Oldham, Lancashire
PSteel	Pressed Steel Ltd, Linwood, Paisley, Renfrewshire
PTL	Positive Traction Ltd, Stonegravels Works, Chesterfield, Derbyshire
PVRA	Plym Valley Railway Association, Marsh Mills, Plympton, Devon
PWR	Pikrose & Co Ltd, Wingrove & Rogers Division, Delta Road, Audenshaw, Greater Manchester (successors to WR)

QRS	Quinton Railway Society, Quainton, Buckinghamshire

R&R	Ransomes & Rapier Ltd, Riverside Works, Ipswich, Suffolk
RADev	RA Developments, Scunthorpe, Lincolnshire
Rail-Ability	Rail-Ability Ltd, Stafford, Staffordshire
Ravenglass	Ravenglass & Eskdale Railway Co Ltd, Ravenglass, Cumbria
Red(F)	Redland Bricks Ltd, Funton Works, near Sittingbourne, Kent
Red(T)	Redland Bricks Ltd, Baltic Road, Tonbridge, Kent
Redstone	Mr Redstone, Penmaenmawr, North Wales
RegentSt	Regent Street Polytechnic, London
Renault	Régie Nationale des Usines Renault, Division Matériel Ferroviaire, Choisy-le-Roi, near Paris, France (from 1945) earlier:- SA de Usines Renault, Boulogne-Billancourt, Hauts de Seine, near Paris
Resco	Resco (Railways) Ltd, Manor Road Industrial Estate, Erith, Greater London
Resita	Uzinele de Fier si Domeniile din Resita Societate Anonima Resita, Resita, Romania (later: Combinatul Metalurgic Resita)
RFSD	R.F.S. Engineering Ltd, Doncaster Works, Hexthorpe Road, Doncaster
RFSK	R.F.S. Engineering Ltd, Kilnhurst Works, Hooton Road, Kilnhurst, South Yorkshire (successors to TH)
RH	Ruston & Hornsby Ltd, Lincoln
RHDR	Romney Hythe & Dymchurch Railway, New Romney, Kent
Rhiw	Rhiw Valley Light Railway (J.Woodruffe), Lower House, Manafon, near Welshpool, Powys, Mid Wales
Rhiwbach	Rhiwbach Quarries Ltd, Rhiwbach Slate Quarry, North Wales
Richard	Establishments B Richard, Saint-Denis-de l' Hôtel, Loire, France
Riley	Ian V.Riley Engineering, Arbour Locomotive Works, Kirkby, Merseyside later at Bury Car Sheds, Bury, Greater Manchester later at Premier Locomotive Works, Sefton Street, Heywood, Greater Manchester

Riordan	Riordan Engineering Ltd, Surbiton, Surrey
RLR	Richmond Light Railway, near Headcorn, Kent
RM	Road Machines (Drayton) Ltd, West Drayton, Middlesex
	later at Iver, Buckinghamshire
RMS	RMS Locotec Ltd, West Yorkshire
RMS(D)	RMS Locotec Ltd, Dewsbury, West Yorkshire
Roanoke	Roanoke Engineering, Grange Hill Industrial Estate, Bratton Fleming, Devon
Robel	Robel & Co, Maschinenfabrik, München, Bayern, Germany
Rosewall	K. Rosewall, Cross Elms Nursery, Bristol
RP	Ruston, Proctor & Co Ltd, Lincoln
RPSI(W)	Railway Preservation Society of Ireland, Whitehead, Co.Antrim, Northern Ireland
RR	Rolls Royce Ltd, Sentinel Works, Harlescott, Shrewsbury, Shropshire
	(successors to Sentinel)
RRS	Rapido Rail Systems, Dudley, West Midlands
RS	Robert Stephenson & Co Ltd, Forth Street, Newcastle-upon-Tyne
	and Darlington, Co.Durham
RSH	Robert Stephenson & Hawthorns Ltd
RSHD	Robert Stephenson & Hawthorns Ltd, Darlington Works, Co.Durham
RSHD/WB	built by RSHD but ordered by WB
RSHN	Robert Stephenson & Hawthorns Ltd, Newcastle-upon-Tyne Works
	(successors to HL)
RSM	Royal Scottish Museum, Chambers Street, Edinburgh
Ruhrthaler	Ruhrthaler Maschinenfabrik Schwarz & Dyckerhoff AG, Mülheim/Ruhr, Germany
Ruislip	Ruislip Lido Railway, Ruislip, Greater London
RVEL	RVEL Ltd, London Road, Derby, Derbyshire
RVM	RVM Engineering, Hastings, East Sussex
RVR	Rother Valley Railway Ltd, Robertsbridge, East Sussex
RWH	R.& W.Hawthorn & Co, Forth Bank Works, Newcastle-upon-Tyne (later HL)

S

S	Sentinel (Shrewsbury) Ltd, Harlescott, Shrewsbury, Shropshire.
	(Diesel locomotives numbered between 10001 and 10183 were actually designed and built by Rolls Royce Ltd, but were fitted with Sentinel worksplates)
Sabero	Hulleras de Sabero y Anexas SA, Sabero, Spain
Sanline	Sanline Systems Ltd, Lucan, Dublin, Ireland
Sara	Sara & Burgess, Penryn, Cornwall
Sartori	Sartori Rides Srl, Montagnana, Italy
SaundersT	T. Saunders
Saxby	Frank Saxby & Co, Guildford Works, Surrey
Saxton	C. Saxton, Cheadle Hulme, Greater Manchester
Scarrott	D.J. Scarrott, Kingsteignton, Newton Abbot, Devon
Schalke	Gewerkschaft Schalke Eisenhütte, Gelsenkirchen-Schalke, Nordrhein-Westfalen, Germany
Schichau	Maschinenbau F.Schichau, Maschinen-und Lokomotivfabrik, Elbing, Germany
	later at Elbtag, Poland
Schöma	Christoph Schöttler Maschinenfabrik GmbH, Diepholz, Niedersachsen, Germany
SchörlingB	Schörling-Brock GmbH, Gehrden, Germany
Schw	L. Schwartzkopff, Berlin, Germany
	(later became Berliner Maschinenbau AG — see BMAG)
Schwingel	Paul Schwingel GmbH, Leverkusen, Germany
ScienceMus	Science Museum, South Kensington, London.
ScottP	Peter Scott, Hillsborough, Northern Ireland
Scul	Sculfort, Zac des fonds St Jaques, Feignies, France
SCW	I.C.I. Ltd, South Central Workshops, Tunstead, Derbyshire
Sdn	Swindon Works, Wiltshire — Great Western Railway / British Railways
SdnCol	Swindon College, Department of Engineering, North Star Avenue, Swindon
SDSI(S)	South Durham Steel & Iron Co Ltd, Stockton Works, Co.Durham
Selhurst	Selhurst Maintenance Depot, Greater London — British Rail, Southern Region
SET	Stored Energy Technology, Litchurch Lane, Derby
SfP	Steaming for Pleasure, Yapton, West Sussex

SGLR	Steeple Grange Light Railway, Steeplehouse Junction, Wirksworth, Derbyshire
S&H	Strachan & Henshaw Ltd, Ashton, Bristol
Shackerstone	Shackerstone Station (The Battlefield Line), Leicestershire
SharmanJ	J. Sharman, Stone, Stoke-on-Trent, Staffordshire
Sharon	Sharon Engineering Ltd, Leek, Staffordshire
—	Shelmerdine & Mulley Ltd, Edgeware Road, Cricklewood, London NW2 (see code "LemonB")
ShepherdFG	F.G. Shepherd, Flow Edge Colliery, Middle Fell, Alston, Cumbria
SherwoodF	Sherwood Forest Railway, Edwinstowe, Mansfield, Nottinghamshire
ShooterA	A. Shooter, Oxfordshire
Siemens	Siemens Bros Ltd, London — possibly agents for :-
S&H(B)	Siemens & Halske, Berlin, Germany (until 1903)
SSW	Siemens-Schuckert-Werke, Berlin, Germany (from 1903)
SkinnerD	D. Skinner, 660 Streetsbrook Road, Solihull, West Midlands
SL	Severn-Lamb UK Ltd, Western Road, Stratford-Upon-Avon, Warwickshire then Alcester, Warwickshire.
SLM	Schweizerische Lokomotiv- und Maschinenfabrik, Winterthur, Switzerland
SM	Southern Motors Ltd, Ringsend, Dublin, Ireland
SMH	Simplex Mechanical Handling Ltd, Elstow Road, Bedford (successor to MR)
SmithEL	E.L. Smith, Garsington, Oxfordshire
SmithN	N. Smith, Heatherslaw Light Railway Co, Heatherslaw Mill, Northumberland
SmithP	Smith, Pengam, South Glamorgan
SMR	Saltburn Miniature Railway, Valley Gardens, Saltburn, North Yorkshire
SolHütte	Federal-Mogul Sollinger Hütte GmbH, Uslar, Germany
SouthCrofty	South Crofty Ltd, Pool, near Camborne, Cornwall
SPA	Specialist Plant Associates Ltd, 23 Podington Airfield, Hinwick, Bedfordshire
Spence	Wm.Spence, Cork Street Foundry, Dublin
SPL	Science Projects Ltd, (Constructors), Hammersmith, London
Spondon	Spondon Power Station, Derbyshire Derbyshire & Nottinghamshire Electric Power Co Ltd
Spoor	NV Spoorijzer, Delft, Netherlands
SRS	Swedish Rail Systems Euroc, Solna, Sweden
SS	Sharp, Stewart & Co Ltd, Atlas Works, Manchester (until 1888) later at Atlas Works, Glasgow
StanhopeT	T.Stanhope, Arthington Station, near Leeds, West Yorkshire
Statfold	Statfold Barn Railway, Tamworth, Staffordshire
StationRoad	Station Road Steam Ltd, Metheringham, Lincolnshire
Steamtown	Steamtown Railway Museum, Warton Road, Carnforth, Lancashire
StewartWP	W.P.Stewart, Washington Sheet Metal Works, Industrial Road, Hertburn Industrial Estate, Washington, Co.Durham
Stoke	Stoke Works, Stoke-on-Trent North Staffordshire Railway
StokesMJ	M.J. Stokes, Little West Garden Railway, Southerndown, Mid Glamorgan
Str	Stratford Works, London Great Eastern Railway / London & North Eastern Railway / British Railways
StrawberryHill	Strawberry Hill Depot, South London (British Rail)
StRollox	St.Rollox Works, Glasgow Caledonian Railway / London, Midland & Scottish Railway / British Railways
STRPS	South Tynedale Railway Preservation Society, Alston Station, Cumbria
Strüver	Ad.Strüver AG, Hamburg, Germany
SUSTRACO	Sustainable Transport Co Ltd, Bristol
SVI	SVI S.p.A., Perugia, Italy
SVR	Severn Valley Railway, Bridgnorth, Shropshire
Swanhaven	Swanhaven, Hull, East Yorkshire
Sweet	Brendon Sweet, Polgooth, Cornwall
Syl	Sylvester Steel Co, Lindsay, Ontario, Canada

T

Tambling	N.J. Tambling, Lappa Valley Railway, St. Newlyn East, near Newquay, Cornwall
TargettR	R.C. & J. Targett, Shrewsbury, Shropshire

TaylorB	B.Taylor, 7 Abbey Road, Shepley, Huddersfield, West Yorkshire
TaylorJ	J. Taylor, The Ford, Woolhope, Herefordshire
TeasdaleS	S. Teasdale, Sleights, near Whitby, North Yorkshire
TDI	Transport Design International Innovations Ltd, Long Marston, Warwickshire
TEE	The Engineering Emporium, Bramcote Works, Bramcote, Warwickshire
TG	T. Green & Son Ltd, Leeds
TH	Thomas Hill (Rotherham) Ltd, Vanguard Works, Kilnhurst, South Yorkshire
Thakeham	Thakeham Tiles Ltd, Thakeham, Sussex
Thomson	Thomson Rail Equipment Ltd, Cinderford, Gloucestershire
Thunes	Thunes Mekaniske Værksted A/S, Skøyen, Norway
Thursley	The Thursley Railway, Hampshire
Thurston	T.J. Thurston, Cove, Farnborough, Hampshire
ThurstonTS	T.S. Thurston
Thwaites	Thwaites Ltd, Leamington Spa, Warwickshire
TK	Oy Tampella Ab, Tampere, Finland
TK & L	Todd, Kitson & Laird, Leeds
TMA	TMA Automation Ltd, Feeds Automated Systems, Jubilee Works, Erdington, Birmingham
TPC	Tractive Power Corporation, North Vancouver, Canada
Trackmobile	Trackmobile Ltd, La Grange, Illinois, USA
	(originally part of the Whiting Corporation [WhC])
TRRF	The Railroad Factory Ltd, Cork, Ireland
TS&S	Track Supplies & Services Ltd, Old Wolverton, Milton Keynes, Buckinghamshire
TU	Task Undertakings Ltd, Birmingham
Tunn	Tunnequip Ltd, Nowhurst Lane, Broadbridge Heath, Horsham, West Sussex
TurnerT	T. Turner, Long Eaton, Derbyshire
TyseleyLW	Tyseley Locomotive Works (Standard Gauge Steam Trust), Tyseley, Birmingham

U

U23A	Uzinele 23 August, Bucuresti, Romania
UCA	UCA, Antwerp, Belgium
UK Loco	UK Loco Ltd, Unit 1, The Heath Works, Main Road, Cropthorne, Worcestershire
Unilok	Unilok locomotives, but whether "(G)" or "(H)" as yet unrecorded.
Unilok(G)	Unilokomotive Ltd, International Division, Mervue Industrial Estate, Galway, Co.Galway, Republic of Ireland
Unilok(H)	Hugo Aeckerle & Co, Hamburg, Germany
	[some early locomotives had frames constructed by Jung]
Unimog	Unimog road/rail locomotives — Mercedes Benz AG, Stuttgart, Germany
UphillJ	John Uphill

V

Vanstone	D. Vanstone, Pixieland Mini-Zoo, Kilkhampton, near Bude, Cornwall
VE	Victor Electrics Ltd, Burscough Bridge, Lancashire
VER	Volks Electric Railway (Magnus Volk), Madeira Drive, Brighton, Sussex
VF	Vulcan Foundry Ltd, Newton-le-Willows, Lancashire
VIW	Vulcan Iron Works, Wilkes-Barre, Pennsylvania, USA
VL	Vickers Ltd, Barrow-in-Furness, Cumbria
Vollert	Hermann Vollert GmbH & Co KG, Maschinenfabrik, Weinsberg/Wurtt, Germany
Volvo	AB Volvo, Gothenburg, Sweden

W

WalkerG	G. Walker, Lakeside Miniature Railway, Marine Lake, Southport, Merseyside
WalkerS	Samuel Walker
Waterfield	James Waterfield, Boston, Lincolnshire
Watson&Haig	Watson & Haig Ltd, Andover, Hampshire
WB	W.G.Bagnall Ltd, Castle Engine Works, Stafford
Wbton	Wilbrighton Wagon Works, Shropshire
Wcb	Whitcomb Locomotive Co, Illinois, U.S.A.
WCI	Wigan Coal & Iron Co Ltd, Kirkless, Wigan, Lancashire
WeaverP	P. Weaver, New Farm, Lacock, near Corsham, Wiltshire
Werk	Werkspoor NV, Utrecht, Holland

WhalleyB	Bruce Whalley, Weston Rail Ltd, Weston Park, Shifnal, Shropshire
WhC	Whiting Corporation, Harvey, Illinois, U.S.A.
WHR(GF)	Welsh Highland Light Railway (1964) Ltd, Gelert's Farm Works, Porthmadog, North Wales
WilliamsH	H. Williams, Barrow-on-Humber, Lincolnshire
WilliamsWJ	W.J. Williams, Blaenau Ffestiniog, North Wales
Wilmott	Wilmott Bros (Plant Services) Ltd, Ilkeston, Derbyshire
WilsonAJ	A.J. Wilson, 6 Trentdale Road, Carlton, Nottingham
Wilton(ICI)	Wilton Works (Imperial Chemical Industries Ltd), Middlesbrough, North Yorkshire
Windhoff	Rheiner Maschinenfabrik Windhoff AG, Rheine, Germany
	later WINDHOFF Bahn-und Anlagentechnik GmbH, Rheine, Germany
Winson	Winson Engineering, Porthmadog (later Penrhyndeudraeth), North Wales
	and at Daventry, Northamptonshire
WkB	Walker Bros (Wigan) Ltd, Wigan, Lancashire
Wkm	D.Wickham & Co Ltd, Ware, Hertfordshire
WkmR	Wickham Rail, Bush Bank, Suckley, Worcestershire (successors to Bruff & Wkm)
WLLR	West Lancashire Light Railway, Hesketh Bank, near Preston, Lancashire
W&LLR	Welshpool & Llanfair Light Railway Preservation Co Ltd,
	Llanfair Caereinion, Powys, Mid Wales
WMD	Waggon & Maschinenbau GmbH, Donauwörth, Germany
Wolverton	Wolverton Works, Buckinghamshire — British Rail Engineering Ltd
Woodings	Woodings Railcar Ltd, Alexandria, Ontario, Canada
Woolwich	Woolwich Arsenal, London
WR	Wingrove & Rogers Ltd, Kirkby, Liverpool (successors to BEV)
WSR	West Somerset Railway, Williton, Somerset
WVanHeiden	W. Van der Heiden, Rotterdam, Netherlands

Y

YE	Yorkshire Engine Co Ltd, Meadow Hall Works, Sheffield
YEC	Yorkshire Engine Company Ltd, Unit 7, Meadow Bank Industrial Estate,
	Harrison Street, Rotherham, South Yorkshire; *and later at*
	Unit A3, Templeborough Enterprise Park, Bowbridge Close, Rotherham,
	South Yorkshire
York	York Works, York, North Yorkshire
	North Eastern Railway/ London & North Eastern Railway/ British Railways
York(BRE)	York Works (British Rail Engineering Ltd), North Yorkshire
Young&Co	J. Young & Co, Leeds

Z

Zagro	ZAGRO Bahn-und Baumaschinen GmbH, Bad Rappenau-Grombach, Germany
Zephir	Zephir S.p.a., via Salvador Allende 85, 41100 Modena, Italy
Zetor	The Zetor Tractor Factory Co, Brno, Czech Republic
Zweiweg	Zweiweg International GmbH & Co KG, Leichlingen, Germany (also Sehnde, Germany)
Zwiehoff	G. Zwiehoff GmbH, Tegernseestrabe 15, 83022 Rosenheim, Germany

0-9

567LG	GCR 567 Locomotive Group, Ruddington, Nottinghamshire
9E	Nine Elms Works, London — London & South Western Railway

CONTRACTORS

Listed below are the Civil Engineering Contractors / Plant Hire specialists who own locomotives for use on tunnelling and sewer contracts, etc. The locomotives are to be found in all parts of the Country but the details of the locomotive fleets are listed under the firm's main depot in the County shown below.

TITLE OF FIRM	COUNTY
Costain Group plc	Warwickshire
Kier Group plc	Hertfordshire
J.Murphy & Sons Ltd	Greater London
Edmund Nuttall Ltd	Greater London
Olympic Aquatic Engineers	Norfolk
Specialist Plant Associates Ltd	Bedfordshire
A.E.Yates Ltd	Greater Manchester

HIRE FLEET OPERATORS

A number of companies are particularly active in the locomotive hire business, maintaining a fleet of locomotives for hire or sale. Such locomotives are hired out to industrial users and main-line "TOC" (Train Operating Companies). As such the locomotives can be seen all over the country, frequently on short-term loan. Hire Fleet Operators are listed in this book in the county of their principal or headquarters address, together with known details of their current fleet. In several cases, however, few or no locomotives are stabled at the addresses given.

Alan Keef Ltd (narrow gauge locomotives)	Herefordshire
Depot Rail Ltd	Lincolnshire
Hunslet Ltd	Staffordshire
MCL Rail Ltd	North Yorkshire
Ed Murray & Sons Ltd	Teesside
Harry Needle Railroad Co Ltd	Derbyshire
Northumbria Rail Ltd	Northumberland
E.G. Steele & Co Ltd	North Lanarkshire
Rail Management Services Ltd t/a RMS Locotec	Derbyshire
Railway Support Services Ltd	Warwickshire

RAILWAY CONTRACTORS

Listed below are the Railway Engineering Contractors / Plant Hire specialists who own locomotives for use on railway contracts etc. The locomotives are to be found in all parts of the Country but the details of the locomotive fleets are listed under the firm's main depot in the County shown below.

Avondale Environmental Services Ltd	Kent
Balfour Beatty Rail Plant Ltd	Nottinghamshire
Land Recovery Ltd	Cumbria
W & D McCulloch	Ayrshire
QTS Rail Ltd	Ayrshire
Quattro Plant Ltd	Greater London
Shovlin Plant Hire Ltd	Greater Manchester
Story Rail Ltd	Cumbria
TXM Plant Ltd	Greater Manchester

SECTION 1 — ENGLAND

BEDFORDSHIRE

INDUSTRIAL SITES

**CRANFIELD SAFETY & ACCIDENT INVESTIGATION CENTRE,
SCHOOL OF ENGINEERING, ACADEMIC OPERATIONS UNIT,
CRANFIELD UNIVERSITY, CRANFIELD**

Gauge 4ft 8½in   www.cranfield.ac.uk

MK43 0AL
SP 938422

69933	(390033)	4w-4wWER	Alstom	2002

**HARMILL SYSTEMS LTD,
UNIT B1, CHERRYCOURT WAY, LEIGHTON BUZZARD**
www.harmill.co.uk

LU7 4UH
SP 940249

New Harmill locomotives under construction or repair occasionally present

SPECIALIST PLANT LTD, PODINGTON

Locomotives present in yard between contracts and in store for third parties.

Gauge 2ft 6in   www.specialistplant.co.uk

SP 948608

SP 543		4wBE		CE	B4056B	1995

Gauge 2ft 0in / 1ft 6in

SP 83		4wBE		CE	5942A	1972
	RR51-069	4wBE		CE	5943	1972
	RR51-071	4wBE		CE	5949D	1972
SP 86		4wBE		CE		1970
			reb	SPA		1995
	RR51 7002	4wBE		CE	B3686D	1990
			reb	CE	B4075.2	1995 a

Gauge 2ft 0in

SP 1014		4wBE		CE		1968
SP 1015		4wBE		CE		1968
RR51 7001		4wBE		CE	B3686C	1990
			reb	CE	B4075.1	1995 a

a   currently at Costain Group plc

PRESERVATION SITES

BEDFORDSHIRE STEAM ENGINE PRESERVATION SOCIETY, CLIFTON

Gauge 2ft 0in   www.bseps.org.uk

Private Site

THE IRON & STEEL RAIL CO LTD No.4	4wVBT	VCG	ForshawJJ		c1998
1 VERACRUZ	0-6-0WT	OC	OK	11009	1925

GREAT WOBURN RAILWAY, PETER SCOTT WOBURN SAFARI PARK, WOBURN, near MILTON KEYNES

Gauge 1ft 8in   www.woburnsafari.co.uk

MK17 9QN

SP 962343

LADY ALEXANDRA	0-4-0DH	s/o	AK		70	2004

LEIGHTON BUZZARD NARROW GAUGE RAILWAY SOCIETY

Locomotives are kept at :-

Pages Park Shed   LU7 4TG   SP 929242
Stonehenge Works   LU7 9LA   SP 941275

Gauge 2ft 6in   www.buzzrail.co.uk

WD 767139	2w-2PMR		Wkm	3282	1943

Gauge 2ft 0in

	GERTRUDE	0-6-0T	OC	AB	1578	1918
4	DOLL	0-6-0T	OC	AB	1641	1919
	SEZELA No.4	0-4-0T	OC	AE	1738	1916
	ELIDIR	0-4-0T	OC	AE	2071	1933
No.3	RISHRA	0-4-0T	OC	BgC	2007	1921
778		4-6-0T	OC	BLW	44656	1917
(1)	CHALONER	0-4-0VBT	VC	DeW		1877
	PENLEE	0-4-0WT	OC reb	Freud ARC	73	1901 c1983
(5)	(HF 2023)	0-8-0T	OC	KraussM	7455	1918
	PETER PAN	0-4-0ST	OC	KS	4256	1922
2	PIXIE	0-4-0ST	OC	KS	4260	1922
"No.2"		0-4-0WT	OC	OK	2544	1907
No.11	P.C.ALLEN	0-4-0WT	OC	OK	5834	1912
	PEDEMOURA	0-6-0WT	OC	OK	10808	1924
"5"	ELF	0-6-0WT	OC	OK	12740	1936
NG 23		4wBE	reb	BD AB	3702	1973 1987
NG 46		4wDH		BD	3698	1973
NG 51		4wDH	reb	AB HAB	720	1987 1996
LR 2182		4wPM		MR	461	1917
LOD	758009	4wDM		MR	8641	1941
(LOD	758220)	4wDM		MR	8745	1942
A.M.W. No.165	FIRE "TRUMPTON"	4wDM		RH	194784	1939
"7" 8986	FALCON	4wDM		OK	8986	1938	a
8	"GOLLUM"	4wDM		RH	217999	1942
"9"	"MADGE"	4wDM		OK	7600	1938
10	HAYDN TAYLOR	4wDM		MR	7956	1945
12		4wPM		MR	6012	1930
13	ARKLE	4wDM		MR	7108	1936
"14"		4wDM		HE	3646	1946
16		4wDM		L	11221	1939
17	DAMREDUB	4wDM		MR	7036	1936
18	FËANOR	4wDM		MR	11003	1956
"19"		4wDM		MR	11298	1965	Dsm
No.21	FESTOON	4wPM		MR	4570	1929
"22"	"FINGOLFIN"	4wDM		LBNGRS	1	1989	b

No.	Name	Type	Builder	Works No.	Year	Notes	
23		4wDM	RH	164346	1932		
24		4wDM	MR	4805	1934		
24		4wDM	MR	11297	1965		
"25"		4wDM	MR	7214	1938		
"26"	YIMKIN	4wDM	RH	203026	1942	a	
26	ANNA	4wDM	MR	8720	1941		
27		4wDH	RH	408430	1957		
28	R.A.F. STANBRIDGE	4wDM	RH	200516	1940	c	d
29	CREEPY YARD No. P 19774	4wDM	HE	6008	1963		
30		4wDM	MR	8695	1941		
31		4wPM	L	4228	1931		
"32"		4wDM	RH	172892	1934		
34	RED RUM	4wDM	MR	7105	1936		
36	"CARAVAN"	4wDM	MR	7129	1936		
No.38	HARRY BARNETT	4wDM	L	37170	1951		
LM 39	T.W.LEWIS	4wDM	RH		1954	e	
"40"	"TRENT"	4wDM	RH	283507	1949		
41	LOD/758054 SOMME	4wDM	HE	2536	1941	f	
"42"	"SARAH"	4wDM	RH	223692	1943		
43		4wDM	MR	10409	1954		
44		4wDM	MR	7933	1941		
"45"		4wDM	MR	21615	1957	Dsm	
"46"		4wDM	RH	209430	1942		
"48"	MCNAMARA	4wDM	HE	4351	1952		
80	BEAUDESERT	4wDH	SMH	101T018	1979		
		reb	AK	59R	1999		
81	PETER WOOD	4wDH	HE	9347	1994		
No.1568		4wPM	FH	1568	1927		
	BLUEBELL	4wDM	FH		1943	g	
–		4wDM	HU	38384	1930		
"REDLANDS"		4wDM	MR	5603	1931	Dsm	
R8		4wDM	MR	5612	1931	Dsm	
No.131		4wDM	MR	5613	1931	Dsm	
–		4wDM	MR	8731	1941	Dsm	
–		4wDM	MR			Dsm	
–		4wDM	RH	187105	1937		
–		4wDM	RH	218016	1943	Dsm	
–		4wDM	RH	229656	1944		
RTT/767182		2w-2PMR	Wkm	2522	1938		
No.5	ISABEL	4wDM	MR	5608	1931	h	+
"6"	MoTeL"	4wDM	MR	5875	1935	i	+
No.8		2w-2DMR	Bg	3539	1959		
		reb	AK		1988	j	+

+	numbered in the coaching stock fleet
a	nameplate is in Arabic
b	built from parts of RH 425798/1958 and RH 444207/1961
c	stored off site
d	carries plate RH 200513
e	either RH 375315 or RH 375316
f	on loan to the Chemin de Fer Froissy-Cappy-Dompierre, Somme, France
g	either FH 2631 or FH 2834
h	converted into mobile compressor for air braking

i   converted into a mobile toilet
j   rebuilt from 4ft 8½in gauge 2w-2DMR, now a bogie unpowered coach

Mrs. MACKINNON
Gauge 1ft 8in   **Private Site**

| – | 4wDM | OK | 6703 | 1936 | Pvd |

Locomotive believed to be preserved by a member of the family at an unknown location.

RAY MASLEN & FRIENDS, Private Railway, ARLESEY
Gauge 2ft 0in   **Private Site**

CLARABEL	4wDMF	HE	4758	1954	
–	4wDM	L	37911	1952	
–	4wDM	RH	441951	1960	
RTT/767187	2w-2PM	Wkm	2559	1939	Dsm
RTT/767094	2w-2PM	Wkm	3033	1941	Dsm

WHIPSNADE WILD ANIMAL PARK LTD, (a ZSL conservation zoo),
THE GREAT WHIPSNADE RAILWAY, WHIPSNADE ZOO, DUNSTABLE   LU6 2LF
Gauge 2ft 6in   www.whipsnadezoo.org   TL 004172

No.2	EXCELSIOR	0-4-2ST	OC	KS	1049	1908
No.4	SUPERIOR	0-6-2T	OC	KS	4034	1920
3		4wDH		BD	3780	1983
No.8	VICTOR	0-6-0DM		JF	4160004	1951
No.9	HECTOR	0-6-0DM		JF	4160005	1951

BERKSHIRE

INDUSTRIAL SITES

FIRST GREATER WESTERN LTD, t/a GREAT WESTERN RAILWAY,
READING TRAINCARE DEPOT, 101 COW LANE, READING   RG1 8FN
(part of the First Group plc)
Gauge 4ft 8½in   www.gwr.com   SU 701741

| – | 4wBE | R/R | Zephir | 2535 | 2014 |
| – | 4wBE | | Zephir | 2643 | 2016 |

PRESERVATION SITES

D. BUCK, near WINDSOR
Locomotives are kept at a private site.
Gauge 5ft 0in   Private Site

91016	LADY PATRICIA	4-6-2	OC	TK		946	1955

Gauge 4ft 8½in

	SWANSCOMBE	0-4-0ST	OC	AB		699	1891
334\G102	SIR VINCENT	4wWT	G	AP		8800	1917
(89-94)	(VIGILANT)	0-4-0ST	OC	HE		287	1883
	HORNPIPE	0-4-0ST	OC	P		1756	1928
–		0-4-0DM		AB		352	1941
	ARMY 9112	4wDMR		Bg		3538	1959
	"TRACKRAT"	2-2wDHR		Group Eng	RAT2001c		1998
–		2w-2PMR		Bance		062	1998
–		2w-2PMR		Geismar	ST/01/21		2001

LEGOLAND WINDSOR PARK LTD, LEGOLAND WINDSOR,
WINKFIELD ROAD, WINDSOR (part of the Merlin Entertainments Group)   SL4 4AY
Gauge 2ft 0in   www.legoland.co.uk   SU 939746

	AMEY	4wDM		MR		7902	1939	OOU
–		4-4-0DH	s/o	SL		663	1995	

NICK WILLIAMS, Private Railway, READING
Gauge 2ft 0in   Private Site

9	JACK	0-4-0T	OC	AB		1871	1925	
2	SEZELA No.2	0-4-0T	OC	AE		1720	1915	
6	SEZELA No.6	0-4-0T	OC	AE		1928	1923	
–		4wDM		FH		2163	1938	
–		4wDM		LB		53225	1962	
	LR 2478	4wPM		MR		1757	1918	
–		4wDM		MR		7512	1938	
–		4wDM		MR		11264	1964	
–		4wDM		RH		277265	1949	
–		4wDM		RH		296091	1949	
–		0-4-0BE		WR		N7661	1974	
RTT/767186		2w-2PM		Wkm		2558	1939	Dsm

WINDSOR STATION LTD, WINDSOR & ETON CENTRAL STATION,
THAMES STREET, WINDSOR   SL4 1PJ
Gauge 4ft 8½in   SU 969773

3041	THE QUEEN	4-2-2		Steamtown	1983
	non-working replica of Sdn 1401 / 1894				

BRISTOL

INDUSTRIAL SITES

COUNTY OF AVON FIRE BRIGADE,
AVONMOUTH FIRE STATION, ST. ANDREWS ROAD, AVONMOUTH
BS11 9HQ

Gauge 4ft 8½in   www.avonfire.gov.uk   **ST 517784**

CK70 CJF	99709 978028-7	4wDH	R/R	_(Mitsubishi C00189 2020	
				(GOS RRC459 2020	

FIRST GREATER WESTERN LTD, t/a GREAT WESTERN RAILWAY,
ST. PHILIP'S MARSH T&RSMD, ALBERT ROAD, BRISTOL
BS2 0GW

(part of the First Group plc)

Gauge 4ft 8½in   www.gwr.com   **ST 608721**

(D3990)	08822 DAVE MILLS	0-6-0DE	Derby		1960
	MUFFIN	4wBE	Niteq	B188	2002
	–	4wBE	SET	1079	2009

LONDON & NORTH WESTERN RAILWAY CO LTD, t/a ARRIVA TRAINCARE,
BARTON HILL DEPOT, DAY'S ROAD, ST. PHILIPS, BRISTOL
BS2 0QS

(Arriva - A DB company)   Preserved locomotives between Mainline duties occasionally present.

Gauge 4ft 8½in   www.arrivatc.com   **ST 604728**

(D3678)	08516	0-6-0DE	Dar	1958

J. MURPHY & SONS LTD, WESSEX WATER,
NORTH BRISTOL RELIEF SEWER CONTRACT

TBM drive at Lawrence Weston ST 547789

Shafts at Henbury ST 566790, ST 567792, ST 567795 and Catbrain, Gloucestershire ST 576803

Gauge 750mm   www.murphygroup.com

–	4wBE	Schöma	7105	2019
–	4wBE	Schöma	7106	2019
–	4wBE	Schöma	7107	2019
–	4wBE	Schöma	7109	2019

PRESERVATION SITES

M SHED, BRISTOL HARBOUR RAILWAY,
WAPPING ROAD, PRINCES WHARF, CITY DOCK, BRISTOL
BS1 4RN

Gauge 4ft 8½in   www.bristolmuseums.org.uk/m-shed   **ST 585722**

I.W. & D. 34	0-6-0ST	OC	AE	1764	1917	
3	0-6-0ST	OC	FW	242	1874	a

HENBURY	0-6-0ST	OC	P		1940	1937	
–	0-4-0DM		RH	418792	1959	a

a   currently stored at Cumberland Road, Bristol ST 570721

BUCKINGHAMSHIRE

INDUSTRIAL SITES

CHILTERN RAILWAYS CO LTD, AYLESBURY DEPOT, LEACH ROAD, AYLESBURY
(A subsidiary of DB Regio, part of the Deutsche Bahn AG Group)   **HP21 8LG**
Gauge 4ft 8½in   www.chilternrailways.co.uk   **SP 816134**

(01509) LESLEY		0-6-0DH	RH	468043	1963

GEMINI RAIL SERVICES UK LTD, WOLVERTON WORKS,
STRATFORD ROAD, WOLVERTON (part of the Gemini Rail Group)   **MK12 5NT**
Gauge 4ft 8½in   www.geminirailgroup.co.uk   **SP 812413**

(D3720)	09009	0-6-0DE	Dar		1959	a
	(TITCHIE)	4wDM	SMH	103GA078	1978	
(18001)	9770 0018001-7	4w-4wBE	CE	B4660.1	2021	
(18004)	9770 0018004-1	4w-4wBE	CE	B4660.4	2022	
(18005)	9770 0018005-8	4w-4wBE	CE	B4660.5	2022	
(18006)	9770 0018006-6	4w-4wBE	CE	B4660.6	2022	
(18007)	9770 0018007-4	4w-4wBE	CE	B4660.7	2022	
(18008)	9770 0018008-2	4w-4wBE	CE	B4660.8	2022	
(18009)	9770 0018009-8	4w-4wBE	CE	B4660.9	2022	
(18010)	9770 0018010-6	4w-4wBE	CE	B4660.10	2023	
(18011)	9770 0018011-4	4w-4wBE	CE	B4660.11	2023	
(18012)	9770 0018012-2	4w-4wBE	CE	B4660.12	2023	
(18013)	9770 0018013-2	4w-4wBE	CE	B4660.13	2023	
(18014)	9770 0018014-0	4w-4wBE	CE	B4660.14	2023	
(18015)	9770 0018015-5	4w-4wBE	CE	B4660.15	2023	

a   property of Railway Support Services Ltd, Wishaw, Warks

PRESERVATION SITES

AMERSHAM TOWN COUNCIL,
OAKFIELD CORNER, CHESHAM ROAD, AMERSHAM   **HP6 5HP**
Gauge 2ft 0in   www.amersham-tc.gov.uk   **SU 963985**

1	0-4-4T	IC	Chesham SG	2021

Replica locomotive

Mrs. BRITT, c/o SUKAMI, 43 WEST STREET, STEEPLE CLAYDON   **MK18 2NS**
Gauge 750mm   SP 693268

–		0-6-0WT	OC	Chrz	2959	1951	Pvd

THE FAWLEY HILL RAILWAY,
FAWLEY HILL, FAWLEY GREEN, near HENLEY-on-THAMES   **(RG9 6JA)**
Gauge 4ft 8½in   www.fawleyhill.co.uk   **Private site**   SU 755861

D2120	(03120)	0-6-0DM		Sdn		1959
–		4wDM		FH	3817	1956
	ERNIE	4wDM		FH	3894	1958
			reb	AB	6930	1988
	R436 XRA	4wDM	R/R	_(Landrover		1998
				(Perm		1998

GULLIVERS LAND LTD, (part of the Gullivers Theme Park group),
LIVINGSTON DRIVE, NEWLANDS, MILTON KEYNES   **MK15 0DT**
Gauge 1ft 3in   www.gulliversfun.co.uk   SP 872399

–		4+6wDE s/o	Gullivers	2005
	a rebuild of	0-6-0+0-6-0DE		1999

MILTON KEYNES MUSEUM,
McCONNELL DRIVE, WOLVERTON, MILTON KEYNES   **MK12 5EL**
Gauge 4ft 8½in   www.miltonkeynesmuseum.org.uk   SP 820404

1009	WOLVERTON	2-2-2	IC	Mercia	1991	a
a	non-working replica					

PENN MEADOW FARM, PAULS HILL, PENN, HIGH WYCOMBE   **HP10 8PD**
Gauge 4ft 8½in   SU 919924

21147	4w-4wRER	MetCam	1949	Dsm

QUAINTON RAILWAY SOCIETY LTD, BUCKINGHAMSHIRE RAILWAY CENTRE,
QUAINTON ROAD STATION, STATION ROAD, QUAINTON,
near AYLESBURY   **HP22 4BY**
Gauge 4ft 8½in   www.bucksrailcentre.org   SP 736189, 739190

3020	CORNWALL	2-2-2	OC	Crewe		1858
6984	OWSDEN HALL	4-6-0	OC	Sdn		1948
6989	WIGHTWICK HALL	4-6-0	OC	Sdn		1948
7200		2-8-2T	OC	Sdn		1934
(7715)	LT 99	0-6-0PT	IC	KS	4450	1930
30585		2-4-0WT	OC	BP	1414	1874
–		0-4-0F	OC	AB	1477	1916
	LAPORTE	0-4-0F	OC	AB	2243	1948

–		4wWT	G	AP	807	1872	
	SYDENHAM	4wWT	G	AP	3567	1895	
No.1	SIR THOMAS	0-6-0T	OC	HC	1334	1918	
	MILLOM	0-4-0ST	OC	HC	1742	1946	
	ARTHUR	0-6-0T	IC	QRS		2022	
	a rebuild of	0-6-0ST	IC	HE	3782	1953	
–		0-6-0ST	IC	HE	3890	1964	
–		0-4-0ST	OC	HL	3717	1928	
	COVENTRY No.1	0-6-0T	IC	NBH	24564	1939	
	BEAR	0-4-0ST	OC	P	614	1896	
		reb		AB	5997	1941	
	(MAY)	0-4-0ST	OC	P	1370	1915	
–		0-4-0T	OC	P	1900	1936	
–		0-4-0ST	OC	P	1903	1936	
	ROKEBY	0-4-0ST	OC	P	2105	1950	a
	GERVASE	0-4-0VBT	VCG	S	6807	1928	
	a rebuild of	0-4-0ST	OC	MW	1472	1900	b
LMS 7165	WILLIAM	4wVBT	VCG	S	9599	1956	
5208		2w-2-2-2w-4-4	12CGR	S	9418	1950	
	"SCOTT"	0-4-0ST	OC	WB	2469	1932	
	CHISLET	0-6-0ST	OC	YE	2498	1951	
D2298		0-6-0DM		_(RSHD	8157	1960	
				(DC	2679	1960	
249	ESL 118A/ESL 118B	4w-4+4-4wRE		BRCW		1932	
		reb		Acton		1961	
T1		4wDM		FH	2102	1937	
	"WALRUS"	0-4-0DM		FH	3271	1949	
	TARMAC	4wDM		FH	3765	1955	
WD 849	ESSO	0-4-0DM		HE	2067	1940	
GWR No.1	OSRAM	0-4-0DM		JF	20067	1933	
	REDLAND	4wDM		KS	4428	1930	
–		0-4-0DE		RH	425477	1959	
		reb		Resco		1979	
(01585)		0-6-0DH		RH	459518	1961	
1139	HILSEA	4wDM		RH	463153	1961	
M51886		2-2w-2w-2DMR		DerbyC&W		1960	
M51899	AYLESBURY COLLEGE	2-2w-2w-2DMR		DerbyC&W		1960	
53028		2w-2-2-2wRER		BRCW		1938	
54233		2w-2-2-2wRER		GRC&W		1939/40	
		reb		Acton		1941	
ARMY 9040		2w-2PMR		Wkm	6963	1955	
B55W	PWM 4316	2w 2PMR		Wkm	7519	1956	
RLC 009037		2w-2PMR		Wkm	8197	1958	Dsm
	TP 53P	2w-2PMR		Wkm	8263	1959	
–		2w-2PMR		Geismar	97/15	1997	

Gauge 3ft 6in

3405	JANICE	4-8-4	OC	NBH	27291	1953	

Gauge 2ft 0in

803		2w-2-2-2wRE		EEDK	803	1931	c

a   actually built in 1948 but plates dated as shown
b   carries plate S 6710
c   built 1931 but originally carried plates dated 1930

SIR JEREMY SULLIVAN, WOTTON LIGHT RAILWAY, near AYLESBURY

Gauge 1ft 3in   Private Site

SANDY	0-6-0T	OC	ESR	301	1996	
PAM	0-4-0DH		AK	52	1996	
POMPEY	4w-4wDH		AK	64	2001	

CAMBRIDGESHIRE

INDUSTRIAL SITES

EUROPEAN METAL RECYCLING LTD, MAYER PARRY RECYCLING,
111 FORDHAM ROAD, SNAILWELL, NEWMARKET   **CB8 7ND**
Gauge 4ft 8½in   www.emrgroup.com   **TL 638678**

ARMY 410	0-4-0DH	NBQ	27645	1958	OOU
468048	0-6-0DH	RH	468048	1963	OOU

GB RAILFREIGHT LTD (part of the Hector Rail Group),
PETERBOROUGH DEPOT, MASKEW AVENUE, PETERBOROUGH   **PE1 2AS**
Gauge 4ft 8½in   www.gbrailfreight.com   **TF 180003**

(D3799)	08632	0-6-0DE	Derby	1959	a

   a   property of Railway Support Services Ltd, Wishaw, Warwickshire

GB RAILFREIGHT LTD (part of the Hector Rail Group),
WHITEMOOR YARD, off HUNDRED ROAD, MARCH   **PE15 8QP**
Gauge 4ft 8½in   www.gbrailfreight.com   **TL 415984**

(D3747)	08580	0-6-0DE	Crewe	1959	a
(D3870)	08703 STEVE BLICK (CONCRETE BOB) SHUNTERSPOT				
		0-6-0DE	Hor	1960	a

   a   property of Railway Support Services Ltd, Wishaw, Warwickshire

PRESERVATION SITES

BARNOLD SUPPLIES & SERVICES LTD, CARRIAGES OF CAMBRIDGE,
CAPABILITY BARNS, HUNTINGDON ROAD, FEN DRAYTON   **CB24 4SD**
Gauge 4ft 8½in   www.carriagesofcambridge.co.uk   **TL 327678**

(3051 Car No.89 289)	4w-4wRER	MetCam	289	1932	

BRAMLEY LINE HERITAGE TRUST, BRAMLEY LINE, WALDERSEA DEPOT, LONG DROVE, FRIDAY BRIDGE, WISBECH

Gauge 4ft 8½in   www.bramleyline.org.uk

(PE14 0NP)
TF 433040

–		0-4-0DE	RH	312989	1952

IMPERIAL WAR MUSEUM, DUXFORD AERODROME

Gauge 2ft 0in   www.iwm.org.uk

CB22 4QR
TL 456456

–		4wPM	MR	1364	1918
–		4wPM	MR	3849	1927

NENE VALLEY RAILWAY LTD

Locomotives are kept at :-

Wansford Depot and Yard	PE8 6LR	TL 091978
Wansford Tunnel		TL 087976
Ferry Meadows Station	PE2 5UU	TL 151970

Gauge 4ft 8½in   www.nvr.org.uk

34081	92 SQUADRON	4-6-2	3C	Bton		1948	
73050	CITY OF PETERBOROUGH	4-6-0	OC	Derby		1954	
	TOBY	0-4-0VBT	OC	Cockerill	1626	1890	
Nr.656		0-6-0T	OC	Frichs	360	1949	
	THE BLUE CIRCLE	2-2-0WT	G	AP	9449	1926	
	DEREK CROUCH	0-6-0ST	IC	HC	1539	1924	
1	THOMAS	0-6-0T	OC	HC	1800	1947	
	JACKS GREEN	0-6-0ST	IC	HE	1953	1939	
75006		0-6-0ST	IC	HE	2855	1943	
1178		2-6-2T	OC	Motala	516	1914	
S.V.J.B	101A	4-6-0	OC	Nohab	2082	1944	
	THE MEG	0-8-0T	OC	Chrz	5485	1961	a
(D53)	45041 ROYAL TANK REGIMENT	1Co-Co1DE		Crewe		1962	
(D)9529	01577 (14029)	0-6-0DH		Sdn		1965	
801		Bo-BoDE		Alco	77120	1950	b
–		0-4-0DH		EEV	D1123	1966	
–		4wPM		FH	2895	1944	
–		4wDM		FH	2896	1944	OOU
323.674-2	SIMONSIDE / SPLUTTER	4wDH		Gmd	4991	1957	b
–		0-4-0DM		RH	304469	1951	
	BARABEL	0-4-0DH		RR	10202	1964	
DL83		0-6-0DH		RR	10271	1967	
55651	143602	4wDMR		_(AB	701	1985	
				(Alex	1784/3	1985	
	rebuilt as	4wDHR		RFSD		1988	
55668	143602	4wDMR		_(AB	672	1986	
				(Alex	1784/4	1986	
	rebuilt as	4wDHR		RFSD		1988	
DR 98500		4wDHR		Plasser	52788	1985	
1212	HELGA	4w-4DMR		EK		1958	b
–		2w-2PMR		Bance	023	1995	
MPP 10188		2w-2BER		Bance	098	2000	b
(960243)	(749) NVR 1612	2w-2PMR		Wkm	(1642	1934?)	c

a   carries plate Chrz 4939/1957
b   property of Northumbria Rail Ltd, Bedlington, Northumberland
c   believed to be incorrect identity

The Night Mail Museum, Ferry Meadows, Nene Park, Peterborough   **PE2 5UU**
Gauge 2ft 0in   www.irps-wl.org.uk   **TL 151970**

807	[211	212]	2w-2-2-2wRE	EEDK	807	1931

S. OWEN, OLD NORTH ROAD STATION, LONGSTOWE, near BOURN   **CB23 2TZ**
Gauge 4ft 8½in   **Private Site**   **TL 313546**

NL 1305	2-2w-2w-2RER	MetCam	1961

C.J. & A.M. PEARMAN, HUNTINGDON AREA
Gauge 1ft 11½in   **Private Site**

–	4wPM	MR	2059	1920

   locomotive stored at a private location

RAILWORLD WILDLIFE HAVEN, (MUSEUM of WORLD RAILWAYS),
OUNDLE ROAD, WOODSTON, PETERBOROUGH   **PE2 9NR**
Gauge 4ft 8½in   www.railworld.org.uk   **TL 188982**

–		4-6-2	4C	Frichs	415	1950
43045	(253022) THE GRAMMAR SCHOOL DONCASTER AD1350					
		Bo-BoDE		Crewe		1977
43060	(254003) COUNTY OF LEICESTERSHIRE					
		Bo-BoDE		Crewe		1977

SHEPRETH WILDLIFE PARK LTD,
THE SIDINGS, STATION ROAD, SHEPRETH, near ROYSTON   **SG8 6PZ**
Gauge 400mm   www.sheprethwildlifepark.co.uk   **TL 393481**

–	4-6wRE	s/o	UK Loco	2007

S. THOMASON, SPRINGFIELD AGRICULTURAL RAILWAY, HUNTINGDON
Gauge 2ft 0in   www.ingr.co.uk   **Private Site**

SP 203	4wBE	CE	B0176A	1974
–	4wBE	WR	3557	1946
12	4wDM	Moës		
112	4wDM	Spoor	112	1952

Gauge 600mm

2	4wDM	Diema	1553	1953

CHESHIRE

INDUSTRIAL SITES

ALSTOM TRANSPORT (part of Alstom Holdings S.A.)
CREWE WORKS, WEST STREET, CREWE   **CW1 3JB**
Gauge 4ft 8½in   www.alstom.com   **SJ 691561**

–		4wDH	R/R	NNM	77501	1980	OOU
1		4wDH	R/R	NNM	82503	1983	a
T2		4wDH	R/R	NNM	83501	1983	a
T5		4wDH	R/R	NNM	83504	1984	a
–		4wDH	R/R	Zephir	1470	1997	b

a   one of these locomotives has been sold to E.G. Steele, Hamilton; one locomotive has been
     reported as scrapped. and the remaining locomotive is OOU
b   property of Depot Rail Ltd, Gainsborough, Lincolnshire

ALSTOM TRANSPORT TECHNOLOGY CENTRE (part of Alstom Holdings S.A.),
LOVEL'S WAY, HALEBANK, WIDNES   **WA8 8WQ**
Gauge 4ft 8½in   www.alstom.com   **SJ 484848**

(D3569)	08454	0-6-0DE		Derby		1958
(D3889)	08721	0-6-0DE		Crewe		1960
–		4wBE	R/R	Zephir	2990	2021

DB CARGO UK LTD, CREWE INTERNATIONAL ELECTRIC MAINTENANCE DEPOT,
VICTORIA AVENUE, CREWE   **CW2 7RL**
Gauge 4ft 8½in   www.uk.dbcargo.com   **SJ 694556**

–	4wBE	R/R	Zephir	2563	2015

FREIGHTLINER GROUP LTD, CREWE BASFORD HALL TRAIN DEPOT
and YARDS, off GRESTY ROAD, CREWE   **CW2 5AA**
(part of the America Genesee & Wyoming Railway Co)
Gauge 4ft 8½in   www.freightliner.co.uk   **SJ 713535**

(18006)	9770 0018006-6	4w-4wBE	CE	B4660.6	2022

LOCOMOTIVE STORAGE LTD,
CREWE DIESEL DEPOT, off NANTWICH ROAD, CREWE   **CW2 6GT**
Gauge 4ft 8½in   www.iconsofsteam.com   **SJ 711542**

5029	NUNNEY CASTLE	4-6-0	4C	Sdn	1936
6024	KING EDWARD I	4-6-0	4C	Sdn	1930
34046	BRAUNTON	4-6-2	3C	Bton	1946
			reb	Elh	1959
(35022)	(HOLLAND-AMERICA LINE)	4-6-2	3C	Elh	1948

			reb	Elh		1956	a	Dsm
(35027)	(PORT LINE)	4-6-2	3C	Elh		1948		
			reb	Elh		1957	a	
45231	THE SHERWOOD FORESTER	4-6-0	OC	AW	1286	1936		
(46100)	6100 ROYAL SCOT	4-6-0	3C	Derby		1930		
			reb	Crewe		1950		
60007	(4498) SIR NIGEL GRESLEY	4-6-2	3C	Don	1863	1937		
60532	BLUE PETER	4-6-2	3C	Don	2023	1948	b	
61306	(1306) MAYFLOWER	4-6-0	OC	NBQ	26207	1948		
70000	BRITANNIA	4-6-2	OC	Crewe		1951		
D213	(40013) ANDANIA	1Co-Co1DE		_(EE	2669	1959		
				(VF	D430	1959		
(D1726	47134 47622) 47841	Co-CoDE		BT	497	1964		
(D1948	47505) 47712 LADY DIANA SPENCER							
		Co-CoDE		BT	610	1966		
(D3598)	08483 BUNGLE	0-6-0DE		Hor		1958		
(D3798)	08631	0-6-0DE		Derby		1959		
D3948	(08780) ZIPPY	0-6-0DE		Derby		1960		
(D6905	37205) 37688	Co-CoDE		_(EE	3383	1963		
	GREAT ROCKS			(EEV	D849	1963		
(D6968	37268) 37401	Co-CoDE		_(EE	3528	1965		
	MARY QUEEN OF SCOTS			(EEV	D957	1965		
(D6974	37274) 37402	Co-CoDE		_(EE	3534	1965		
				(EEV	D963	1965		
(D8118)	20118	Bo-BoDE		_(EE	3024	1962		
	SALTBURN-BY-THE-SEA			(RSHD	8276	1962		
(D8132)	20132	Bo-BoDE		_(EE	3603	1966		
				(EEV	D1002	1966		
D9000	(55022)	Co-CoDE		_(EE	2905	1960		
	ROYAL SCOTS GREY			(VF	D557	1960		
55022	(960 014 977873) FLORA	2-2w-2w-2DMR		PSteel		1960		
(E6001)	73001	Bo-BoDE/RE		Elh		1962	c	
55544	142003	2w-2DMR		_(BRE(D)		1985		
				(Leyland R5.10		1985		
	rebuilt as	2w-2DHR		HAB		1990		
55594	142003	2-2wDMR		_(BRE(D)		1985		
				(Leyland R5.09		1985		
	rebuilt as	2-2wDHR		HAB		1990		
99288	3050 (288 3051) BERYL	4w-4wRER		MetCam		1932		
99291	3050 (291 3052) MABEL	4w-4wRER		MetCam		1932		

a currently off site for restoration
b currently at the Severn Valley Railway, Bridgnorth, Shropshire
c currently at Ecclesbourne Valley Railway, Wirksworth, Derbyshire

LONDON & NORTH WESTERN RAILWAY COMPANY LTD, t/a ARRIVA TRAINCARE, CREWE CARRIAGE WORKS, off WESTON ROAD, CREWE   CW1 6NE

Mainline and preserved vehicles usually present.

Gauge 4ft 8½in   www.arrivatc.com   SJ 715538

(D3884	08717) 09204	0-6-0DE	Crewe	1960
(D4036)	08868	0-6-0DE	Dar	1960

AVANTI WEST COAST, THE TALENT ACADEMY, TATTON HOUSE,
WESTMERE DRIVE, CREWE BUSINESS PARK, CREWE   **CW1 6ZD**
Gauge 4ft 8½in   www.avantiwestcoast.co.uk   **SJ 719548**

| 390033 | 69133 | | 4w-4wWER | Alstom | | 2002 | |

PRESERVATION SITES

CREWE HERITAGE TRUST LTD,
CREWE HERITAGE CENTRE, VERNON WAY, CREWE   **CW1 2DB**
Gauge 4ft 8½in   www.crewehc.org   **ST 708553**

2013	PRINCE GEORGE	4-4-0	IC	Keyte-Smith		2023	u/c
(D2073)	03073	0-6-0DM		Don		1959	
(D6808)	37108	Co-CoDE		_(EE	3237	1962	
				(VF	D762	1962	
W43018	253009	Bo-BoDE		Crewe		1976	
43081		Bo-BoDE		Crewe		1978	
M49002	THE CREWE HERITAGE CENTRE 1987 - 2017						
		Bo-BoWE		Derby		1977	
M49006		Bo-BoWE		Derby		1977	
87035	ROBERT BURNS	Bo-BoWE		Crewe		1974	
90050	(90150)	Bo-BoWE		BRE(C)		1990	
91120	(91020)	Bo-BoWE		BRE(C)		1990	
–		4wDH		S	10007	1959	

DUKE OF WESTMINSTER, EATON HALL RAILWAY,
EATON HALL, ECCLESTON, near CHESTER   **Private Site**
Private location with no public access, except on advertised open days
Gauge 1ft 3in   www.grosvenor.com   **SJ 386612**

KATIE	0-4-0T	OC	_(Thursley		1994	
			(FMB		1994	
SIBELL	4wDH		HE	9331	1994	

GULLIVERS WORLD LTD, (part of the Gullivers Theme Park group),
GULLY'S EXPRESS RAILROAD, SHACKLETON CLOSE, WARRINGTON **WA5 9YZ**
Gauge 1ft 3in   www.gulliversfun.co.uk   **SJ 590900**

NEVILLE	0-6-0+6wDE	s/o	Meridian		1989

HIGH LEGH MINIATURE RAILWAY, HIGH LEGH GARDEN CENTRE,
HALLIWELL'S BROW, HIGH LEGH, KNUTSFORD   **WA16 0QW**
Gauge 1ft 3in   www.highleghminiaturerailway.co.uk   **SJ 700836**

"BLACK SMOKE"	2-4-2PM	s/o	SmithEL	c1956	Pvd

PAUL WALLEY, c/o CAMBI UK LTD,
CONGLETON TECHNOLOGY PARK, RADNOR PARK INDUSTRIAL ESTATE,
FIRST AVENUE, BACK LANE, CONGLETON   **CW12 4XJ**

Gauge **4ft 8½in**   **Private Site**   **SJ 846637**

2859		2-8-0	OC	Sdn	2765	1918	
	"IVOR"	0-4-0ST	OC	P	1555	1920	

Gauge **2ft 0in**

3		0-8-0T	OC	Hen	14928	1917	
3010		0-6-0T	OC	KS	3010	1916	

Locomotives stored off site at an undisclosed private location

CORNWALL

INDUSTRIAL SITES

DRILLSERVE LTD, PLANT YARD, ROSCROGGAN MILL, near CAMBORNE TR14 0BA
Locomotives for resale are occasionally present.   www.drillserve.com   **SW 648418**

FIRST GREATER WESTERN LTD t/a GREAT WESTERN RAILWAY,
LONG ROCK TRAIN DEPOT, LONGROCK, PENZANCE   **TR20 8HX**
(part of the First Group plc)
Gauge **4ft 8½in**   www.gwr.com   **SW 492312**

(D3812)	08645	ST. PIRIAN	0-6-0DE		Hor	1959	**M**

IMERYS MINERALS LTD, IMERYS CLAY, ROCKS WORKS, near BUGLE   **PL26 8PJ**
Gauge **4ft 8½in**   www.imerys.com   **SX 024587**

(D3920)	08752	0-6-0DE		Crewe	1960	a

    a   property of Railway Support Services Ltd, Wishaw, Warwickshire

MPOWER KERNOW CIC, ST. BLAZEY ROUNDHOUSE,
THE RAILWAY YARD, BLAZEY ROAD, ST. BLAZEY, PAR   **PL24 2HY**
Gauge **4ft 8½in**   www.mpowerkernow.org   **SX 073537**

	PETER	0-4-0DM	JF	22928	1940	a
	"BRIAN"	4wDM	RH	443642	1960	

    a   carries works plate JF 4000001/1945

SUMMERCOURT SCRAPYARD LTD, SCRAP METAL MERCHANTS & VEHICLE
DISMANTLERS, TREFULLOCK, SUMMERCOURT, NEWQUAY   **TR8 5BY**
Gauge **5ft 0in**   **SW 897563**

1103		2-8-0	OC	LO	141	1943	Pvd

T. WARE & SONS,
HEATHER BANK, UNITED ROAD, CARHARRACK, near REDRUTH
Gauge 4ft 8½in   www.twareandsons.co.uk

TR16 5HT
SW 741417

TO 9362	4wDM	RH	349041	1953	Dsm

WHEAL JANE GROUP, WHEAL JANE EARTH SCIENCE PARK,
WHEAL JANE, BALDHU, near TRURO
Gauge 1ft 10in   www.wheal-jane.co.uk

TR3 6EH
SW 777422

–	4wBE	CE			Pvd

PRESERVATION SITES

BODMIN & WENFORD RAILWAY plc
Locomotives are kept at :-

	Bodmin General Station	PL31 1AQ	SX 074664
	Bodmin	PL30 4BB	SX 110640

Gauge 4ft 8½in   www.bodminrailway.co.uk

4612		0-6-0PT	IC	Sdn			1942
5552		2-6-2T	OC	Sdn			1928
5553		2-6-2T	OC	Sdn			1928
6412		0-6-0PT	IC	Sdn			1934
6435		0-6-0PT	IC	Sdn			1937
30120	(120)	4-4-0	IC	9E			1899
30587	(3298)	2-4-0WT	OC	BP		1412	1874
	JUDY	0-4-0ST	OC	WB		2572	1937
75178		0-6-0ST	IC	WB		2766	1944
No.19		0-4-0ST	OC	WB		2962	1950
	ALFRED	0-4-0ST	OC	WB		3058	1953
(D442)	50042 TRIUMPH	Co-CoDE		_(EE		3812	1968
				(EEV		D1183	1968
(D1787)	47306	Co-CoDE		BT		549	1964
D3452		0-6-0DE		Dar			1957
D3489	COLONEL TOMLINE	0-6-0DE		Dar			1958
(D3559)	08444	0-6-0DE		Derby			1958
(D6527)	33110	Bo-BoDE		BRCW	DEL119		1960
(D6842)	37142	Co-CoDE		_(EE		3317	1963
				(EEV		D816	1963
121020	W55020	2-2w-2w-2DMR		PSteel			1960

CORNWALL COUNTY COUNCIL, GEEVOR TIN MINES MUSEUM,
PENDEEN, near ST.JUST
Gauge 1ft 6in   www.geevor.com

TR19 7EW
SW 375346

–	4wBE	CE	B1501	1977	
–	4wBE	CE	B1592A	1978	
–	4wBE	CE	B1592B	1978	

13		4wBE	CE	B1851A	1978	
B3606A		4wBE	CE	B3606A	1989	
–		4wBE	CE	B3606B	1989	
–		4wBE	CE	B3606C	1989	
59		4wBE	CE	B3606D	1989	
1		0-4-0BE	Geevor			
2	80	0-4-0BE	Geevor			
4		0-4-0BE	Geevor			Pvd
11		0-4-0BE	Geevor			
13		0-4-0BE	Geevor			
15		0-4-0BE	Geevor			
19		0-4-0BE	Geevor			
6		0-4-0BE	Geevor			
16		0-4-0BE	Geevor			
–		4wBE	WR			
3		4wBE	WR			

The Geevor-built locomotives are based on and use parts of, WR or CE locomotives.

Two of the un-numbered CE locomotives are numbered 2 and 8, but which works numbers these pair up with is unknown.

CORNWALL at WAR MUSEUM, NOTTLES PARK, DAVIDSTOW, near CAMELFORD

PL32 9YF

Gauge 2ft 0in   www.cornwallatwarmuseum.co.uk   **SX 139862**

–	4wDM	MR	8882	1944	

THE DELABOLE SLATE COMPANY LTD, PENGELLY, DELABOLE, CAMELFORD

PL33 9AZ

Gauge 1ft 11in   www.delaboleslate.co.uk   **SX 074836**

No.2	4wDM	MR	3739	1925	

GREAT WESTERN RAILWAY YARD, near ST. AGNES

TR5 0PD

Gauge 4ft 8½in   **SW 721492**

YARD No.5200	4wDM	FH	3776	1956	Pvd

HELSTON RAILWAY PRESERVATION COMPANY LTD, HELSTON RAILWAY, PROSPIDNICK HALT, SITHNEY

TR13 0RY

Gauge 4ft 8½in   www.helstonrailway.co.uk   **SW 647305, 645309**

–		0-6-0ST	IC	P	2000	1942
2	WILLIAM MURDOCH	0-4-0ST	OC	P	2100	1949
	KINGSWOOD	0-4-0DM		AB	446	1959
97649		0-4-0DM		RH	327974	1954
–		0-4-0DM		RH	395305	1956
(W50413)	W56168	2-2w-2w-2DMR		PRoyal	B38848	1958
(M51616)		2-2w-2w-2DHR		DerbyC&W		1959

M51622		2-2w-2w-2DHR		DerbyC&W		1959	a
DB 965079	68/016	2w-2PMR		Wkm	7594	1957	

a   currently at Vincent Engineering, Henstridge, Somerset

KING EDWARD MINE LTD,
KING EDWARD MINE MUSEUM, TROON, near CAMBORNE   TR14 9HW
Gauge 1ft 10in   www.kingedwardmine.co.uk   SW 664389

60	4wBE		SouthCrofty		1992

LAPPA VALLEY STEAM RAILWAY & COUNTRY LEISURE PARK,
BENNY HALT, ST.NEWLYN EAST, near NEWQUAY   TR8 5LX
Gauge 1ft 3in   www.lappavalley.co.uk   SW 838574, 839564

No.2	MUFFIN	0-6-0	OC	Berwyn		1967	
			reb	Tambling		1991	
	RUBY	0-4-2T	OC	ESR	302	1997	
	ELLIE	0-4-2T	OC	ESR	331	2006	
No.1	ZEBEDEE	0-6-2T	OC	SL	7434	1974	
	rebuilt as	0-6-4T	OC	Tambling		1990	
	CITY OF TRURO	4w-4wDH		AK	113	2023	
D5905	CITY OF DERBY	4w-4wDM		BrownJ		1995	
4	ARTHUR	4wDM		L	20698	1942	
			reb	MalyanG		2011	
3	GLADIATOR	4w-4DH		Minirail		c1960	OOU

LAUNCESTON STEAM RAILWAY, ST. THOMAS ROAD, LAUNCESTON   PL15 8DA
Gauge 600mm   www.launcestonsr.co.uk   SX 328850

	LILIAN	0-4-0STT	OC	HE	317	1883	
	VELINHELI	0-4-0ST	OC	HE	409	1886	a
	COVERTCOAT	0-4-0STT	OC	HE	679	1898	
	DOROTHEA	0-4-0ST	OC	HE	763	1901	
89		2w-2VBT	VC	ParmenterC		2004	
38		2w-2-2-2wRE		EEDK	761	1930	
42		2w-2-2-2wRE		EEDK	806	1930	Dsm
–		4wDM		FH	1896	1935	Dsm
–		4wDM		MR	5646	1933	
	"ELECTRIC DILLY"	2w-2BER		Bowman		c1986	
	rebuilt as	4wBE		Bowman			
–		4w-4DER		Bowman		2003	
2	"THE DILLY"	4wPER		Bowman		2003	
	rebuilt as	2w-2DER		Bowman		2004	
–		4wPE		ShooterA		2014	

a   currently at Ffestiniog Railway Co, Boston Lodge, Gwynedd

MOSELEY N.G. INDUSTRIAL TRAMWAY & MUSEUM, MOSELEY HERITAGE MUSEUM, TUMBLY DOWN FARM, TOLGUS MOUNT, REDRUTH

Visitors by appointment only or on advertised open days.

Gauge 2ft 0in SW 686428

	THE LADY D	4wDM		MR	8934	1944
	(SMELTER)	4wDM	s/o	RH	229647	1943
15		4wBE		CE	B3132A	1984
	"BINCLEAVES"	4wBE		GB	2345	1951
	CATHODE	4wBE		GB	2960	1959
	ANODE	4wBE		GB	420172	1969
	"LITTLE GEORGE"	0-4-0BE		WR	1298	1938
20	DIODE	0-4-0BE		WR	L1021	1983

Gauge 1ft 11in

–		4wBE		?		
			reb	SouthCrofty		

Gauge 1ft 10in

3		4wBE		CE	5739	1970
–		4wBE		(CE	B2944D 1982)?	
–		4wBE		(CE	B2930B 1981)?	
–		4wBE		CE		Dsm
1	MACCA	4wBE		CE		
			reb	SouthCrofty		
6		4wBE		CE		
			reb	SouthCrofty		
(66)	LEWIS	4wBE		SouthCrofty	1996	
–		4wBE		(SouthCrofty?)		

Gauge 1ft 6in

–		0-4-0BE		WR
15		4wBE		(Geevor?)

Gauge 1ft 4in

–		4wWE		Saxton	1988

PARADISE RAILWAY - JUNGLE EXPRESS, PARADISE PARK, 16 TRELISSICK ROAD, HAYLE

TR27 4HB

Gauge 1ft 3in www.paradisepark.org.uk SW 555365

No.3	ZEBEDEE	4wDM		L	10180	1938

POLDARK MINE, TRENEAR, WENDRON, HELSTON

TR13 0ER

Gauge 4ft 8½in www.poldarkmine.org.uk SW 682315

–	0-4-0ST	OC	P	1530	1919

Gauge 900mm

–	0-4-0WT	OC	OK	5102	1912

ROSEVALE MINING HISTORICAL SOCIETY, ZENNOR

Gauge 2ft 0in   www.rosevalemine.com   SW 458380

		0-4-0BE		WR		
–			reb	SouthCrofty		

JOHN SPENCELEY, TREVAYLOR FARM
Locomotives stored at private location with no public access.

Gauge 2ft 0in   Private Site

		0-4-0VBT	VC	Roanoke		2004
–	(LOD/758221)	4wDM		MR	8886	1944

ST. AUSTELL CHINA CLAY MUSEUM LTD, WHEAL MARTYN MUSEUM & COUNTRY PARK, WHEAL MARTYN, CARTHEW, near ST. AUSTELL   PL26 8XG

Gauge 4ft 6in   www.wheal-martyn.com   SX 004555

	LEE MOOR No.1	0-4-0ST	OC	P	783	1899

Gauge 2ft 6in

		4wDM		RH	244558	1946
–						

BRENDON SWEET, GREAT POLGOOTH TIN MINE, POLGOOTH

Gauge 1ft 11½in   Private Site

		4wPM		Sweet
–				

CUMBRIA

INDUSTRIAL SITES

AYLE COLLIERY CO LTD, QUARRY DRIFT, ALSTON
The site of the colliery is actually located in Northumberland, though the original
Ayle East Drift and related workings were located in Cumbria. See Northumberland entry for details.

CUMBERLAND COUNCIL, THE PORT OF WORKINGTON, PRINCE OF WALES DOCK, WORKINGTON   CA14 2JH

Gauge 4ft 8½in   www.portofworkington.co.uk   NX 993294

		0-6-0DH		HE	8976	1979
–				HE	8976	1979
			reb	HE	9306	1992
			reb	HE	6706	2000
–		0-6-0DH		HE	8977	1979
			reb	HE	9307	1992
			reb	HE	6707	2000

DIRECT RAIL SERVICES LTD,
KINGMOOR DEPOT, ETTERBY ROAD, ETTERBY, CARLISLE   CA3 9NZ
Gauge 4ft 8½in   www.directrailservices.com   NY 385575

Locomotives from British Nuclear Fuels plc, Sellafield occasionally present for repair/overhaul.

EGREMONT MINING CO LTD, FLORENCE IRON ORE MINE, EGREMONT   CA22 2NR
Gauge 2ft 6in (Closed) www.florenceartscentre.com   NY 018103

7		4wBE		WR	6218	1961	OOU

HONISTER SLATE MINE LTD,
HONISTER SLATE QUARRY, HONISTER PASS, BORROWDALE   CA12 5XN
Gauge 2ft 0in   www.honister.com   NY 225136

–		0-4-0DM	s/o	Bg	3236	1947	Pvd
–		4wBE		PWR	AO.296V.02	1992	a
–		4wBE		PWR	AO.296V.03	1992	

a   currently stored off site

LAND RECOVERY LTD,
BLACKDYKE ROAD, KINGSTOWN, CARLISLE   CA3 0PJ
Road-rail vehicles are present in this yard between use on contracts.
Gauge 4ft 8½in   www.landrecovery.co.uk   NY 389594

W 031	99709 977011-4	4wDM	R/R	_(Unimog160572	1990
				(Zweiweg 1291	1990
W 020	99709 977012-2	4wDM	R/R	_(Unimog166228	c1991
				(Rail-Ability	c1991
W 061	99709 970010-6	4wDM	R/R	_(Unimog195311	2000
				(Zweiweg 1850	2000

LOCOMOTIVE SERVICES GROUP,
UPPERBY DEPOT, off ST. NICHOLAS BRIDGES, CARLISLE   (CA2 4AA)
Gauge 4ft 8½in   Private site   NY 410544

preserved steam and diesel locomotives usually present

MINISTRY OF DEFENCE, DEFENCE MUNITIONS,
LONGTOWN DEPOT, near CARLISLE
See Section 6 for details.

NUCLEAR DECOMMISSIONING AGENCY,
SELLAFIELD SITE, SELLAFIELD, near SEASCALE   CA20 1PG
(operated by Sellafield Ltd, a subsidiary of Nuclear Management Partners Ltd)
Gauge 4ft 8½in   NY 025034

1		4w-4wBE		CE	B4659/1	2021

2		4w-4wBE		CE	B4659/2	2021	
3		4wBE		CE	B4668	2024	
B.N.F.L.1	PB 1056.860	0-4-0DH		HE	7426	1982	
			reb	HAB	6480	1997	OOU
3		0-4-0DH		HE	7406	1977	
			reb	HE	9200	1983	
			reb	HAB	6478	1997	OOU
4		0-4-0DH		HE	7427	1982	
			reb	HE	9384	2014	
No.6	H0344	4wDH	R/R	_(Minilok	158	1991	
				(YEC	L105	1992	

"Pile No.2", Building B12, Windscale, Sellafield Site, Sellafield

Gauge 2ft 5½in   NY 030043

PV 7	2w-2BE	WR	

F.G.SHEPHERD, FLOW EDGE SCRAP YARD, MIDDLE FELL, ALSTON

Locomotives stored at unknown location

Gauge 2ft 0in   NY 734442

3	0-4-0BE	WR
9	0-4-0BE	WR

STORY RAIL LTD, BURGH ROAD INDUSTRIAL ESTATE, CARLISLE   CA2 7NA

Road-rail vehicles are present in this yard between use on contracts.

Gauge 4ft 8½in   www.storycontracting.com/rail   NY 377562

SR 924	99709 977013-0	4wDM	R/R	_(Unimog130580	1986	
				(Zweiweg 1143	1986	

TATA STEEL EUROPE, SHAPFELL LIMESTONE QUARRIES, SHAP, PENRITH

(part of the Tata Group)   CA10 3QG

Gauge 4ft 8½in   www.tatasteeleurope.com   NY 571134

272	GROSMONT	6wDE	GECT	5470	1978
278		6wDE	GECT	5468	1977
(No.403)		0-6-0DH	HE	7543	1978
(404)		0-6-0DH	HE	8978	1979

PRESERVATION SITES

BASSENTHWAITE LAKE RAILWAY STATION, BASSENTHWAITE LAKE, near COCKERMOUTH

CA13 9YL

Gauge 4ft 8½in   www.basslakestation.co.uk   NY 198309

241-010		4-8-2	OC	Longcross	c2016	+

     +   replica locomotive

BLENNERHASSET WATERMILL, near ASPATRIA
Private site
Gauge 2ft 0in
NY 184419

–		4wDM		FH	3756	1955
–		4wDM		RH	273525	1949
–		0-4-0BE		WR	(4149	1949?)
			reb	ShepherdFG		
NVR No.1		4wBE	s/o	ShepherdFG		c1973
			reb	‡		1988
41		4wBER		HallamP		

‡   rebuilt by S.Frogley, Nent Valley Rly

Gauge 1ft 3in

–		4wPH		CurleA		c1985	Dsm

EDEN VALLEY RAILWAY TRUST, WARCOP STATION, WARCOP
CA16 6PR
Gauge 4ft 8½in   www.evr-cumbria.org.uk
NY 753156

(D1654 47070 47620 47835) 47799	Co-CoDE		Crewe		1965
(D6742) 37042	Co-CoDE		_(EE	3034	1962
			(VF	D696	1962
ND 3815	0-4-0DM		HE	2389	1941
21	0-4-0DH		JF	4220045	1967
DARLINGTON	0-6-0DH		_(RSHD	8343	1962
			(WB		1962
2	0-6-0DH		S	10111	1963
		reb	TH		1987
		reb	HAB	6479	1997
(ARMY 244 88 EQ)	0-4-0DH		TH	130c	1963
	a rebuild of	0-4-0DM	JF	22971	1942
(227)	0-4-0DM		_(VF	5262	1945
			(DC	2181	1945
205009 60108	4-4wDER		Afd/Elh		1957
2315 61798	4w-4wRER		Afd/Elh		1961
2315 61799	4w-4wRER		Afd/Elh		1961
2311 61804	4w-4wRER		Afd/Elh		1961
2311 61805	4w-4wRER		Afd/Elh		1961
9003 (S68003 931093)	4w-4RE/BER		Afd/Elh		1960
(9005) S68005 931095	4w-4RE/BER		Afd/Elh		1960
9010 68010 (931090)	4w-4RE/BER		Afd/Elh		1961
(68045)	2w-2PMR		Wkm	730	1932	DsmT
DE 320468 (950042 753)	2w-2PMR		Wkm	1724	1934

JOHN FOWLER & CO (LEEDS) LTD,
OLD HALL FARM, WEAR BRIDGE, BOUTH, ULVERSTON
LA12 8JA
Gauge 4ft 8½in   www.oldhallfarmbouth.com   **Private Site**
SD 325855

"ASKHAM HALL"		0-4-0ST	OC	AE	1772	1917

Gauge 3ft 0in

No.14 THORNHILL		2-4-0T	OC	BP	2028	1880
1 DROMAD		0-4-2ST	OC	KS	3024	1916
	rebuilt as	0-4-2T	OC	AK		1993

Gauge 2ft 0in

ANNIE		0-4-2T	OC	BoothR		1997

Gauge 1ft 3in

24		2-6-2	OC	Fairbourne	No.4	1990	
–		4-6-2	OC	KraussM	8473	1929	Dsm
–		4wBE		GB	2782	1957	
–	(QUARRYMAN)	4wPM		MH	2	1926	

Railway vehicles under restoration usually present.

LAKESIDE & HAVERTHWAITE RAILWAY CO LTD, HAVERTHWAITE   LA12 8AL
Gauge 4ft 8½in   www.lakesiderailway.co.uk   SD 349843

42073		2-6-4T	OC	Bton		1950
42085		2-6-4T	OC	Bton		1951
46441		2-6-0	OC	Crewe		1950
	CARRON No.14	0-6-0T	OC	AB	1245	1911
1	DAVID	0-4-0ST	OC	AB	2333	1953
	"REPULSE"	0-6-0ST	IC	HE	3698	1950
	PRINCESS	0-6-0ST	OC	WB	2682	1942
	VICTOR	0-6-0ST	OC	WB	2996	1951
(D2072	03072)	0-6-0DM		Don		1959
(D2117)		0-6-0DM		Sdn		1959
(D8314)	20214 AUSTIN MAHER - CHAIRMAN LAKESIDE & HAVERTHWAITE RAILWAY 1970-2006					
		Bo-BoDE		_(EE	3695	1967
				(EEV	D1090	1967
(7120)	AD601	0-6-0DE		Derby		1945
	"RACHEL"	4wPM		MR	2098	1924
(M52071)		2-2w-2w-2DMR		BRCW		1962
(M52077)		2-2w-2w-2DMR		BRCW		1961

NATIONAL TRUST, FORCE CRAG MINE, near BRAITHWAITE
Gauge 1ft 10in   www.nationaltrust.org.uk   NY 199216

–	4wBE	WR	2489	1943

NATURAL ENGLAND, UNIT 2, KIRKBRIDE AIRFIELD, KIRKBRIDE, WIGTON
Gauge 2ft 0in   NT 222551

–	4wDM	MR	9231	1947

H. POTTS, APPLEBY
Gauge 2ft 0in   Private Site

5	4wBE	WR

RAVENGLASS & ESKDALE RAILWAY CO LTD, RAVENGLASS   CA18 1SW
Gauge 1ft 3¼in   www.ravenglass-railway.co.uk   SD 086967, NY 137000

No.9	RIVER MITE	2-8-2	OC	Clarkson	4669	1966

No.	Name	Wheel	Cyl	Builder	No.	Date	Notes
No.3	RIVER IRT	0-8-2	OC	DB	3	1894	
			reb	Ravenglass		1927	
	RIVER ESK	2-8-2	OC	DP	21104	1923	
	WHILLAN BECK	4-6-2	OC	KraussM	8457	1929	
No.10	NORTHERN ROCK	2-6-2	OC	Ravenglass	10	1976	
I.C.L.9	CYRIL	4wDM		L	4404	1932	
			reb	Ravenglass		1986	
–		4wDM		L	40009	1954	a DsmT
	LES	4wDM		LB	51721	1960	
	PERKINS	4w-4DM		Ravenglass		1933	
	a rebuild of	4wPM		MH	NG39A	1929	
	LADY WAKEFIELD	4w-4wDH		Ravenglass		1980	
	ANITA	4wDM		RH	277273	1949	b
I.C.L. No.11	DOUGLAS FERRIERA	4w-4wDH		TMA	28800	2005	
136	(126 RC1)	4w-4wDHR		Crow		1975	c Dsm
137	(RC 2)	4w-4wDHR		Ravenglass		1983	c Dsm

a   converted to a flat wagon
b   in use as a non-self propelled flail mower
c   converted from coaching stock; since reverted to coaching stock

RAVENGLASS RAILWAY MUSEUM TRUST, RAVENGLASS RAILWAY MUSEUM, RAVENGLASS

CA18 1SW

Gauge 1ft 3in   www.ravenglassrailwaymuseum.co.uk   **SD 086967**

No.	Name	Wheel	Cyl	Builder	No.	Date	Notes
	ELLA	0-6-0T	OC	DB	2	1881	Dsm
	KATIE	0-4-0T	OC	DB	4	1896	
			reb	StationRoad		2016	
	LITTLE GIANT	4-4-2	OC	BL	10	1905	
	SYNOLDA	4-4-2	OC	BL	30	1912	
No.1	BLACOLVESLEY	4-4-4PM	s/o	BL		1909	a
I.C.L.No.1		4-4wPM		Ravenglass		1925	
	(SCOOTER)	2-2wPMR		Ravenglass		1971	Dsm

a   currently located at an unknown private location

SOUTH TYNEDALE RAILWAY PRESERVATION SOCIETY, ALSTON STATION, ALSTON

CA9 3JB

Gauge 4ft 8½in   www.south-tynedale-railway.org.uk   **NY 717467**

No.	Name	Wheel	Cyl	Builder	No.	Date	Notes
	(KITTY)	4wBE		CE	B4427D	2006	

Gauge 2ft 0in

No.	Name	Wheel	Cyl	Builder	No.	Date	Notes
10	NAKLO	0-6-0WTT	OC	Chrz	3459	1957	
	GREEN DRAGON	0-4-2T	OC	HE	1859	1937	
	BARBER	0-6-2ST	OC	TG	441	1908	
	CARLISLE	4wBE		CE	B4427B	2006	
			reb	AK	96R	2015	
			reb	SBR	271	2016	
	NEWCASTLE	4wBE		CE	B4427C	2006	
			reb	SBR	272	2017	
NG 25		4wBE		BD	3704	1973	
			reb	AB	6526	1987	
	(2103/35)	4wBEF		_(EE	2519	1958	
				(Bg	3500	1958	a Dsm

4	NAWORTH STR 04	0-6-0DMF		HC	DM819	1952		
–		0-6-0DMF		HC	DM1169	1960	b	Dsm
1247	OLD RUSTY STR 18	0-6-0DMF		HC	DM1247	1961		
9	(2403/50)	0-4-0DMF		HE	4109	1952		
			reb	STRPS		1992		
4110	2403/51	0-4-0DMF		HE	4110	1953	c	Pvd
11	CUMBRIA	4wDM		HE	6646	1967		
21		4wDM		Moës				
51		2w-2PMR		Wkm		193x	d	DsmT

a   chassis used as a frame for a mobile crane
b   frame only.
c   plinthed outside the Hub Museum, Station Road, Alston (NY 717467)
d   converted to weedkiller wagon

STAINMORE RAILWAY COMPANY,
KIRKBY STEPHEN EAST STATION, KIRKBY STEPHEN   CA17 4LA

Gauge 4ft 8½in   www.kirkbystepheneast.co.uk   **NY 769075**

No.910		2-4-0	IC	Ghd		1875	
68009		0-6-0ST	IC	HE	3825	1954	
	F.C. TINGEY	0-4-0ST	OC	P	2084	1948	
	LYTHAM ST.ANNE'S	0-4-0ST	OC	P	2111	1949	
	ELIZABETH	4wDH		FH	3958	1961	
	STANTON No.50	0-6-0DE		YE	2670	1958	
(PWM 2222 B154W)		2w-2PMR		Wkm	4139	1947	
	DB 965095	2w-2PMR		Wkm	7610	1957	
–		2w-2PMR		Wkm			DsmT

THRELKELD QUARRY & MINING MUSEUM,
THRELKELD QUARRY, near KESWICK   CA12 4TT

Locomotives are also kept at Hilltop Quarry - NY 320123

Gauge 3ft 2¼in   www.threlkeldquarryandminingmuseum.co.uk   **NY 326245, 328244**

–		4wDM		RH	320573	1951	Dsm

Gauge 2ft 6in

–		0-4-0DM		HE	2248	1940	Dsm
–		0-4-0DM		HE	2267	1940	Dsm

Gauge 2ft 0in

	SIRTOM	0-4-0ST	OC	WB	2135	1925	
–		0-4-0DMF		HC	DM752	1949	Dsm
–		0-4-0DMF		HE	3149	1945	a
	SILVERBAND	4wDM		HE	3595	1948	
–		4wDM		LB	51651	1960	
3	K 11143	4wDM		MR	7191	1937	Dsm
–		4wDM		MR	8627	1941	
–		4wDM		RH	217993	1943	
	ND 6456	4wDM		RH	221626	1943	
–		4wDM		RH	223744	1944	
	ND 6440	4wDM		RH	242918	1947	

2	B173106	AA 1	L.R. 10227	BR(E) No.1067			
				4wBE	Red(T)	1980	b
7	TAMAR			4wBE	WR		
	LLECHWEDD			4wBE	WR	C6766	1963
–				4wBE	WR	P7624	1975

a   carries worksplate HE 2254/1940
b   carries worksplate MR 461/1917

DERBYSHIRE

INDUSTRIAL SITES

ABELLIO EAST MIDLANDS LTD, t/a EAST MIDLANDS RAILWAY, DERBY ETCHES PARK TRACTION & ROLLING STOCK DEPOT, DEADMANS LANE, DERBY

DE24 8BS

Gauge 4ft 8½in   www.eastmidlandsrailway.co.uk   **SK 368349**

–	4wBE		Zephir	2849	2019
–	4wBE	R/R	Zephir	3029	2023

ALLSOP (PLANT & HAULAGE) LTD, HOLLY MOUNT, HEANOR ROAD, SMALLEY, ILKESTON

DE7 6DW

Gauge 4ft 8½in   **SK 412451**

44 MITCHELL	0-6-0DH		HE	7396	1974
		reb	Wilmott		2000

ALSTOM TRANSPORT (part of Alstom Holdings S.A.), DERBY WORKS, LITCHURCH LANE, DERBY

DE24 8AD

Gauge 9ft 10in   www.alstom.com   **SK 364345**

(TRAVERSER No.4)	0-4-0WE		DerbyC&W	1985

Gauge 4ft 8½in

–	4wDH	R/R	Zephir	2677	2016
–	4wDH	R/R	Zephir	2733	2017

BREEDON CEMENT LTD, HOPE CEMENT WORKS, PINDALE ROAD, HOPE

S33 6RP

Gauge 4ft 8½in   www.breedongroup.com   **SK 167823**

(D4047) 08879		0-6-0DE		Dar		1961	a
(D8075 20075) 20309	Bo-BoDE		_(EE	2981	1961	a
			(RSHD	8233	1961	a
(D8168 20168) 2	Bo-BoDE		_(EE	3639	1966	
SIR GEORGE EARLE			(EEV	D1038	1966	a
4	0-6-0DH		AB	616	1977	
		reb	HN(R)	DH L102	2008	a

5		0-6-0DH		AB	613	1977		
	reb			AB		1986		
	reb			HN(R)	DH L101	2008	a	
01570	BLUE JOHN	B-B DH		HAB	773	1990		OOU
	PEVERIL	0-6-0DH		S	10087	1963		
		reb		AB	6140	1989		OOU
	DERWENT	0-6-0DH		S	10156	1963		
		reb		AB	6004	1988		OOU
	62	6wDH		TH	V316	1987	a	OOU

a   property of Harry Needle Railroad Co Ltd, Derbyshire

Freightliner Maintenance Ltd, Hope Wagon Works

Gauge 4ft 8½in   www.freightliner.co.uk   **SK 165826**

P720	P720 OVR 99709 949063-6	4wDH	R/R	_(Benford ES09AH504 (Rexquote 1078	1997	
		reb		Shovlin	c2020	a

a   converted from road/rail dumper; property of Shovlin Plant Hire Ltd, West Gorton

PETER BRIDDON, LOCOMOTIVE ENGINEER, GEOFFREY BRIDDON BUILDING, STATION ROAD, DARLEY DALE, near MATLOCK   DE4 2EQ

Gauge 4ft 8½in   www.petebriddon.co.uk   SK 273625

    Locomotives for overhaul or repair usually present

BROOKSIDE ENGINE CO LTD, BROOKSIDE WORKS, NEW ROAD, WHALEY BRIDGE   SK23 7JG

Gauge 2ft 0in   Private Site - no visitors without prior permission   **Private Site**

–		0-6-0WT	OC	OK	9239	1921	
7		4wBE		WR		1973	a
	"PANTHER WAGON"	2-2wPMR		Brookside		2017	

a   one of WR N7606, WR N7620 or WR N7621

CEMEX UK OPERATIONS LTD, DOVE HOLES QUARRY, DALE ROAD, DOVE HOLES, SMALLDALE, BUXTON   SK17 8BH

Gauge 4ft 8½in   www.cemex.co.uk   SK 091769

    shunting may be carried out using privately owned or mainline locomotives

W.H. DAVIS LTD, LANGWITH ROAD, LANGWITH JUNCTION   NG20 9RN

Gauge 4ft 8½in   www.whdavis.co.uk   SK 529683

DWAYNE THE TRAIN	0-4-0DH	HC	D1387	1967	
"CHEEDALE"	4wDH	TH	284V	1979	a

a   property of Andrew Briddon, Darley Dale, Derbyshire

DONFABS & CONSILLIA LTD,
THE OLD IRON WAREHOUSE, THE WHARF, SHARDLOW, near DERBY   DE72 2GH

www.trackgeometry.co.uk   SK 442303

New Consillia railcars under construction or repair occasionally present

ELMEC SOLUTIONS LTD, WHITELEY ROAD, RIPLEY   DE5 3QL

Gauge 4ft 8½in   www.elmecsolutions.com   SK 404493

0051	E15 RRV	99709 977014-8	4wDM	R/R	_(Unimog 208529	2006
	(AE06 PFN)				(Zweiweg	2006

        Other road/rail vehicles usually present at this location

        Vehicles may also be found at CK Rail Ltd, Efficiency Works, Burley Close,
        Turnoaks Business Park, Birdholme, Chesterfield, Derbyshire S40 2UB SK 385688

EP INDUSTRIES LTD, PYE BRIDGE INDUSTRIAL ESTATE,
MAIN ROAD, PYE BRIDGE, ALFRETON   DE55 4NX

Gauge 4ft 8½in   www.epindustries.co.uk   SK 435528

992	DANNY		4wDH	HE	9388	2019
		a rebuild of	4wDM	Wkm	11622	1986

INDEPENDENT RAIL ENGINEERING LTD and POSITIVE TRACTION LTD,
STONEGRAVELS WORKS, STONEGRAVELS LANE, SHEFFIELD ROAD,
CHESTERFIELD   www.independentrail.co.uk   S41 7JH

This site is also associated with **Rail Management Services Ltd** and **Eastern Rail Services Ltd**

Gauge 4ft 8½in   www.positivetraction.co.uk   SK 383724

(D3538)	08423	H 011 14	0-6-0DE	Derby		1958	a	
(D3740)	08573		0-6-0DE	Crewe		1959	a	
(D3922)	08754	H 041	0-6-0DE	Hor		1961	a	
(D4042)	08874		0-6-0DE	Dar		1960	a	OOU
(D4115)	08885	H 042 18	0-6-0DE	Hor		1962	a	OOU
(D4166)	08936	H 075	0-6-0DE	Dar		1962	a	OOU

Gauge 3ft 0in – (locomotives for storage)

LM 363		4wDH	DunEW	LM363	c1984	a	OOU
(LM 370)		4wDH	DunEW			a	OOU
H 048	BERTIE	0-6-0DM	RH	281290	1949	a	OOU

       a   property of Rail Management Services Ltd, Chesterfield, Derbyshire

       vehicles for overhaul, re-construction and repair occasionally present

       This site is strictly private. No visits are permitted

LEANDER ARCHITECTURAL, FLETCHER FOUNDRY,
HALLSTEADS CLOSE, DOVE HOLES, near BUXTON   SK17 8BP

Locomotives usually present for restoration or overhaul.
[visitors welcome – but first phone T. McAvoy on 01298 814941]

Gauge 2ft 0in   www.leanderarchitectural.co.uk   SK 076784

87009		4wDM	MR	4572	1929

22			4wDM	MR	8756	1942	
–			4wDM	RH	264252	1952	a
FAULD			4wDM	RH	444208	1961	

a   currently stored off site

LORAM UK LTD, RTC BUSINESS PARK, LONDON ROAD, DERBY   DE24 8UP
Locomotives for repair usually present.

Gauge 4ft 8½in   www.loram.com   SK 364349

(D3924)	08756	H 039	0-6-0DE	Hor		1961	a
(D4039)	08871	H 074	0-6-0DE	Dar		1960	a
922			4w-4wDH	CNES	0003	2010	
			refurbished	HE	9379	2012	

a   property of Rail Management Services Ltd, Chesterfield, Derbyshire

HARRY NEEDLE RAILROAD COMPANY LTD, BARROW HILL ROUNDHOUSE, CAMPBELL DRIVE, BARROW HILL, STAVELEY   S43 2PR
Gauge 4ft 8½in   www.harry-needle.co.uk   SK 412754

D2853	02003		0-4-0DH	YE		2812	1960
D2996	(07012)		0-6-0DE	RH		480697	1962
(D3543)	08428		0-6-0DE	Derby			1958
(D3666)	09002		0-6-0DE	Dar			1959
(D3689)	08527		0-6-0DE	Dar			1959
(D3738)	08571		0-6-0DE	Crewe			1959
(D3852)	08685		0-6-0DE	Hor			1959
(D3910)	08742		0-6-0DE	Crewe			1960
(D3933	08765)		0-6-0DE	Hor			1961
(D3950)	08782		0-6-0DE	Derby			1960
(D3954)	08786		0-6-0DE	Derby			1960
(D3966)	08798		0-6-0DE	Derby			1960
(D3992	08824)	IEMD 01	0-6-0DE	Derby			1960
D4092			0-6-0DE	Dar			1962
(D4186)	08956		0-6-0DE	Dar			1962
(D8047	20047)	20301	Bo-BoDE	_(EE		2769	1959
				(VF		D494	1959
(D8066)	20066	(82)	Bo-BoDE	_(EE		2972	1961
				(RSHD		8224	1961
(D8084	20084)	20302	Bo-BoDE	_(EE		2990	1961
				(RSHD		8242	1961
D8096	(20096)		Bo-BoDE	_(EE		3002	1961
				(RSHD		8254	1961
(D8102	20102)	20311	Bo-BoDE	_(EE		3008	1961
				(RSHD		8260	1961
D8107	(20107)		Bo-BoDE	_(EE		3013	1961
				(RSHD		8265	1961
(D8117	20117)	20314	Bo-BoDE	_(EE		3023	1962
				(RSHD		8275	1962
(D8120	20120)	20304	Bo-BoDE	_(EE		3026	1962
				(RSHD		8278	1962
(D8319	20219	20906) 3	Bo-BoDE	_(EE		3700	1968
				(EEV		D1095	1968

L127	BILL		0-6-0DH	EEV	D1199	1967
		reb		YEC	L127	1996
L149	BEN		0-6-0DH	EEV	D1200	1967
		reb		YEC	L149	1996
10			0-6-0DH	EEV	D1228	1967
–			0-6-0DH	GECT	5365	1972
(11)	"VALIANT"		0-6-0DH	RR	10213	1964
		reb		TH		1988
01562	MS 6475 CHUG		0-6-0DH	TH	257V	1975
		reb		RMS	LWO2918	2006

See also entry for **Barrow Hill Engine Shed Society** in preservation section

RAIL MANAGEMENT SERVICES LTD, t/a RMS LOCOTEC, CHESTERFIELD

(part of the Proviso Holdings Group)   Administration address only.
The following is a FLEET LIST of locomotives owned by this contractor.

Gauge 4ft 8½in   www.rmslocotec.com

(D3378) 08308 08e		0-6-0BE	PTL		2023	
	a rebuild of	0-6-0DE	Derby		1957	a
(D3538) 08423) H 011 14 LOCO 2		0-6-0DE	Derby		1958	b
(D3685) 08523 H 061		0-6-0DE	Don		1958	c
(D3740) 08573		0-6-0DE	Crewe		1959	b
(D3755) 08588 H 047		0-6-0DE	Crewe		1959	d
(D3780) 08613 H 064		0-6-0DE	Derby		1959	e
(D3789) 08622 H 028 19		0-6-0DE	Derby		1959	f
(D3815) 08648 H 065		0-6-0DE	Hor		1959	g
(D3922) 08754 H 041		0-6-0DE	Hor		1961	h
(D3924) 08756 H 059		0-6-0DE	Hor		1961	i
(D3930) 08762 H 067		0-6-0DE	Hor		1961	d
(D3956) 08788 H 068		0-6-0DE	Derby		1960	e
(D3977) 08809		0-6-0DE	Derby		1960	f
(D4015) 08847		0-6-0DE	Hor		1961	e
(D4039) 08871 H 074		0-6-0DE	Dar		1960	i
(D4042) 08874		0-6-0DE	Dar		1960	b
(D4115) 08885 H 042 18		0-6-0DE	Hor		1962	b
(D4166) 08936 H 075		0-6-0DE	Dar		1962	b
(653) H 050		0-6-0DE	_(EE	2150	1956	
			(VF	D340	1956	h
(01573) H 006 15		0-6-0DH	HE	6294	1965	h
(H 032)		0-6-0DH	HE	7541	1976	h
H 055 VINCENT DE RIVAZ		4wDH	S	10037	1960	k
H 003 BRIAN LARK CME		4wDH	S	10070	1961	k
H 057 01582		4wDH	S	10177	1964	h
TNS 107		2-2wDMR	Robel			
			56.27-10-AG38		1983	l

a	currently at Barrow Hill Engine Shed Society, Staveley, Derbyshire
b	currently at Independent Rail Engineering Ltd, Chesterfield, Derbyshire
c	currently Electro-Motive Diesel Ltd, Longport, Staffordshire
d	currently at Alstom Transport, Ilford Depot, Greater London
e	currently at PD Ports Teesport, Tees Dock
f	currently at Heidelberg Cement Group, t/a Hanson Cement, Ketton Works, Rutland
g	currently at Abellio Scotrail Ltd, Inverness T&RSMD, Highland, Scotland
h	currently at Eastern Rail Services Ltd, Great Yarmouth, Norfolk

i   currently at Loram UK Ltd, Derby, Derbyshire
k   currently at EDF Energy plc, Heysham Power Stations, Lancashire
l   currently at Flixborough Wharf Ltd, Scunthorpe, Lincolnshire

Gauge 3ft 0in

LM 363		4wDH	DunEW	LM363	c1984	a
(LM 370)		4wDH	DunEW			a
H 048	BERTIE	0-6-0DM	RH	281290	1949	a

a   currently at Independent Rail Engineering Ltd, Chesterfield, Derbyshire

TARMAC plc - A CRH Company,
TUNSTEAD QUARRY, GREAT ROCKS, TUNSTEAD, BUXTON   **SK17 8TG**

Gauge 4ft 8½in   www.tarmac.com/tunstead   **SK 101743, 097755**

PATRICK D. DUGGAN	0-6-0DH	HE	9386	2015
	rebuild of	EEV	D1226	1967 a
GRAHAM LEE JNR	0-6-0DH	HE	9387	2015
	rebuild of	EEV	_(D1249	1968
		(3947	1968 a
HIGH PEAK	6w-6wDH	Vollert	02/013	2003

a   property of Hunslet Ltd, Barton-under-Needwood, Staffordshire

WALKER & PARTNERS LTD, INKERSALL ROAD INDUSTRIAL ESTATE,
STEPHENSON ROAD, STAVELEY, CHESTERFIELD   **S43 3JN**

Dealers yard, with locomotives for resale occasionally present.

Gauge 2ft 0in   **SK 436743**

1524	2143	4wBEF	CE	B3611	1989

S.E. WARD ENGINEERING LTD, STATION ROAD, KILLAMARSH

Gauge 2ft 0in   **SK 448810**

Locomotives for restoration or overhaul occasionally present

PRESERVATION SITES

BARROW HILL ENGINE SHED SOCIETY, BARROW HILL ROUNDHOUSE RAILWAY
CENTRE, CAMPBELL DRIVE, BARROW HILL, STAVELEY   **S43 2PR**

Site also encompasses some commercial operations including **Rampart Engineering Ltd**, and **Harry Needle Railroad Company Ltd**, along with other short term contractors.

Gauge 4ft 8½in   www.barrowhill.org   **SK 414755**

158A		2-4-0	IC	Derby		1866
(41000)	1000	4-4-0	3C	Derby		1902
41708		0-6-0T	IC	Derby		1880
(62660)	No.506 BUTLER-HENDERSON	4-4-0	IC	Gorton		1920
(65567)	No.8217	0-6-0	IC	Str		1905
68006		0-6-0ST	IC	HE	3192	1944
			reb	HE	3888	1964
	HENRY	0-4-0ST	OC	HL	2491	1901

VULCAN	0-4-0ST	OC	VF	3272	1918

(D67) 45118

THE ROYAL ARTILLERYMAN	1Co-Co1DE		Crewe		1962

(D86) 45105 1Co-Co1DE Crewe 1961

(D100) 45060 SHERWOOD FORESTER 1Co-Co1DE Crewe 1961

(D147) 46010 1Co-Co1DE Derby 1961

(D2066)	03066	0-6-0DM	Don		1959	
D2868		0-4-0DH	YE	2851	1961	
(D3378)	08308 08e	0-6-0BE	PTL		2023	
	a rebuild of	0-6-0DE	Derby		1957	a
(D5209)	25059	Bo-BoDE	Derby		1963	
(D5300)	26007	Bo-BoDE	BRCW	DEL45	1958	
(D5386	27103) 27066	Bo-BoDE	BRCW	DEL229	1962	
D5910		Bo-BoDE	BabyDeltic		2013	
	a rebuild of	Co-CoDE	_(EE	3337	1963	
(D6859	37159 37372)		(EES	8390	1963	b
(D6521)	33108	Bo-BoDE	BRCW	DEL113	1960	
(E3003)	81002	Bo-BoWE	_(BRCW		1960	
			(BTH	1085	1960	
E3035	(83012)	Bo-BoWE	_(EE	2941	1960	
			(VF	E277	1960	
(E3054)	82008	Bo-BoWE	_(BP	7892	1961	
			(AEI/MV	1029	1961	
(E3061)	85006 (85101)	Bo-BoWE	Don		1961	
3	HARRY	0-4-0DM	_(RSHN	7922	1957	
			(DC	2589	1957	
(3918	62321)	4w-4wRER	York(BRE)		1970	
1499	62364	4w-4wRER	York(BRE)		1971	
–	(99709 909133-9)	2w-2PMR	Geismar ST/04/01		2004	
Q270 GAN	(70108491)	4wDM	R/R	Unimog 132889	c1986	

a   property of Rail Management Services Ltd, Chesterfield, Derbyshire
b   carries worksplate EEV 4004/D1281

Deltic Preservation Society

Gauge 4ft 8½in   www.thedps.co.uk   SK 413756

(D9009)	55009	ALYCIDON	Co-CoDE	_(EE	2914	1961
				(VF	D566	1961
D9015	(55015)	TULYAR	Co-CoDE	_(EE	2920	1961
				(VF	D572	1961
(D9019)	55019		Co-CoDE	_(EE	2924	1961
	ROYAL HIGHLAND FUSILIER			(VF	D576	1961

ANDREW BRIDDON, GEOFFREY BRIDDON BUILDING,
STATION ROAD, DARLEY DALE, MATLOCK   DE4 2EQ

Gauge 4ft 8½in   www.andrewbriddonlocos.co.uk   SK 273625

D2128	(03128 03901)	0-6-0DH	BriddonA		2011	
	a rebuild of	0-6-0DM	Sdn		1960	
(D2994)	07010	0-6-0DE	RH	480695	1962	a
(D9524)	14901	0-6-0DH	Sdn		1964	
–		0-4-0DH	AB	499	1965	a
(64)		0-6-0DE	BT	803	1978	
	"PLUTO"	4wDM	FH	3777	1956	

–		0-6-0DH		EEV	D1197	1967	
–		0-6-0DH		GECT	5395	1974	a
	LUDWIG MOND	6wDE		GECT	5578	1980	
	(GRACE)	0-4-0DH		HC	D1345	1970	
LH 005		0-6-0DH		HE	6295	1965	
873	JOHN BOY	0-6-0DH		AB	512	1966	b
11		0-6-0DH		EEV-AEI	3994	1970	
			rep	YEC	L180	2000	
No.503		0-6-0DH		HE	6614	1965	a
	LOUISE	0-6-0DH		HE	6950	1967	a
01531	H4323	0-6-0DH		HE	7018	1971	
			reb	HAB	6576	1999	b
No.5		0-4-0DH		HE	7161	1970	a
–		0-6-0DH		HE	7276	1972	a
(875)		0-4-0DH		HE	9222	1984	
–		0-4-0DH		HE	9225	1984	
			rep	YEC	L173	2001	a
	(CORONATION)	0-4-0DH		NBQ	27097	1953	
107		0-6-0DH		NBQ	27932	1959	
–		0-4-0DE		RH	421439	1958	a
	(ST MIRREN)	0-4-0DE		RH	423658	1958	a
D2	ARMY 610	0-8-0DH		S	10143	1963	a
	"TOM"	0-6-0DH		S	10180	1964	b
RS 8		0-4-0DH		SCW		1960	
	a rebuild of	0-4-0ST	OC	AE	1913	1923	
–		4wDH		TH	144V	1964	
	(CHEEDALE)	4wDH		TH	284V	1979	c
	CHARLIE	4wDH		TH	265V	1976	d
B.S.C.2		0-4-0DE		YE	2480	1950	b
	"JACK"	0-4-0DH		YE	2679	1962	a
H 051	(LIBBY)	0-6-0DH		YE	2940	1965	
2	JAMES	0-4-0DH		YE	2675	1961	
372		0-6-0DE		YE	2760	1959	a
2895		0-6-0DH		YE	2895	1964	
No.305		0-4-0DH		YE	2952	1965	
9120		4wDHR		BD	3709	1975	a
DR 98307		4w-4wDHR		Geismar	825	1998	
–		2w-2DMR		Wkm	(6607	1953?)	Dsm

a   privately owned
b   currently at Peak Rail plc, Rowsley, Derbyshire
c   currently at W.H. Davis Ltd, Langwith Junction, Derbyshire
d   currently at Ecclesbourne Valley Railway, Wirksworth, Derbyshire

BUXTON PAVILLION GARDENS RAILWAY,
PAVILLION GARDENS, ST. JOHNS ROAD, BUXTON
Gauge 1ft 0¼in   www.paviliongardens.co.uk

SK17 6XN
SK 055734

	EDWARD MILNER	0-6-0DH	s/o	AK	60	2000

CRICH TRAMWAY VILLAGE, CRICH, near MATLOCK

Gauge 4ft 8½in   www.tramway.co.uk

DE4 5DP
SK 345549

47		0-4-0VBTram	BP	2464	1885	
–		0-4-0VBTram	BP	2734	1886	a Dsm
–		4wWE	EEDK	717	1927	
–		4wDM	RH	223741	1944	b
	G.M.J.	4wDM	RH	326058	1952	c
058	(53.0692-3)	2w-2DMR	SolHütte	7808	1978	
	JOHN GARDNER	reb	SolHütte	692	1999	

a   currently stored at Clay Cross store
b   rebuilt in 1963 from 600mm gauge
c   rebuilt in 1969 from 3ft 3in gauge

Gauge 1000mm

–	4wDM	RH	373363	1954	Dsm

Gauge 1ft 2½in (1ft 3in tramway gauge)

46	0-4-0VBTram	OC	Pritchard		1941

Cliff Quarry, Wake Bridge Station

Gauge 2ft 0in   www.pdmhs.co.uk

SK 341555

4	4wBE	WR	3492	1946	a

a   property of Peak District Mines Historical Society

TERRY GIBSON, DERBY

Gauge 1ft 3in

Private Site

	KING GEORGE	4-4-2	OC	BL	22	1915
6100	ROYAL SCOT	4-6-0	OC	Carland	[6100]	1950

GULLIVERS KINGDOM LTD, MATLOCK BATH

Gauge 1ft 9½in   www.gulliverskingdomresort.co.uk

DE4 3PG
SK 289578

OLD TIMER	4w-4RE	s/o	Sharon		1988

MIDLAND RAILWAY – BUTTERLEY (MIDLAND RAILWAY TRUST LTD), BUTTERLEY HILL, RIPLEY

DE5 3QZ

Locomotives are kept at :-

Butterley SK 403520
Swanwick Junction SK 412519
Hammersmith SK 397519
Diesel Locomotive Shed, Swanwick Junction SK 414519

Gauge 4ft 8½in   www.midlandrailway-butterley.co.uk

(47327	16410) 23		0-6-0T	IC	NBH	23406	1926	
(47357)	16440		0-6-0T	IC	NBQ	23436	1926	
		reb			Derby		1973	
(47445)			0-6-0T	IC	HE	1529	1927	Dsm
(47564)			0-6-0T	IC	HE	1580	1928	a Dsm
73129			4-6-0	OC	Derby		1956	
	"STANTON No.24"		0-4-0CT	OC	AB	1875	1925	

No.2	PN 8292	0-4-0F	OC	AB	2008	1935	
	LORD PHIL	0-6-0ST	IC	HE	2868	1943	
			reb	HE	3883	1963	
–		0-4-0ST	OC	HL	2918	1912	
	F.D. & E. Co. No.3	0-4-0ST	OC	HL	3597	1926	
	"GLADYS"	0-4-0ST	OC	Mkm	109	1894	
	"OSWALD"	0-4-0ST	OC	NW	454	1894	
	WHITEHEAD	0-4-0ST	OC	P	1163	1908	b
	VICTORY	0-4-0ST	OC	P	1547	1919	
No.15		0-4-0ST	OC	RSHN	7063	1942	
CASTLE DONINGTON POWER STATION 1		0-4-0ST	OC	RSHN	7817	1954	
D4	(44004) GREAT GABLE	1Co-Co1DE		Derby		1959	
(D40)	45133	1Co-Co1DE		Derby		1961	
(D212)	40012 AUREOL	1Co-Co1DE		_(EE	2668	1959	
				(VF	D429	1959	
(D1500)	47401 NORTH EASTERN	Co-CoDE		BT	342	1962	
D1516	(47417)	Co-CoDE		BT	358	1963	
(D1619	47564) 47761	Co-CoDE		Crewe		1964	
D2138		0-6-0DM		Sdn		1960	
(D2858)		0-4-0DH		YE	2817	1960	
(D3401)	08331	0-6-0DE		Derby		1957	
(D3757)	08590 RED LION	0-6-0DE		Crewe		1959	
(D5522)	31418 BOADICEA	A1A-A1A DE		BT	121	1959	
(D5526)	31108	A1A-A1A DE		BT	125	1959	
(D5814	31414) 31514	A1A-A1ADE		BT	315	1961	
D7671	(25321)	Bo-BoDE		Derby		1967	
(D)8001	(20001) VULCAN PIONEER	Bo-BoDE		_(EE	2348	1957	
				(VF	D376	1957	
(D8007)	20007	Bo-BoDE		_(EE	2354	1957	
				(VF	D382	1957	c d
(D8142)	20142	Bo-BoDE		_(EE	3614	1966	
	SIR JOHN BETJEMAN			(EEV	D1013	1966	c
(D8189)	20189	Bo-BoDE		_(EE	3670	1966	
				(EEV	D1065	1966	c
(D8305)	20205	Bo-BoDE		_(EE	3686	1967	
				(EEV	D1081	1967	c d
(D8327)	20227	Bo-BoDE		_(EE	3685	1967	
	SHERLOCK HOLMES			(EEV	D1080	1967	c
43048	(253024) TCB MILLAR MBE	Bo-BoDE		Crewe		1977	
43089	(254017) RIO WARRIOR	Bo-BoDE		Crewe		1978	
43159	(254036)	Bo-BoDE		Crewe		1981	
12077		0-6-0DE		Derby		1950	
(E)27000	(ELECTRA)	Co-CoWE		Gorton	1065	1953	
No.4		0-4-0DM		AB	416	1957	
	rebuilt	0-4-0DH		AB		1980	
441		0-4-0DH		AB	441	1959	
	ALBERT	0-6-0DM		HC	D1114	1958	
16038	ANDY	0-4-0DM		JF	16038	1923	
–		2-2wDM		Mercury	5337	1927	
"RS 12"		4wDM		KC		c1924	+
"RS 9"		4wDM		MR	2024	1920	
–		0-4-0DE		RH	384139	1955	
E50019		2-2w-2w-2DMR		DerbyC&W		1957	

51118		2-2w-2w-2DMR	GRC&W		1957	
M51591	(M55966)	2-2w-2w-2DHR	DerbyC&W		1959	
51669		2-2w-2w-2DMR	DerbyC&W		1960	
M51849		2-2w-2w-2DMR	DerbyC&W		1960	
E50015	(55929 977775)	2-2w-2w-2DMR	DerbyC&W		1956	
(M 51610)	M 55967	2-2w-2w-2DHR	DerbyC&W		1959	
M51625	(M55976)	2-2w-2w-2DHR	DerbyC&W		1959	
M51907	LO 262	2-2w-2w-2DMR	DerbyC&W		1960	
55513	141113	4wDHR	_(BRE(D)		1984	
			(Leyland R4.012		1984	
		reb	AB	761	1989	
55533	141113	4wDHR	_(BRE(D)		1984	
			(Leyland R4.011		1984	
		reb	AB	760	1989	
55552	142011	2-2wDMR	_(BRE(D)		1986	
			(Leyland R5.30		1986	
		rebuilt as	2-2wDHR	AB		1991
55602	142011	2w-2DMR	_(BRE(D)		1986	
			(Leyland R5.29		1986	
		rebuilt as	2w-2DHR	AB		1991
55554	142013	2-2wDMR	_(BRE(D)		1986	
			(Leyland R5.34		1986	
		rebuilt as	2-2wDHR	AB		1991
55604	142013	2w-2DMR	_(BRE(D)		1986	
			(Leyland R5.33		1986	
		rebuilt as	2w-2DHR	AB		1991
	(99709 909296-4)	2w-2PMR	Bance	084/00	2000	
F540 YCK		4wDM	R/R	_(Multicar 16305	1988	
			(Perm		1988	
PWM 3949		2w-2PMR	Wkm	6934	1955	DsmT
DX 68062	(TR34 DB 965566 PT52P)	2w-2PMR	Wkm	8272	1959	

a   currently at Riley & Sons (E) Ltd, Heywood, Greater Manchester
b   based here, but visits other locations
c   property of Class 20189 Ltd, based here for maintenance
d   currently at Severn Valley Railway Co Ltd, Kidderminster, Worcestershire
+   may have been fitted with worksplate and minor parts from MR 460

Golden Valley Light Railway
Gauge 2ft 0in   www.gvlr.org.uk   **SK 412519, 414519**

	JOAN	0-4-2IST	OC	Civil	No.1	1997
2		0-4-0WT	OC	OK	7529	1914
BD 3753	DARCY	4wDH		BD	3753	1980
Dtz	10249	4wDM		Dtz	10248	1931
–		0-6-0DMF		HC	DM1117	1957
AD 34		4wDH		HE	7009	1971
–		4wDM		HE	7178	1971
–		4wDM		L	3742	1931
			reb	FMB		1993
–		4wDM		L	10994	1939
–		4wDM		LB	53726	1963
LOD 758228	G.V.L.R. 15 TUBBY	4wDM		MR	8667	1941
(T2)	PIONEER	4wDM		MR	8739	1942
LOD 758028	LO 3009	4wDM		MR	8855	1943
	HOLWELL CASTLE	4wDM		MR	11177	1961
15		4wDM		MR	11246	1963

	CAMPBELL BRICKWORKS	4wDM		MR	60S364	1968	
AD 41	(LOD 758263) LYDDIA	4wDM		RH	191646	1938	
	BERRY HILL	4wDM		RH	222068	1943	
(L203N)	U84	4wDM		RH	7002/0567/6	1967	
	ELLISON	4wDH		SMH	101T020	1979	
40SD529		4wDM		SMH	40SD529	1984	
NG 24		4wBE		BD	3703	1973	
			reb	AB		1986	
19	BABY JAYNE	0-4-0BE		WR		198x	

Princess Royal Class Locomotive Trust, Swanwick Junction
Gauge 4ft 8½in   www.westshedmuseumcom.wordpress.com   **SK 410520**

46203	PRINCESS MARGARET ROSE	4-6-2	4C	Crewe	253	1935	
(46233)	6233						
	DUCHESS OF SUTHERLAND	4-6-2	4C	Crewe		1938	
80098		2-6-4T	OC	Bton		1954	
92212		2-10-0	OC	Sdn		1959	
	GEORGE	0-4-0ST	OC	RSHN	7214	1945	

Gauge 1ft 9in

6201	(PRINCESS ELIZABETH)	4-6-2DH	s/o	HC	D611	1938	
6203	PRINCESS MARGARET ROSE	4-6-2DH	s/o	HC	D612	1938	
	GEORGINA	0-4-0ST	OC	NBRES	(001)	2016	

PEAK DISTRICT MINES HISTORICAL SOCIETY, PEAK DISTRICT MINING MUSEUM, TEMPLE MINE, TEMPLE ROAD, MATLOCK BATH (DE4 3NR)
Gauge 1ft 5in   www.peakdistrictleadminingmuseum.co.uk   **SK 293583**

–	4wBE		GB	1445	1936	

PEAK RAIL plc,
ROWSLEY SOUTH STATION, off HARRISON WAY, near MATLOCK   DE4 2LF
Gauge 4ft 8½in   www.peakrail.co.uk   **SK 262640**

5224		2-8-0T	OC	Sdn		1924	
6634		0-6-2T	IC	Sdn		1928	
	WALESWOOD	0-4-0ST	OC	HC	750	1906	
S.100		0-6-0T	OC	HC	1822	1949	
	CATHRYN	0-6-0T	OC	HC	1884	1955	
(RRM 192)		0-6-0ST	OC	HL	3138	1915	
(No.16)		0-6-0ST	OC	HL	3837	1934	
	WESTMINSTER	0-6-0ST	OC	P	1378	1914	
	OLWEN	0-4-0ST	OC	RSHN	7058	1942	
WD 150	ROYAL PIONEER	0-6-0ST	IC	RSHN	7136	1944	
			reb	HE	3892	1969	
No.2	"KENT No.2"	0-4-0ST	OC	WB	2842	1946	
D8	(44008) PENYGHENT	1Co-Co1DE		Derby		1959	
(D172)	46035 (97403) (IXION)	1Co-Co1DE		Derby		1962	
(D429)	50029 RENOWN	Co-CoDE		_(EE	3799	1968	
				(EEV	D1170	1968	
(D430)	50030 REPULSE	Co-CoDE		_(EE	3800	1968	
				(EEV	D1171	1968	

D1501	(47402) (GATESHEAD)	Co-CoDE	BT	343	1962	
(D3998)	08830	0-6-0DE	Derby		1960	
(D6852	37152) 37310	Co-CoDE	_(EE	3327	1963	
	BRITISH STEEL RAVENSCRAIG		(EEV	D826	1963	
D7659	(25309 25909)	Bo-BoDE	BP	8069	1966	
873	JOHN BOY "BAGNALL"	0-6-0DH	AB	512	1966	a
No.5		0-4-0DM	Bg	3027	1939	
E1	CASTLEFIELD	0-6-0DM	HC	D1199	1960	
01531	H4323	0-6-0DH	HE	7018	1971	
		reb	HAB	6576	1999	
	"TOM"	0-6-0DH	S	10180	1964	a
–		0-4-0DH	TH	146C	1964	
	a rebuild of	0-4-0DM	JF	4210018	1950	
B.S.C.2		0-4-0DE	YE	2480	1950	a
M51950		2-2w-2w-2DMR	DerbyC&W		1961	
W52062		2-2w-2w-2DMR	DerbyC&W		1961	
(3905)	62266	4w-4wRER	York		1969	
(DX 68006	DB 965950 MPP 0006)	2w-2PMR	Wkm	10646	1972	
	"TONY"	4wDMR	Wkm	9688	1965	

a   property of Andrew Briddon, Darley Dale, Derbyshire

Gauge 1ft 11½in

109		2-6-2+2-6-2T 4C	BP	6919	1939

Heritage Shunters Trust, Rowsley South   DE4 2LF
Gauge 4ft 8½in   www.heritageshunterstrust.com   SK 261640

(D2027	03027)	0-6-0DM	Sdn		1958
(D2099)	03099	0-6-0DM	Don		1960
(D2113)	03113	0-6-0DM	Don		1960
D2139		0-6-0DM	Sdn		1960
(D2180)	03180	0-6-0DM	Sdn		1962
D2199		0-6-0DM	Sdn		1961
D2205		0-6-0DM	_(VF	D212	1953
			(DC	2486	1953
(D2229)		0-6-0DM	_(VF	D278	1955
			(DC	2552	1955
D2272	2272 "ALFIE"	0-6-0DM	_(RSHD	7914	1958
			(DC	2616	1958
D2284		0-6-0DM	_(RSHD	8102	1960
			(DC	2661	1960
(D2289)	3945	0-6-0DM	_(RSHD	8122	1960
			(DC	2669	1960
D2337	"DOROTHY"	0-6-0DM	_(RSHD	8196	1961
			(DC	2718	1961
D2420	(06003 97804)	0-4-0DM	AB	435	1959
D2587		0-6-0DM	HE	5636	1959
		reb	HE	7180	1969
D2854		0-4-0DH	YE	2813	1960
(D2866)		0-4-0DH	YE	2849	1961
D2953		0-4-0DM	AB	395	1955
(D2985)	07001	0-6-0DE	RH	480686	1962
(D3023)	08016	0-6-0DE	Derby		1953
(D3665)	09001	0-6-0DE	Dar		1959
PWM 650	97650	0-6-0DE	RH	312990	1952

PWM 654	(97654)		0-6-0DE	RH	431761	1959
319284	FARADAY		0-4-0DM	RH	319284	1952
	BIGGA		0-4-0DH	TH	102C	1960
		a rebuild of	0-4-0DM	JF	4200019	1947

Ashover Light Railway Society, Rowsley South
Gauge 2ft 0in www.alrs.org.uk

DE4 2LF
SK 261642

–		4wVBT	VC	Jaywick		1939
8917	NCB 24 LINBY	4wDHF		HE	8917	1980
–		4wDM		RH	260712	1948
85051		4wDM		RH	404967	1957
–		4wDM		RH	487963	1963
No.1	SPONDON	4wBE		Spondon		1926

PLEASLEY PIT TRUST,
PLEASLEY COLLIERY, PIT LANE, PLEASLEY, MANSFIELD
Gauge 3ft 0in www.pleasleypittrust.org.uk

NG19 7PH
SK 498643

E15	P6	4wBEF	CE	B3045B	1983

PRIVATE SITE, SWADLINCOTE
Gauge 4ft 8½in

Private Site

DERBYSHIRE STONE No.2	4wDM	KC	1470	1925

THE STANTON COLLECTION, WOOD HOUSE FARM, ASHBOURNE
Gauge 1ft 3in www.stantoncollections.co.uk

DE6 3GS
SK 206382

INVICTA	4wPH	Maxitrack		1989

STEEPLE GRANGE LIGHT RAILWAY, STEEPLEHOUSE JUNCTION, WIRKSWORTH
Gauge 1ft 6in www.sglr.co.uk

SK 288554

ZM32	No.11 HORWICH	4wDM	RH	416214	1957	
551		4wBE	BEV	551	1924	a
No. 4	(LIZZIE)	4wDM	ClayCross		1973	b
No.9	4008 ULLR	4wDM	FH	4008	1963	
No.10		4wDM	L	37366	1951	
4	5A	4wBE	GB	1326	1933	
No.6		4wBE	GB	2493	1953	
No.3	GREENBAT	4wBE	GB	6061	1961	
No.15		4wBE	CE	5431	1968	
No.17		4wBE	CE	5942B	1972	
No.13	TOM	4wBE	CE	5965B	1973	
No.16	PEGGY	4wBE	CE	B0109B	1973	
No.18	HAZEL	4wBE	CE	B0111C	1973	
No.14	PETER	4wBE	CE	B0922B	1975	
12	GREENSBURG	4wBE	Greensburg 2368		1950	
4	"COSTAIN"	4wBE	Plymouth			a

No.2 (HUDSON)	2-2wPMR		SGLR		1988	
ML-2-17	4wBH		Tunn		1980	
		reb	Tunn		1996	c
–	4wBH		Tunn		1980	
		reb	Tunn		1996	c

a	currently under renovation elsewhere
b	constructed from parts supplied by Lister
c	two of Tunn TQ121 to TQ126

WYVERNRAIL plc, ECCLESBOURNE VALLEY RAILWAY, WIRKSWORTH STATION, STATION ROAD, WIRKSWORTH

Gauge 4ft 8½in   www.e-v-r.com

DE4 4FB
SK 289541

Number(s)	Name	Wheel	Gear	Builder	Works No	Date	Note
80080		2-6-4T	OC	Bton		1954	
	HENRY ELLISON	0-4-0ST	OC	AB	2217	1947	
No.3	BRIAN HARRISON	0-4-0ST	OC	AB	2360	1954	
68012		0-6-0ST	IC	WB	2746	1944	
(D1842) 47192		Co-CoDE		Crewe		1965	
(D5609 31186) 31601	DEVON DIESEL SOCIETY	A1A-A1ADE		BT	209	1960	
(D6514) 33103	SWORDFISH	Bo-BoDE		BRCW	DEL106	1960	
D9525		0-6-0DH		Sdn		1965	
D9537	ERIC	0-6-0DH		Sdn		1965	
58022		Co-CoDE		Don		1984	Dsm
(E6001) 73001		Bo-BoDE/RE		Elh		1962	
(E6022 73116) 73210	SELHURST	Bo-BoDE/RE		_(EE	3584	1966	
				(EEV	E354	1966	
402803		0-4-0DE		RH	402803	1956	
	L.J. BREEZE	6wDH		RR	10275	1969	
RRM 134	"TOM"	4wDH		TH	188C	1967	
	a rebuild of	4wVBT	VCG	S	9597	1955	
	CHARLIE	4wDH		TH	265V	1976	a
E50170 (53170)		2-2w-2w-2DMR		MetCam		1957	
E50253 (53253)		2-2w-2w-2DMR		MetCam		1957	
(E50599) E53599 L990		2-2w-2w-2DMR		DerbyC&W		1958	
W51073 L594		2-2w-2w-2DMR		GRC&W		1959	
E51505		2-2w-2w-2DMR		MetCam		1959	
(51567) 977854		2-2w-2w-2DMR		DerbyC&W		1959	
W55006		2-2w-2w-2DMR		GRC&W		1958	
(960302 977975 55027)		2-2w-2w-2DMR		PSteel		1960	
960303 977976 (55031)		2-2w-2w-2DMR		PSteel		1960	
W55034 121034		2-2w-2w-2DMR		PSteel		1960	
M79018 975007)		2-2w-2w-2DMR		DerbyC&W		1954	
M79900 (975010) IRIS		2-2w-2w-2DMR		DerbyC&W		1956	b
68500 (S61269) 9101		4w-4RER		Afd/Elh		1959	
			reb	Elh		1983	c
(68506 S61292 9107)		4w-4RER		Afd/Elh		1957	
			reb	Elh		1984	c
7501		4w-4wRER		MetCam		1979	
	L263 MNU	4wDMR	R/R	_(Multicar		1993	
				(Perm		1993	
	R447 XRA	4wDM	R/R	_(Landrover	146643	1998	
				(Hy-Rail		1998	

296	(R482 JGG)		4wDM	R/R	_(Landrover	1998
					(Perm	1998
	MURIEL		2w-2DER		EVRA	2000
	99709 901209-5		2w-2PMR		Lesmac LMA001	

a   property of Andrew Briddon, Darley Dale, Derbyshire
b   currently at the Great Central Railway, Loughborough, Leicestershire
c   unpowered hauled coaching stock

Gauge 2ft 0in — Stone Line

| | LESLEY THE LISTER | 4wDM | | L | 26288 | 1944 |

DEVON

INDUSTRIAL SITES

BABCOCK INTERNATIONAL GROUP plc,
DEVONPORT ROYAL DOCKYARD, DEVONPORT
(PL1 4SG)

Gauge 4ft 8½in   www.babcockinternational.com   SX 449558

RTU 1	DENNIS	4wDM		CE	B4314A	2000
RTU 2	HENRY	4wDM		CE	B4314B	2000
			reb	CE	B4622	2016

FIRST GREATER WESTERN LTD, t/a GREAT WESTERN RAILWAY,
LAIRA DEPOT, EMBANKMENT ROAD, PLYMOUTH
PL4 9JN

(part of the First Group plc)

Gauge 4ft 8½in   www.gwr.com   SX 504556

(D3808)	08641	FRED	0-6-0DE		Hor	1959	M
(D3811)	08644	LAIRA DIESEL DEPOT 50 YEARS 1962 - 2012					
			0-6-0DE		Hor	1959	M
(D4004)	08836		0-6-0DE		Derby	1960	M
	–		4wBE		Windhoff 101005675/50	2008	

LEAKY FINDERS LTD,
GREAT WESTERN YARD, STATION ROAD, HELE, EXETER
EX5 4PW

Gauge 4ft 8½in   www.leakyfindersltd.co.uk   SS 995022

	1638		0-6-0PT	IC	Sdn		1951
	9629		0-6-0PT	IC	Sdn		1945
(30541)	541		0-6-0	IC	Elh		1939
	47	CARNARVON	0-6-0ST	IC	K	5474	1934

locomotives for overhaul/restoration usually present

MINISTRY OF DEFENCE, OKEHAMPTON RANGES TARGET RAILWAY, OKEHAMPTON
Gauge 2ft 6in SX 586932

RTT/767138	CAPTAIN		2w-2PM	Wkm	3284	1943

PRESERVATION SITES

ASHBRITTLE LIGHT RAILWAY, HOLE FARM, HOCKWORTHY
Gauge 2ft 0in Private Site

YARD No.1073	EARL OF MORTAIN	4wDH	HE	7450	1976
34	SO 34	LR 10756	4wDM	MR	5879	1935
25		4wDM	RH	375694	1954

BICTON WOODLAND RAILWAY,
BICTON PARK BOTANICAL GARDENS, EAST BUDLEIGH
Gauge 1ft 6in   www.bictongardens.co.uk

EX9 7BJ
SY 074862

	SIR WALTER RALEIGH	0-4-0DH	s/o	AK	61	2000
	CLINTON	4wDM		HE	2290	1941
B.W.R. 2	BICTON	4wDM		RH	213839	1942
		reb	4wDH	s/o	AK	75R	2007

BIDEFORD HERITAGE RAILWAY CENTRE C.I.C.,
BIDEFORD STATION, BIDEFORD HILL, BIDEFORD
Gauge 4ft 8½in   www.bidefordrailway.co.uk

EX39 4BB
SS 457262

DS 1170	KINGSLEY	4wDM	FH	3832	1957

C. BURGES, EXETER & TEIGN VALLEY RAILWAY, CHRISTOW STATION,
SHELDON LANE, DODDISCOMBSLEIGH, EXETER
Gauge 4ft 8½in   www.teignrail.co.uk

EX6 7YT
SX 839868

	PERSEUS	0-4-0DM	_(VF	D98	1949
			(DC	2269	1949
DX 90011		4wDMR	Matisa	PV6 620	1967
(DX 68004) (PWM)2831 (B)194(W)		2w-2PMR	Wkm	5009	1949

'BYGONES' VICTORIAN EXHIBITION STREET AND RAILWAY MUSEUM,
FORE STREET, ST.MARYCHURCH, TORQUAY
Gauge 4ft 8½in   www.bygones.co.uk

TQ1 4PR
SX 922658

No.5	PATRICIA	0-4-0ST	OC	HC	1632	1929

COMBE MARTIN WILDLIFE AND DINOSAUR PARK, HIGHER LEIGH MANOR, LEIGH ROAD, COMBE MARTIN, near ILFRACOMBE   EX34 0NG

Gauge 1ft 3in   www.cmwdp.co.uk   **SS 600452**

–		2-8-0PH	s/o	SL	70.5.87	1987

DART VALLEY RAILWAY plc, DARTMOUTH STEAM RAILWAY & RIVERBOAT CO, DARTMOUTH STEAM RAILWAY

Locomotives are kept at :-

Churston	TQ5 0LL	SX 896564
Kingswear	TQ6 0AA	SX 884515
Park Siding, Paignton		SX 889606

Gauge 4ft 8½in   www.dartmouthrailriver.co.uk

2873		2-8-0	OC	Sdn	2779	1918	
3803		2-8-0	OC	Sdn		1939	
4277	HERCULES	2-8-0T	OC	Sdn	2857	1920	
5239	GOLIATH	2-8-0T	OC	Sdn		1924	
5542		2-6-2T	OC	Sdn		1928	
7827	LYDHAM MANOR	4-6-0	OC	Sdn		1950	
75014	BRAVEHEART	4-6-0	OC	Sdn		1951	
2253	OMAHA	2-8-0	OC	BLW	69496	1944	
D2192	TITAN	0-6-0DM		Sdn		1961	
(D2371)	03371	0-6-0DM		Sdn		1958	
D3014		0-6-0DE		Derby		1952	
(D6767)	37067 (37703)	Co-CoDE		_(EE	3059	1962	
				(VF	D721	1962	
(D)6975	(37275)	Co-CoDE		_(EE	3535	1965	
				(EEV	D964	1965	
(DB 966030)	CE 68200	2w-2DMR		Plasser	419	1975	
	99709 901185-7	2w-2PMR		Geismar	M44/135	2012	

DARTMOOR RAILWAY ASSOCIATION, STATION ROAD, OKEHAMPTON   EX20 1EJ

Locomotives are kept at :-   Meldon Quarry   EX20 4LT   SX 567927

Gauge 4ft 8½in   www.dartmoor-railway-association.org

(S.103)		0-6-0T	OC	HC	1864	1952
D4167	(08937)	0-6-0DE		Dar		1962

DEVON RAILWAY CENTRE, BICKLEIGH MILL, CADLEIGH STATION, TIVERTON   EX16 8RG

Gauge 4ft 8½in   www.devonrailwaycentre.co.uk   **SS 938076**

	"BORIS"	0-4-0DM		Bg	3357	1952

Gauge 2ft 0in

No.14	REBECCA	0-4-0WT	OC	OK	5744	1912	
	"PLANET"	4wDM		FH	2201	1939	
–		4wDM		L	34025	1949	a
	IVOR	4wDM	s/o	MR	8877	1944	
	EDEN	4wDM		MR	20058	1949	
	SIR TOM	4wDM		MR	40S273	1966	

	HORATIO	4wDM	s/o	RH	217967	1942	
	"RUSTON"	4wDM		RH	418770	1957	
1300	S14	0-4-0BE		WR		c1950	a

a   currently under restoration off site

EXMOOR STEAM RAILWAY, CAPE OF GOOD HOPE FARM, BRATTON FLEMING

Locomotives under construction and repair are usually present.   **Private Site**

Gauge 2ft 0in   **SS 662383**

	77	2-6-2+2-6-2T 4C	Hano	10629	1928
	115	2-6-2+2-6-2T 4C	BP	6925	1937
	135	2-8-2 OC	AFB	2685	1952
	–	0-6-0WT+T OC	OK	7122	1914
	1	4wDH	BD	3783	1984
	2	4wDH	BD	3784	1984
LOD 758035		4wDM	MR	8856	1944

Gauge 1ft 0¼in

	(EXMOOR RANGER)	0-4-4-0	Meyer	ESR	298		u/c
	DENZIL	0-4-2T+T	OC	ESR	299	1995	
	LORNA DOONE	0-4-2T	OC	ESR	330	2006	
	–	4wDH		HE	9333	1994	
	DOUGAL	4wDH		HE	9334	1994	

C. GROVE, THE TAMAR BELLE HERITAGE GROUP,
BERE FERRERS STATION, YELVERTON   PL20 7LT

Gauge 4ft 8½in   www.tamarbelle.co.uk   **SX 452635**

	HILDA	0-4-0ST	OC	P	1963	1938
PD&SWJR 3	A.S. HARRIS	0-4-0DM		HE	2642	1941
PD&SWJR 4	EARL OF MOUNT EDGCUMBE					
		0-4-0DM		HE	3133	1944
PD&SWJR 5	LORD ST. LEVAN	0-4-0DM		HE	3395	1946

Gauge 2ft 0in

	–	4wDM	RH	186318	1937

LYNTON & BARNSTAPLE RAILWAY TRUST,
WOODY BAY STATION, MARTINHOE CROSS, near PARRACOMBE   EX31 4RA

Gauge 2ft 0in   www.lynton-rail.co.uk   **SS 682464**

E 762	LYN	2-4-2T	OC	AK	92	2017	
	AXE	0-6-0T	OC	Gartell		2006	a
	CHARLES WYTOCK	4-4-0T	OC	WB	2819	1946	
D2393	PILTON	0-6-0DM		Bg/DC	2393	1952	
D6652		4wDH		HE	6652	1965	
	–	4wDH		HE	6660	1965	

a   locomotive built with new frames, incorporating parts from KS 2451

THE MILKY WAY ADVENTURE PARK, THE MILKY WAY RAILWAY, DOWNLAND FARM, near CLOVELLY

Gauge 2ft 0in   www.themilkyway.co.uk

EX39 5RY
SS 327229

–		4w-4wDH		SL	23	1973

D. MOORE, BRATTON FLEMING

Gauge 2ft 0in

Private Site

"CHRIS"		4wDH		HLH	001	1996

THE MORWELLHAM & TAMAR VALLEY TRUST, MORWELLHAM, MORWELLHAM QUAY, near TAVISTOCK

Gauge 2ft 0in   www.morwellham-quay.co.uk

PL19 8JL
SX 448699

–			4wBE	AK	67	2003	a
–			4wBE	AK	111	2021	a
1	GEORGE	S259	4wBE	WR	H7197	1968	
(2)	BERTHA		4wBE	WR	6298	1960	
5494 (3)	CHARLOTTE		4wBE	WR	G7124	1967	
4	LUDO		4wBE	WR	6769	1964	
(5)	WILLIAM		4wBE	WR	C6770	1964	
(6)	MARY		4wBE	WR	5665	1957	
No.7	HAREWOOD		4wBE	WR	D6800	1964	
8			4wBE	WR			Dsm

a   fitted with equipment for multiple working

PLYM VALLEY RAILWAY COMPANY LTD, MARSH MILLS STATION, COYPOOL ROAD, PLYMPTON, PLYMOUTH

Gauge 4ft 8½in   www.plymrail.co.uk

PL7 4NW
SX 520571

705		0-4-0ST	OC	AB	2047	1937	
	ALBERT	0-4-0ST	OC	AB	2248	1948	
5374	VANGUARD	0-6-0T	OC	Chrz	5374	1959	
	BYFIELD No.2	0-6-0ST	OC	WB	2655	1942	
–		0-4-0F	OC	WB	3121	1957	
(D3002	11 DULCOTE) 13002	0-6-0DE		Derby		1952	
(D5613)	31190	A1A-A1ADE		BT	213	1960	
429	RIVER ANNAN	0-6-0DH		RH	466618	1961	
10077		4wDH		S	10077	1961	
–		4wDH		TH	125V	1963	
(50222	53222 RDB977693) 901002 IRIS II	2-2w-2w-2DMR		MetCam		1957	
(50338	53338 RDB977694) 901002 IRIS II	2-2w-2w-2DMR		MetCam		1958	
W51365	T304	2-2w-2w-2DMR		PSteel		1960	
W51407	T304	2-2w-2w-2DMR		PSteel		1960	
55564	142023	2-2wDMR		_(BRE(D)		1986	
				(Leyland R5.52		1986	
	rebuilt as	2-2wDHR		RFSD		1989	
55614	142023	2w-2DMR		_(BRE(D)		1986	
				(Leyland R5.51		1986	
	rebuilt as	2w-2DHR		RFSD		1989	

55659	143618 (143018)		2-2wDMR	_(AB	685	1986
				(Alex	1784/35	1986
		rebuilt as	2-2wDHR	RFSD		1988
55684	143618 (143018)		2w-2DMR	_(AB	686	1986
				(Alex	1784/36	1986
		rebuilt as	2w-2DHR	RFSD		1988
–			2w-2PMR	Wkm	3366	1943 Dsm
–			2w-2PMR	Wkm	4154	1949 Dsm
–			2w-2PMR	Wkm	4992	1949
		reb	PVRA		1989	DsmT
–			2w-2PMR	Wkm	5002	1949
		reb	PVRA		1989	DsmT
–			2w-2PMR	Wkm	7139	1955

PRIVATE OWNER, WHITCHURCH, TAVISTOCK
Gauge 2ft 0in   Private Site

–		4wDM	L	11410	1939
	SIMON	4wDM	MR	7126	1936

PRIVATE SITE, Near PLYMOUTH
Gauge 4ft 8½in   Private Site

(PWM 3762 B1W)	2w-2PMR	Wkm	6641	1953

SOUTH DEVON RAILWAY TRUST, THE PRIMROSE LINE
Locomotives are kept at :-

Buckfastleigh	TQ11 0QZ	SX 747663
Staverton Bridge		SX 785638
Totnes Littlehempston		SX 804613

Gauge 7ft 0¼in   www.southdevonrailway.org

151	TINY	0-4-0VBWT VCG	Sara		1868

Gauge 4ft 8½in

1420		0-4-2T	IC	Sdn		1933
3205		0-6-0	IC	Sdn		1946
4160		2-6-2T	OC	Sdn		1948
5526		2-6-2T	OC	Sdn		1928
5786		0-6-0PT	IC	Sdn		1930
	GLENDOWER	0-6-0ST	IC	HE	3810	1954
	LADY ANGELA	0-4-0ST	OC	P	1690	1926
1	ASHLEY	0-4-0ST	OC	P	2031	1942
(D402)	50002 SUPERB	Co-CoDE		_(EE	3771	1967
				(EEV	D1142	1967
D2246	(11216)	0-6-0DM		_(RSHN	7865	1956
				(DC	2578	1956
D2271		0-6-0DM		_(RSHD	7913	1958
				(DC	2615	1958
D3721	(09010)	0-6-0DE		Dar		1959
D6501	(33002)	Bo-BoDE		BRCW	DEL93	1960
(D)6737	(37037 37321)	Co-CoDE		_(EE	2900	1961
				(VF	D616	1961

D7541	(25191)		Bo-BoDE	Derby		1965
D7535	(25185) HERCULES		Bo-BoDE	Derby		1965
D7612	(25262, 25901)		Bo-BoDE	Derby		1966
(51352)			2-2w-2w-2DMR	PSteel		1960
51376			2-2w-2w-2DMR	PSteel		1960
W55000	(122 100)		2-2w-2w-2DMR	GRC&W		1958
	MFP No.4		0-4-0DM	JF	4210141	1958
2745			0-6-0DE	YE	2745	1960
DR 98210			4wDMR	Plasser	52765A	1985
	–		2w-2PMR	Wkm	4149	1947 Dsm
(PWM 3767)	DS 3321 ADRIAN		2w-2PMR	Wkm	6646	1953
(PWM 3773)	B 12 W		2w-2PMR	Wkm	6652	1953
	–		2w-2PMR	Wkm	8198	1958
THE ADDICK	BERNARD WICKHAM		2w-2PMR	Wkm	11717	1976
XLR 8123	S134 TBC		4wDM	R/R	_(Landrover	1998
					(Perm	1998

Gauge 4ft 6in

LEE MOOR No.2	0-4-0ST	OC	P	784	1899

TARKA VALLEY RAILWAY C.I.O.,
TORRINGTON STATION, off STATION HILL, TORRINGTON

EX38 8JD

Gauge 4ft 8½in   www.tarkavalleyrailway.org   **SS 479198**

No.1	PROGRESS	0-4-0DM	JF	4000001	1945
544998	TORRINGTON CAVALIER	0-4-0DE	RH	544998	1969
55644	143617 (143003)	2-2wDMR	_(AB	674	1985
			(Alex	1784/5	1985
	rebuilt as	2-2wDHR	RFSD		1988
55683	143617 (143017) FOUNDER MEMBER & CHAIRMAN ROD GARNER 1943 - 2020				
		2w-2DMR	_(AB	712	1985
			(Alex	1784/34	1985
	rebuilt as	2w-2DHR	RFSD		1988

TIVERTON MUSEUM SOCIETY,
TIVERTON MUSEUM, ST. ANDREW STREET, TIVERTON

EX16 6PJ

Gauge 4ft 8½in   www.tivertonmuseum.org.uk   **SS 955124**

1442	0-4-2T	IC	Sdn	1935

MARTIN TURNER,
c/o BRIGHTLYCOTT COTTAGE, SHIRWELL ROAD, BARNSTAPLE

Private Site

Gauge 4ft 8½in   **SS 571354**

–	4wDM		FH	2893	1944

WORLD OF COUNTRY LIFE,
WEST DOWN LANE, SANDY BAY, near EXMOUTH

EX8 5BY

Gauge 1ft 6in   www.worldofcountrylife.co.uk   **SY 035808**

–	4-2-2	OC	RegentSt	1898
SIR FRANCIS DRAKE	4-6-0	OC	Scarrott	1988

DORSET

INDUSTRIAL SITES

ALASKA ENVIRONMENTAL CONTRACTING LTD,
STOKEFORD FARM, EAST STOKE, WAREHAM   **BH20 6AL**
Locomotive may be present in yard between contracts.
Gauge 2ft 0in   www.alaska.ltd.uk   **SY 873872**

–	4wDM	MR	8614	1941

MINISTRY OF DEFENCE, LULWORTH RANGES, EAST LULWORTH
See Section 6 for details.   **SY 863822**

VITACRESS SALADS LTD, WATERCRESS GROWERS,
DODDINGS FARM, BERE REGIS
Gauge 1ft 6in   www.vitacress.com   **SY 852934**

VITACRESS	4wPH	Jesty	1948

PRESERVATION SITES

BUTCOMBE BREWERY LTD,
AVON CAUSEWAY HOTEL, HURN, near BOURNEMOUTH   **BH23 6AS**
Gauge 4ft 8½in   www.butcombe.com   **SZ 136976**

–	0-4-0DM	JF	22871	1939

G.A. & L.M. FELDWICK, 22A ROPERS LANE, WAREHAM   **BH20 4QT**
Gauge 2ft 0in   **SY 921875**

Private location with locomotives occasionally present. Visitors by appointment only.

JEFF LEAKE, 154 WAREHAM ROAD, LYTCHETT MATRAVERS   Private Site
Gauge 1ft 3in   SY 945950

–	4w-4BER	Leake	1992
–	0-4-0BE	Leake	1992
–	0-4-0DM	Leake	1992

MICKY FINN RAILWAY, near POOLE
Gauge 2ft 0in   Private site

8	4wDM	L	28039	1945
–	4wDM	MR	26007	1964

MOORS VALLEY RAILWAY, MOORS VALLEY COUNTRY PARK, HORTON ROAD, ASHLEY HEATH, RINGWOOD

Gauge 2ft 0in   www.moorsvalleyrailway.co.uk

BH24 2ET
SZ 104061

EMMET		0-4-0T		Moors Valley 20	1995	
a rebuild of		0-4-0DM		OK	21160	1938	a

a   currently at the Old Kiln Light Railway, near Farnham, Surrey

Gauge 1ft 3in

KATIE		0-4-2T	OC	HaylockJ		
a rebuild of		0-4-0T	OC	FMB	002	1997

THE NORTH DORSET RAILWAY TRUST, NORTH DORSET RAILWAY, SHILLINGSTONE STATION, STATION ROAD, SHILLINGSTONE

Gauge 4ft 8½in   www.northdorsetrailway.co.uk

DT11 0SA
ST 825116

(62-669)	30075	0-6-0T	OC	DDak	669	1960
(62-521)	30076	0-6-0T	OC	DDak	521	1954
(D1)	"ASHDOWN"	0-6-0DM		HC	D1186	1959
		reb		HE	8526	1977
DS 1169	LITTLE EVA	4wDM		RH	305302	1951
TP57P		2w-2PMR		Wkm	8267	1959

PRIVATE OWNER, PRIVATE LOCATION

Gauge 4ft 8½in

Private Site

47160	CUNARDER	0-6-0T	OC	HE	1690	1931
92207	MORNING STAR	2-10-0	OC	Sdn		1959

SWANAGE RAILWAY SOCIETY

Locomotives are kept at :-

Swanage Station   BH19 1HB   SZ 026789
Corfe Castle   BH20 5EJ   SY 962821
Furzebrook   (BH20 5AR)   SY 932841
Harmans Cross   BH19 3EB   SY 979800
Herston Railway Works   BH19 1AH   SZ 018792
Norden   BH20 5DW   SY 955829, 957827

Gauge 4ft 8½in   www.swanagerailway.co.uk

No.563		4-4-0	OC	9E	1893	
(30053)	53	0-4-4T	IC	9E	1905	a
(31625)	5 "JAMES"	2-6-0	OC	Afd	1929	
31806	(806)	2-6-0	OC	Afd	1928	
	rebuild of	2-6-4T	OC	Bton	1926	
31874		2-6-0	OC	Woolwich	1925	a
34028	EDDYSTONE	4-6-2	3C	Bton	1946	
		reb		Elh	1958	
34070	MANSTON	4-6-2	3C	Bton	1947	
34072	257 SQUADRON	4-6-2	3C	Bton	1948	
D3551	08436 BEIGHTON	0-6-0DE		Derby	1958	
D3591	(08476)	0-6-0DE		Crewe	1958	
D6515	(33012) LT. JENNY LEWIS RN	Bo-BoDE		BRCW	DEL107	1960

(D6528)	33111	Bo-BoDE	BRCW	DEL120	1960	
W51356		2-2w-2w-2DMR	PSteel		1960	
W51388		2-2w-2w-2DMR	PSteel		1960	
51392	(117701)	2-2w-2w-2DMR	PSteel		1960	
W55028	(960.012 L128 977860)	2-2w-2w-2DMR	PSteel		1960	
2054	BERYL	4wPM	FH	2054	1938	
	MAY	0-4-0DM	JF	4210132	1957	
TRS 802	99709 909271-7	2w-2PMR	Geismar	ST/01/14	2001	
TRS 804	99709 901063-8	2w-2PMR	Geismar	ST/03/31	2003	
	99709 909140-4	2w-2PMR	Geismar	ST/04/09	2004	
TRS 801	99709 909194-1	2w-2PMR	Geismar	ST/09/01	2009	Dsm
(PWM 4302)	68056	2w-2PMR	Wkm	7505	1956	

a   currently at Herston Railway Works, Swanage

Purbeck Mineral & Mining Group,
Corfe Castle Goods Shed, Corfe Castle   BH20 5EJ
Gauge 2ft 8in   www.purbeckminingmuseum.org   SY 962821

(SECUNDUS)	0-6-0WT	OC	B&S		1874

Purbeck Mining Museum, Purbeck Park, Corfe Castle   BH20 5DW
Gauge 2ft 0in   SY 957828

(SNAPPER)		4wDM	RH	283871	1950
(NORDEN)		4wDM	RH	392117	1956
–		4wDM	MR	8994	1946
	a rebuild of	4wDM	MR	8618	1941

P.C. VALLINS, near BLANDFORD
Gauge 2ft 0in   Private Site

–	4wDM	L	9256	1937
–	4wPM	L	18557	1942
–	4wDM	LB	51917	1960

DURHAM

Our definition of "County Durham" includes those locations which were situated within the part of the former County of Tyne & Wear which is South of the River Tyne and now comprises a number of unitary authorities. Locomotives which are located within the County Boroughs of Hartlepool and Stockton-on-Tees are listed under a separate heading of Teesside.

INDUSTRIAL SITES

GRANGE IRON COMPANY LTD,
56 BROOMSIDE LANE, CARRVILLE, DURHAM   DH1 2QT
Gauge 2ft 0in   NZ 309439

new minetubs under construction usually present

HARGREAVES SERVICES plc, WEST TERRACE, ESH WINNING   DH7 9PT
Gauge 2ft 6in   www.hsgplc.co.uk   NZ 194422

29	(12)		4wBEF		CE	B1575*	1978
				reb	CE	B3377	1987

*   one of CE B1575C or CE B1575E
Private site - locomotive cannot be viewed from any public location

HITACHI RAIL EUROPE, TRAIN MANUFACTURING FACILITY, IEP MERCHANT PARK, MILLENNIUM WAY, HEIGHINGTON, NEWTON AYCLIFFE   DL5 6UG
Gauge 4ft 8½in   www.hitachirail.com   NZ 266221

(D3830)	08663	ST. SILAS	0-6-0DE		Hor		1959	a
CRAB 1	NAY 1000075		4wBE	R/R	Zephir	2630	2016	
–			4wDH	R/R	Zephir	2631	2016	
	NAY 1000185		4wBE	R/R	Zephir	2632	2016	
CRAB 2	NAY 1000371		4wBE	R/R	Zephir	2678	2016	

new multiple unit type rolling stock under construction/repair usually present
a   property of Railway Support Services Ltd, Wishaw, Warwickshire

JOSHUA PETER MARSH, t/a RAILWAY SIGNS, FROSTERLEY
Gauge 2ft 0in   Private Site

No.2	4wBE	MarshJP	2020

NBR ENGINEERING SERVICES LTD, UNIT 6, 2 BARTON STREET, DARLINGTON   DL1 2LP
Gauge 3ft 0in   www.nbres.co.uk   NZ 300155

KETTERING FURNACE No.3	0-4-0ST	OC	BH	859	1885

Gauge 1ft 8in

STEAMPLEX		4wVBT	VCG	AK	93R	2013
	rebuild of	4wDM		MR	5877	1935

locomotives under construction, refurbishment, overhaul or repair usually present

TYNE & WEAR FIRE & RESCUE SERVICE, BARMSTON MERE TRAINING CENTRE, NISSAN WAY, WASHINGTON   SR5 3QY
Gauge 4ft 8½in   www.twfire.gov.uk   NZ 329570

3721	4w-4wRER	MetCam	1983	OOU

used as a static training aid

U.K. MINING VENTURES LTD, STANHOPE
Rogerley Mine
Gauge 2ft 0in   www.ukminingventures.com   Private Site

1291	2848	4wBE	CE	(5858	1971?)	a

a   battery box carries worksplate CE B0476

Diana Maria Mine
Gauge 2ft 0in   Private Site

1	ROSIE	0-4-0BE	WR		
2		0-4-0BE	WR	G7174	1967

PRESERVATION SITES

BEAMISH – THE LIVING MUSEUM OF THE NORTH,
BEAMISH, Near STANLEY   DH9 0RG
Gauge 4ft 8½in   www.beamish.org.uk   NZ 215548, 215545, 218543, 221543

	"PUFFING BILLY"	4wG	VC	AK	71	2006
E No.1		2-4-0VBCT	OC	BH	897	1887
	"STEAM ELEPHANT"	6wG	VC	_(Dorothea		2001
				(AK		2001
–		0-4-0VBT	VC	HW		1871
			reb	Wilton(ICI)		1984	a
17		0-4-0VBT	OC	HW	33	1873
18		0-4-0T	OC	Lewin	683	1877
	NEWCASTLE	0-6-0ST	IC	MW	1532	1901
No.15	ROKER	0-4-0CT	OC	RSHN	7006	1940
No. 5		0-4-0ST	OC	SDSI(S)		1900
2		4wWE		Siemens	455	1908

    a   carries plate T.H. Head, Engineer, 50 Cannon St, London 1871

Gauge 2ft 0in

	SAMSON	0-4-0WTG	OC	Beamish	BM2	2014
No.2		0-4-0WT	OC	KS	721	1901
	L.R. 3098	4wPM		MR	1377	1918

Gauge 1ft 11½in

	GLYDER	0-4-0WT	OC	AB	1994	1931
	EDWARD SHOLTO	0-4-0ST	OC	HE	996	1909
	ASHOVER	4wDM		FH	3307	1948

BOWES RAILWAY CO LTD,
SPRINGWELL ROAD, SPRINGWELL, GATESHEAD   NE9 7QJ
Gauge 4ft 8½in   www.bowesrailway.uk   NZ 285589

No.22	No.85	0-4-0ST	OC	AB	2274	1949
	W.S.T.	0-4-0ST	OC	AB	2361	1954
101		4wDM		FH	3922	1959
–		0-4-0DH		HE	6263	1964
	PERKY	0-4-0DE		RH	395294	1956
	PINKY	0-4-0DE		RH	416210	1957
	REDHEUGH	4wDM		RH	476140	1963
			reb	Wilton(ICI)		1982

No.1	9307/110	4wBE	CE	5921	1972
No.2	20/270/34	4wBE	CE	B3060	1983

Gauge 2ft 6in

B03		4w-4wDHF	_(HE	8515	1981
			(AB	651	1981
No.8		4wBE	CE	B1840	1978

Gauge 2ft 0in

20.123.945	SER No.2476	4wBEF	_(EE	2476	1958
	PLANT No.2207/456		(RSHN	7980	1958
–		0-6-0DMF	HC	DM842	1954

CLASS G5 LOCOMOTIVE COMPANY LTD, UNIT 8 S, HACKWORTH INDUSTRIAL PARK, SHILDON DL4 1HS
Gauge 4ft 8½in   www.g5locomotiveltd.co.uk   NZ 224255

"1759"	0-4-4T	IC	GNS/G5 Loco Co	u/c

HOPETOWN DARLINGTON - DARLINGTON RAILWAY HERITAGE QUARTER
Encompassing the premises of Darlington Railway Museum, the A1 Steam Locomotive Trust, Darlington Railway Preservation Society and the North Eastern Locomotive Preservation Group, along with historic buildings located on site, celebrating 200 years of the Stockton & Darlington Railway.

Darlington Railway Museum,
North Road Station, Station Road, Hopetown, Darlington DL3 6ST
Gauge 4ft 8½in   www.hopetowndarlington.co.uk   NZ 288157

25	DERWENT	0-6-0	OC	Kitching		1845
No.1463		2-4-0	IC	Dar		1885
No.39		0-6-0T	OC	RSHD	6947	1938
(D6898)	37198	Co-CoDE		_(EE	3376	1964
				(EES	8419	1964

Gauge 1ft 0¼in

No.1	2-2-2	IC	Leatham	No.1	c1845

A1 Steam Locomotive Trust,
Darlington Locomotive Works, Bonomi Way, Darlington DL3 0PY
Gauge 4ft 8½in   www.a1steam.com   NZ 289157

60163	TORNADO	4-6-2	3C	Darlington	2195	2008	a
2007	PRINCE OF WALES	2-8-2	3C	Darlington		2015	u/c

a   currently at Locomotive Maintenance Services Ltd, Loughborough, Leicestershire

Darlington Railway Preservation Society,
North Road Locomotive Works (1861 Shed), Whessoe Road, Darlington DL3 0QT
Gauge 4ft 8½in   www.drps.synthasite.com   NZ 286160

	NORTHERN GAS BOARD No.1	0-4-0ST	OC	P	2142	1953
	PATONS	0-4-0F	OC	WB	2898	1948
185	DAVID PAYNE	0-4-0DM		JF	4110006	1950
–		4wDM		RH	279591	1949
–		0-4-0DE		RH	312988	1952

		0-4-0DM	_(RSHN	7925	1959
			(DC	2592	1959
1		4wWE	GEC	(1928?)	
(DX 68022	DB 965096 DE 320514)	2w-2PMR	Wkm	7611	1957
(DX 68003)	68/007 DB965951 MPP 0007	2w-2PMR	Wkm	10647	1972

Gauge 1ft 8in

		4wDM	RH	476124	1962

North Eastern Locomotive Preservation Group,
North Road Locomotive Works (1861 Shed), Whessoe Road, Darlington   DL3 0QT
Gauge 4ft 8½in   www.nelpg.org.uk   NZ 286160

69023	JOEM	0-6-0T	IC	Dar	2151	1951

KIRK MERRINGTON PRIMARY SCHOOL,
SOUTH VIEW, KIRK MERRINGTON, near SPENNYMOOR   DL16 7JB
Gauge 4ft 8½in   www.kirkmerrington.durham.sch.uk   NZ 262310

55586	142045		2-2wDMR	_(BRE(D)		1986
				(Leyland R5.96		1986
		rebuilt as	2-2wDHR	RFSD		1988
55636	142045		2w-2DMR	_(BRE(D)		1986
				(Leyland R5.97		1986
		rebuilt as	2w-2DHR	RFSD		1988

KYNREN, FLATTS FARM, TORONTO, BISHOP AUCKLAND   DL14 7SF
(operated by Eleven Arches charity)
Gauge 5ft 0in   www.11arches.com   NZ 211307

	LOCOMOTION	0-4-0BE	s/o		c2016

LOCOMOTION – THE NATIONAL RAILWAY MUSEUM AT SHILDON,
DALE ROAD INDUSTRIAL ESTATE, SHILDON   DL4 2RE
Gauge 4ft 8½in   www.locomotion.org.uk   NZ 239255

	LOCOMOTION	0-4-0	VC	RS	1	1825
No.1	ROCKET	0-2-2	OC	RS	19	1829
	"SANS PAREIL"	0-4-0	VC	Hackworth		1829
–		0-4-0	VC	GS(H)		c1849
No.66	AEROLITE	2-2-4T	IC	Ghd		1869
	ROCKET	0-2-2	OC	FlourMill		2010
		rebuild of		LocoEnt	No.2	1979
	SANS PAREIL	0-4-0	VC	BRE(S)		1980
790	HARDWICKE	2-4-0	IC	Crewe	3286	1892
1621		4-4-0	IC	Ghd		1893
34051	WINSTON CHURCHILL	4-6-2	3C	Bton		1946
(45000)	5000	4-6-0	OC	Crewe	216	1934
49395		0-8-0	IC	Crewe	5662	1921
(63460)	901	0-8-0	3C	Dar		1919
	IMPERIAL No.1	0-4-0F	OC	AB	2373	1956
	JUNO	0-6-0ST	IC	HE	3850	1958

No.77	NORWOOD		0-6-0ST	OC	RSHN	7412	1948	
	DELTIC		Co-CoDE		EEDK	2007	1955	
D200	(40122)		1Co-Co1DE		_(EE	2367	1957	
					(VF	D395	1957	
D2090	(03090)		0-6-0DM		Don		1960	
(D3079	08064) 13079		0-6-0DE		Dar		1954	
(D4141)	08911 MATEY		0-6-0DE		Hor		1962	
(D5500)	31018		A1A-A1ADE		BT	71	1957	
41001	(43000) 252001		Bo-BoDE		Crewe		1972	
43102	(43302) THE JOURNEY SHRINKER - 148.5MPH THE WORLDS FASTEST DIESEL TRAIN							
	(254024)		Bo-BoDE		Crewe		1978	
E5001	(71001) 89403		Bo-BoRE/WE		Don		1959	
(26500)	No.1		Bo-BoWE/RE		BE		1905	
(DS 75)	75S		4wRE		_(Siemens	6	1898	
					(HC		1898	a
(BEL 2)	No.1		4wBE		Stoke		1917	
–			4wPM		MR	4217	1931	
H 001			4wDH		S	10003	1959	
55542	142001		2-2wDMR		_(BRE(D)		1985	
					(Leyland R5.002		1985	
		rebuilt as	2-2wDHR		HAB		1990	
55592	142001		2w-2DMR		_(BRE(D)		1985	
					(Leyland R5.001		1985	
		rebuilt as	2w-2DHR		HAB		1990	
2090	S10656		4w-4wRER		_(Lancing		1937	
					(Elh		1937	
3131	11179		4w-4RER		EE/Elh		1938	
4308	61275		4w-4wRER		Afd/Elh		1959	
65217	306 017		4w-4wWER		MetCam		1949	
APT-E	PC1/TC1/TC2/PC2	4w-4w-4-4w-4wArticGTE			Derby		1972	
–			4wBE	R/R	Zagro	80969	2022	b
(960209)			2w-2PMR		Wkm	899	1933	

a   carries worksplate Siemens 7/1898
b   property of SCT Rail, Brough, East Yorkshire

Gauge 3ft 6in

390		4-8-0	OC	SS	4150	1896	

Gauge 3ft 0in

No. 14	No.4 9306/108	0-6-0DMF		HC	DM1274	1961	

Gauge 1ft 3in

1	LOCOMOTION 1825	0-4-0	VC	(Dar?)		1875

Soho Engine Shed, Soho Street, Shildon   DL4 1PQ
Gauge 4ft 8½in   (open during special event days)   NZ 233257

"NELSON"	(BRADYLL)	0-6-0	OC	(Hackworth?)	c1835	a

a   possibly built by Fossick & Hackworth, Stockton-on-Tees

SUNDERLAND CITY COUNCIL, TYNE & WEAR JOINT MUSEUMS SERVICE, WASHINGTON "F" PIT MUSEUM, WASHINGTON NEW TOWN NE37 1BN

Gauge 2ft 0in www.mysunderland.co.uk/washington-f-pit NZ 303574

–	0-4-0DMF	RH	392157	1956	

TANFIELD RAILWAY PRESERVATION SOCIETY, MARLEY HILL, near STANLEY NE16 5ET

Gauge 4ft 8½in www.tanfield-railway.co.uk NZ 207573

–		0-6-0ST	OC	AB	1015	1904	
No.6		0-4-2ST	OC	AB	1193	1910	
(No.32)	STANLEY	0-4-0ST	OC	AB	1659	1920	
	WELLINGTON	0-4-0ST	OC	BH	266	1873	
	IRWELL	0-4-0ST	OC	HC	1672	1937	
38		0-6-0T	OC	HC	1823	1949	
–		0-4-0ST	OC	HL	2711	1907	
		reb		DL		1956	
No.2		0-4-0ST	OC	HL	2859	1911	
L.& H.C. 14		0-4-0ST	OC	HL	3056	1914	
		reb		‡		1980	
	STAGSHAW	0-6-0ST	OC	HL	3513	1927	
	COAL PRODUCTS No.3	0-6-0ST	OC	HL	3575	1923	
–		0-6-0F	OC	HL	3746	1929	
No.3	TWIZELL	0-6-0T	IC	RS	2730	1891	
–		0-4-0CT	OC	RSHN	7007	1940	
	LYSAGHT'S	0-6-0ST	OC	RSHD	7035	1940	
49		0-6-0ST	IC	RSHN	7098	1943	
	SIR CECIL A.COCHRANE	0-4-0ST	OC	RSHN	7409	1948	
No.44	9103/44	0-6-0ST	OC	RSHN	7760	1953	
38		0-6-0ST	OC	RSHN	7763	1954	
21		0-4-0ST	OC	RSHN	7796	1954	
–		0-6-0ST	OC	RSHN	7800	1954	
No.48		0-6-0ST	OC	RSHN	7944	1957	
(No.3)		0-4-0ST	OC	RWH	2009	1884	
(No.4)		4wVBT	VCG	S	9559	1953	
No.20		0-6-0ST	IC	WB	2779	1945	
14		0-4-0DE		AW	D21	1933	
No.2		0-4-0DE		AW	D22	1933	
–		4wDM		FH	3716	1955	
	(2111-125)	0-6-0DH		HE	6612	1965	
No.6		0-6-0DH		JF	4240010	1960	
T.I.C.No.35		0-4-0DE		RH	418600	1958	
–		0-6-0DM		RSHN	7697	1953	
–		0-4-0DM		RSHN	6980	1940	
–		0-6-0DM		RSHN	7746	1954	
–		0-4-0DM		RSHN	7901	1958	
(9)		Bo-BoWE		AEG	1565	1913	
–		4wWE		GB	2509	1955	Dsm
2	DEREK SHEPHERD	Bo-BoWE		HL	3872	1936	a
	rebuilt as	Bo-BoBE		Riley		1993	
	(KEARSLEY No.3)	Bo-BoWE		RSHN	7078	1944	

E10		4wWE		Siemens	862	1913
MPP 10157		2w-2BER		Bance	097	2000
–		2w-2DHR		Bg	3565	1962

a   carries plate RSH 3872
‡   rebuilt by Clark Hawthorn Ltd, Northumberland Engine Works, Wallsend-on-Tyne

Gauge 3ft 6in

(M2)		4-6-2	OC	RSHD	7430	1951

Gauge 2ft 6in

–		4wBEF		CE	B1886B	1980

Gauge 2ft 0in

4	DM1067	0-6-0DMF		HC	DM1067	1959
–		4wDM		HE	2577	1942
2	AYLE	4wDM		HE	2607	1942
No.1		4wDHF		HE	7332	1974
–		4wDM		LB	53162	1962
–		4wDM		LB	54781	1965
–		4wDM		RH	244487	1946
–		4wDM		RH	323587	1952
No.2		4wBEF		CE	B3141B	1984
–		4wBEF		_(EE	2848	1960
				(RSHN	8201	1960
–		0-4-0BE		WR		1972
–		0-4-0BE		WR		

THORPE LIGHT RAILWAY, WHORLTON, near BARNARD CASTLE
Private site with occasional public open days   **Private Site**
Gauge 1ft 3in   www.thorpelightrailway.co.uk   **NZ 106146**

"BESSIE"	4wDM		Eddy/Knowell		2002	
WENDY	4-4wDM		ColebySim		1972	
–	2-8-0DH	s/o	SL	73.35	1973	
reb	2-6-0DH	s/o			1994	
reb	2-8-0DH	s/o	MRWRS		c2009	a

a   property of MRW Railways Ltd, Sheffield, South Yorkshire

WEARDALE RAILWAY LTD, (a subsidiary of The Auckland Trust)
WOLSINGHAM RAILWAY CENTRE, off DURHAM ROAD, WOLSINGHAM   DL13 3JD
(Operated by Weardale Railway Heritage Services Ltd, a subsidiary of Weardale Railway Trust)
Locomotives may also be found at :- Stanhope Station, Bondisle, Stanhope DL13 2YS   NY 998387
No public access to Wolsingham Depot.
Gauge 4ft 8½in   www.weardale-railway.org.uk   **NZ 081370**

No.40	68692 9312/40	0-6-0T	OC	RSHN	7765	1954	
(D5637	31213) 31465	A1A-A1ADE		BT	237	1960	
(D5684	31256) 31459	A1A-A1ADE		BT	285	1961	
(D5817)	31285	A1A-A1ADE		BT	318	1961	
–		0-6-0DH		RR	10187	1964	
–		4wDH		RR	10232	1965	OOU
M50980		2-2w-2w-2DMR		DerbyC&W		1959	

M51572		2-2w-2w-2DMR	DerbyC&W		1960	OOU
M52054		2-2w-2w-2DMR	DerbyC&W		1960	
E55012		2-2w-2w-2DMR	GRC&W		1959	
55728 142078		2-2wDMR	_(BRE(D)		1987	
			(Leyland R5.186		1987	
	rebuilt as	2-2wDHR	AB		1990	
55774 142078		2w-2DMR	_(BRE(D)		1987	
			(Leyland R5.161		1987	
	rebuilt as	2w-2DHR	AB		1990	
(DX 68003 DB 965331)		2w-2PMR	Wkm	10179	1968	
(DX 68080 DB 965993) POINTLESS 3		2w-2DMR	Wkm	10706	1974	DsmT
DX 68090 (DB 966035)		2w-2DMR	Wkm	10843	1975	

ESSEX

INDUSTRIAL SITES

HALTERMANN CARLESS UK LTD, HARWICH REFINERY,
REFINERY ROAD, PARKESTON, HARWICH　　　　　　　　　　　　**CO12 4SS**
Gauge 4ft 8½in　　　www.haltermann-carless.com　　　　　　**TM 232323**

Q240 JBV	4wDM	R/R	_(Unimog 092692 1982	OOU
			(Zweiweg 12/993 1982	

QINETIQ LTD,
ENVIRONMENTAL TEST CENTRE, FOULNESS ISLAND, SHOEBURYNESS
Gauge 2ft 6in　　　www.qinetiq.com　　　　　　　　　　　**TQ 980889**

RAMBO	4wBEF	CE	B0483	1976	
TERMINATOR	4wBEF	CE	B0483	1976	

PRESERVATION SITES

THE LITTLE BRAXTED RAILWAY, BRAXTED BAKERY,
HOMEFIELD HOUSE, LITTLE BRAXTED LANE, WITHAM　　　　**CM8 3ET**
Gauge 2ft 0in　　　Private site　　　　　　　　　　　　**TL 838140**

103	0-4-0T	OC	Hen	16045	1918
09735	4wDH		AK	18	1985
–	4wDM		RH	296111	1950

Private site - visitors by appointment only

BRITISH POSTAL MUSEUM, STORE & ARCHIVE, DEBDEN
Gauge 2ft 0in www.postalmuseum.org

IG10 3UF
TQ 444962

–	4w Atmospheric Car		1861	a

a Cut into two halves, Inventory no.s 31.85/1 and 31.85/2

CHELMSFORD CITY COUNCIL, CHELMSFORD CITY MUSEUM, OAKLANDS PARK, MOULSHAM STREET, CHELMSFORD
Gauge 2ft 8½in www.chelmsford.gov.uk/museums

CM2 9AQ
TL 705069

8	4wRER	BE	1898
	reb	BE	c1911

COLNE VALLEY RAILWAY PRESERVATION SOCIETY LTD, CASTLE HEDINGHAM STATION, YELDHAM ROAD, CASTLE HEDINGHAM, HALSTEAD
Gauge 4ft 8½in www.colnevalleyrailway.co.uk

CO9 3DZ
TL 774362

35010	BLUE STAR		4-6-2	3C	Elh		1942
			reb		Elh		1957
45163			4-6-0	OC	AW	1204	1935
45293			4-6-0	OC	AW	1348	1936
2138	SWORDFISH		0-6-0ST	OC	AB	2138	1941
1875	BARRINGTON		0-4-0ST	OC	AE	1875	1921
WD 190			0-6-0ST	IC	HE	3790	1952
WD 200	No.24		0-6-0ST	IC	HE	3800	1953
–			0-4-0ST	OC	HL	3715	1928
No.60	JUPITER		0-6-0ST	IC	RSHN	7671	1950
(D2046)			0-6-0DM		Don		1958
			reb		HE	6644	1967
D2041			0-6-0DM		Sdn		1959
D2184			0-6-0DM		Sdn		1962
43023	(253011)		Bo-BoDE		Crewe		1976
43071	(254008)		Bo-BoDE		Crewe		1977
43073	(254009)	NEVILLE HILL HST DEPOT - 42 YEARS					
			Bo-BoDE		Crewe		1977
43082	(254014)	RAILWAY CHILDREN - FIGHTING FOR STREET CHILDREN					
			Bo-BoDE		Crewe		1978
43165	(253042)		Bo-BoDE		Crewe		1981
W51339			2-2w-2w-2DMR		PSteel		1960
W51382			2-2w-2w-2DMR		PSteel		1960
W55033	(977826 T003)		2-2w-2w-2DMR		PSteel		1960
–			4wDM		FH	3147	1947
(5)	YARD No.12228		0-4-0DH		HE	6975	1968
	HENRY		4wDM		Lake&Elliot		c1924
			reb		FordTTC		1997
–			0-4-0DM		RH	281266	1950
887			4wDM		RH	394009	1955
–			4wDM	R/R	Unilok(G)	2109	1980
1	DOUG TOTTMAN		Bo-BoBE		RSHN	7284	1945
			reb		Kearsley		1982

–	2w-2PMR	Wkm		1946	1935
(9023)	2w-2PMR	Wkm		8087	1958

CRAVEN HERITAGE TRAINS LTD, EPPING SIGNALBOX, EPPING   CM16 4HW
Gauge 4ft 8½in   www.cravensheritagetrains.co.uk   TL 460013

(L11)	4w-4wRE	MetCam		1931/1932
	reb	Acton		1964

ROGER CRAVEN, PRIVATE LOCATION
Gauge 3ft 0in   Private Site

–		4wDMF	RH	418803	1957

Gauge 750mm

5	BETSY	4wDH	HE	8829	1979
16		4wDH	HE	9079	1984
15		4wDH	HE	9080	1984
(11)		4wDM	Moës		
27		4wDM	Moës		
4		4wDM	Moës		
(3)		4wDMF	RH	375693	1954

Gauge 650mm

2		0-4-0DM	RH	305326	1952

Gauge 600mm

–		0-4-0DM	Dtz	47069	1950
1		4wDM	Diema	1407	1951
–		4wDM	Jung	7649	1937
17		4wDM	Moës		
–		4wDM	Moës		
–		4wDM	Moës		

EAST ANGLIAN RAILWAY MUSEUM,
CHAPPEL & WAKES COLNE STATION, STATION ROAD, WAKES COLNE CO6 2DS
Gauge 4ft 8½in   www.earm.co.uk   TL 898289

(69621 7999)	9621	0-6-2T	IC	Str		1924	
No.11	STOREFIELD	0-4-0ST	OC	AB	1047	1905	
1		0-6-0T	IC	EARM		2008	
	rebuild of	0-6-0ST	IC	RSHN	7031	1941	
	JEFFREY	0-4-0ST	OC	P	2039	1943	
	(JUBILEE)	0-4-0ST	OC	WB	2542	1936	
	LAMPORT No.3	0-6-0ST	OC	WB	2670	1942	
D2279	(11249)	0-6-0DM		_(RSHD	8097	1960	
				(DC	2656	1960	
E51213		2-2w-2w-2DMR		MetCam		1959	
E79963		2w-2DMR		WMD	1268	1958	
A.M.W.No.144	JOHN PEEL	0-4-0DM		AB	333	1938	
7		0-4-0DH	s/o	JF	4220039	1965	a

–		4wPM	MR	2029	1920	
WD 72229		0-4-0DM	_(VF	5265	1945	
		(DC	2184	1945		
	reb	YEC	L120	1993		
XOH 2299 (HCT 008)		4wDH	Perm	008	1988	
(TR 37) (PWM 2797)		2w-2PMR	Wkm	6896	1954	b

a   fitted with GER tram bodywork
b   currently off site

EPPING ONGAR RAILWAY

Locomotives are kept at :-

Ongar Goods Yard, Ongar   CM5 9AB   TL 551034
North Weald Station, North Weald Bassett   CM16 6BT   TL 496036

Gauge 5ft 0in   www.eorailway.co.uk

(1008) DRACULA CASTLE	4-6-2	OC	LO		157	1948	

Gauge 4ft 8½in

4141	2-6-2T	OC	Sdn		1946	
4953 PITCHFORD HALL	4-6-0	OC	Sdn		1929	
(5521) L 150	2-6-2T	OC	Sdn		1927	
1	0-4-4T	IC	Neasden	3	1898	
ISABEL	0-6-0ST	OC	HL	3437	1919	
56	0-6-0ST	IC	RSHN	7667	1950	
63 CORBY	0-6-0ST	IC	RSHN	7761	1954	
(D22) 45132	1Co-Co1DE		Derby		1961	
(D1606 47029) 47635 JIMMY MILNE	Co-CoDE		Crewe		1964	
(D2119) 03119	0-6-0DM		Sdn		1959	
D2170 (03170)	0-6-0DM		Sdn		1960	
(D3180) 08114 (13180) GOTHAM	0-6-0DE		Derby		1955	
(D5557 31139, 31538) 31438	A1A-A1A DE		BT	156	1959	
D6729 (37029)	Co-CoDE		_(EE	2892	1961	
			(VF	D608	1961	
–	4wDM		RH	398616	1956	
51342	2-2w-2w-2DMR		PSteel		1960	
51384	2-2w-2w-2DMR		PSteel		1960	
60110 205205	4-4wDER		Afd/Elh		1957	
1031 (1085)	2-2w-2w-2RER		MetCam		1959	
PM002 BADGER	4wDHR		Perm	T002	1988	
(DX 68086) No.1	2w-2DMR		Wkm	10841	1975	

"GIFFORDS", COXTIE GREEN ROAD, PILGRIMS HATCH, BRENTWOOD   CM14 5RP

Gauge 5ft 0in   **Private Site**   **TQ 562959**

792 HEN	0-6-0T	OC	TK	373	1927	

GLENDALE FORGE, MONK STREET, near THAXTED   CM6 2NR

Gauge 2ft 0in   **(Closed)**   **TL 612287**

145 C.P.HUNTINGTON	4w-2-4wPH	s/o	Chance		
			76-50145-24	1976	
ROCKET	0-2-2+4wPH	s/o	FRgroup4	1970	

| | 4wDM | RH | | 1942 | a Dsm |

a   either RH 217973 / 1942, or RH 213853 / 1942

HIGH BARN HERITAGE, HIGH BARN ROAD, HALSTEAD   CO9 1RR
Gauge 4ft 8½in   www.highbarnheritage.co.uk   TL 800769

YD No.43	4wDM	RH	221639	1943

Private site with no public access

MANGAPPS RAILWAY MUSEUM, SOUTHMINSTER ROAD, BURNHAM-ON-CROUCH   CM0 8QG
Gauge 4ft 8½in   www.mangapps.co.uk   TQ 944980

Identity			Type		Builder	Works No.	Date	Note
TOTO			0-4-0ST	OC	AB	1619	1919	
MINNIE			0-6-0ST	OC	FW	358	1878	
"BROOKFIELD"			0-6-0PT	OC	WB	2613	1940	
(D1778	47183	47793) 47579	JAMES NIGHTALL G.C.					
			Co-CoDE		BT	540	1964	
(D2018	03018)		0-6-0DM		Sdn		1958	
(D2020)	03020	F134L	0-6-0DM		Sdn		1958	a
(D2081)	03081	LUCIE	0-6-0DM		Don		1960	
(D2089)	03089		0-6-0DM		Don		1960	
D2158	(03158)	MARGARET ANN	0-6-0DM		Sdn		1960	
(D2203)	11103		0-6-0DM		_(VF	D145	1952	
					(DC	2400	1952	
D2325			0-6-0DM		_(RSHD	8184	1961	
					(DC	2706	1961	
(D2399)	03399		0-6-0DM		Don		1961	
(D5523)	31105	RADIO CAROLINE	A1A-A1ADE		BT	122	1959	
(D5660)	31233		A1A-A1ADE		BT	260	1960	
(D6530)	33018		Bo-BoDE		BRCW	DEL122	1960	a
ELLAND No.1			0-4-0DM		HC	D1153	1959	
–			4wDM	R/R	S&H	7502	1966	
11104			0-6-0DM		_(VF	D78	1948	
					(DC	2252	1948	
(226)			0-4-0DM		_(VF	5261	1945	
					(DC	2180	1945	
1030			2-2w-2w-2RER		MetCam		1960	
22624			2w-2-2-2wRER		GRC&W	O/2228	1938	
			reb		GRC&W		1950	a
W51381			2-2w-2w-2DMR		PSteel		1960	b
PWM 2786	A14W	(TR36)	2w-2PMR		Wkm	6885	1954	
LLPW 01	PWM 3951	PW2 MAISIE	2w-2PMR		Wkm	6936	1955	
3700-84			2w-2PMR		Woodings	A466	c1980	

a   currently under restoration at Sonic Rail Services Ltd, Burnham-on-Crouch
b   converted to non-powered hauled stock

Gauge 3ft 6in

8	4wRER	ACCars		1949

SOUTHEND-ON-SEA CITY COUNCIL, SOUTHEND BOROUGH PARKS DEPARTMENT, CENTRAL NURSERY, BARLING ROAD, SOUTHEND-ON-SEA

Gauge 3ft 6in   www.southend.gov.uk   **TQ 915873**

7		4wRER	ACCars	1949
11		4wRER	ACCars	1949

SOUTHEND-ON-SEA CITY COUNCIL, SOUTHEND PIER RAILWAY, SOUTHEND-ON-SEA

SS1 1EE

Gauge 3ft 0in   www.southendpier.co.uk   **TQ 884850**

A	SIR JOHN BETJEMAN	4w-4wDH	SL	SE4	1986	
B	SIR WILLIAM HEYGATE	4w-4wDH	SL	SE4	1986	OOU
	SIR DAVID AMESS	(4w)BER	SL		2021	a
	WILLIAM BRADLEY	(4w)BER	SL		2021	a
1835		4wBER	Castleline		1996	

a   power unit within multiple coach unit

SOUTHEND PIER RAILWAY MUSEUM, SOUTHEND-ON-SEA

SS1 1EE

Gauge 3ft 6in   www.southendpiermuseum.co.uk   **TQ 884850**

22		4wRER	ACCars	1949
6		2-2wRER	BE	1890

THE OLD FOUNDRY, HIGH STREET, LEIGH-ON-SEA

SS9 1RP

Gauge 3ft 6in   (Closed)   **TQ 889891**

21		4wRER	ACCars	1949

WALTHAM ABBEY ROYAL GUNPOWDER MILLS CO LTD, BEAULIEU DRIVE, WALTHAM ABBEY

EN9 1JY

Gauge 3ft 0in   www.royalgunpowdermills.com   **TL 376013**

BB 307	2w-2BE	GB	6099	1964

Gauge 2ft 6in

DH 888 (ND 10392)	4wDH	BD	3755	1981
–	4wDH	HE	8828	1979
–	4wDH	Ruhrthaler	3920	1969

Gauge 1ft 6in

BUDLEIGH	4wDM	RH	235624	1945

GLOUCESTERSHIRE

INDUSTRIAL SITES

FIRE SERVICE COLLEGE LTD, LONDON ROAD, MORETON-IN-MARSH   GL56 0RH
Vehicles are used for static training purposes.
Gauge 4ft 8½in   www.fireservicecollege.ac.uk   SP 216329

| 64681 | 508212 (508133) | 4w-4wRER | York(BRE) | 1980 |
| 64724 | 508212 (508133) | 4w-4wRER | York(BRE) | 1980 |

THE FLOUR MILL LTD, BREAM, Forest of Dean
Locomotives for repair and restoration are usually present.
Gauge 4ft 8½in   www.theflourmill.com   SO 604067

| (5521) | LT 150 | 2-6-2T | OC | Sdn | | 1927 | a |
| | "WILLY" | 0-4-0WT | OC | KS | 3063 | 1918 | b |

Preserved vehicles currently present :

1450		0-4-2T	IC	Sdn		1935	
5538		2-6-2T	OC	Sdn		1928	
9642		0-6-0PT	IC	Sdn		1946	a
No.229		0-4-0ST	OC	N	2119	1876	
No.8		0-6-0T	OC	AB	1296	1912	
(No.17)	BANNOCKBURN	0-6-0T	OC	AB	1338	1913	
–		0-4-0ST	OC	AE	1498	1906	
No.3		0-4-0WT	OC	EB	37	1898	
	THE KING	0-4-0WT	OC	EB	(48?)	1906	
No.20	JENNIFER	0-6-0T	OC	HC	1731	1942	
1982	RING HAW	0-6-0ST	IC	HE	1982	1940	
–		0-6-0ST	IC	HE	3183	1944	

Gauge 2ft 6in

| No.1 | CHEVALLIER | 0-6-2T | OC | MW | 1877 | 1915 | c |

Gauge 2ft 0in

| | JANET | 4wDM | | RH | 504546 | 1963 | d |

a	based here, but visits other locations
b	currently at Swindon & Cricklade Railway Society, Wiltshire
c	currently operating in Romania
d	currently located elsewhere

HITACHI RAIL EUROPE LTD,
STOKE GIFFORD IEP DEPOT, LITTLE STOKE, BRISTOL   BS34 7QG
Gauge 4ft 8½in   www.hitachirail.com   ST 612801

| – | | 4wBE | | _(Express | E604 | 2016 |
| | | | | (Hegenscheidt | | 2016 |

PRESERVATION SITES

AVON VALLEY RAILWAY CO LTD, BITTON STEAM CENTRE,
BITTON STATION, BATH ROAD, BITTON   BS30 6HD
Gauge 4ft 8½in   www.avonvalleyrailway.org   ST 670705

44123		0-6-0	IC	Crewe	5658	1925	
TKh 4015	KAREL	0-6-0T	OC	Chrz	4015	1954	a
	EDWIN HULSE	0-6-0ST	OC	AE	1798	1918	
No.7	WIMBLEBURY	0-6-0ST	IC	HE	3839	1956	
	LITTLETON No 5	0-6-0ST	IC	MW	2018	1922	
(No.9)		0-6-0T	OC	RSHN	7151	1944	
–		4wVBT	VCG	S	7492	1928	
(D3429)	08359	0-6-0DE		Crewe		1958	
D3668	(09004)	0-6-0DE		Dar		1959	b
(D4103)	09015 ROB	0-6-0DE		Hor		1961	b
D4118	(08888)	0-6-0DE		Hor		1962	b
(D4157)	08927	0-6-0DE		Hor		1962	c
(D)5518	(31101)	A1A-A1ADE		BT	89	1958	
70043	GRUMPY	0-4-0DM		AB	358	1941	
–		4wDM		RH	235519	1945	
	RH 252823	4wDM		RH	252823	1947	d Dsm
(B8W)	PWM 3769	2w-2PMR		Wkm	6648	1953	

    a   carries plate Chrz 4939/1957
    b   property of Valley Rail Preservation
    c   property of Railway Support Services Ltd, Wishaw, Warwickshire
    d   converted to unpowered weed killing unit

DEAN FOREST RAILWAY CO LTD, LYDNEY, Forest of Dean
Locomotives are kept at :–   Norchard Steam Centre GL15 4ET SO 629044
                                     Lydney Riverside Station GL15 5EW SO 634025
Gauge 4ft 8½in   www.deanforestrailway.co.uk

5541		2-6-2T	OC	Sdn		1928
9681		0-6-0PT	IC	Sdn		1949
9682		0-6-0PT	IC	Sdn		1949
2		0-4-0ST	OC	AB	2221	1946
–		0-6-0ST	IC	HE	2411	1941
	GUNBY	0-6-0ST	IC	HE	2413	1941
WD 132	SAPPER	0-6-0ST	IC	HE	3885	1964
	a rebuild of			HE	3163	1944
	WILBERT REV. W. AWDRY	0-6-0ST	IC	HE	3806	1953
63.000.432	FRED WARRIOR	0-6-0ST	IC	HE	3823	1954
(RRM 28	No.65)	0-6-0ST	IC	HE	3889	1964
6	USKMOUTH 1	0-4-0ST	OC	P	2147	1952

WD 152	RENNES		0-6-0ST	IC	RSHN	7139	1944	
				reb	HE	3880	1961	
D3937	(08769)	GLADYS	0-6-0DE		Derby		1960	
(D5548)	31130		A1A-A1ADE		BT	147	1959	
D9521			0-6-0DH		Sdn		1964	
D9555			0-6-0DH		Sdn		1965	
	DON CORBETT		0-4-0DH		HE	5622	1960	
–			0-4-0DH		HE	6688	1968	
–			0-4-0DM		JF	4210127	1957	
E50619	(53619)	B 962	2-2w-2w-2DMR		DerbyC&W		1958	
M51566			2-2w-2w-2DMR		DerbyC&W		1959	
M51914			2-2w-2w-2DMR		DerbyC&W		1960	
(DS 3057)			4wPMR		Wkm	4254	1947	
	99709 909138-8		2w-2PMR		Geismar ST/04/07		2004	
105639	R135 NAU 99709 976014-9	4wDMR	R/R	_(Landrover		1998	
					(Perm		1998	

BRIAN FAULKNER
Locomotives kept at private location.

Gauge 2ft 0in   **Private Site**

–	4wDM	FaulknerB	2022	a		
–	4wDM	FaulknerB	2022	a		
–	4wPM	L	3834	1931		
–	4wDM	L	8022	1936		
–	4wDM	L	33650	1949		
–	4wDM	LB	56371	1970		
–	2w-2DM	StokesMJ	1986	b	Dsm	
–	4wBE	CE				

a   constructed by Brian Faulkner, using Lister parts
b   converted to non-powered passenger coach

GLOUCESTERSHIRE WARWICKSHIRE STEAM RAILWAY plc

Locomotives are kept at :-	Toddington Goods Yard	GL54 5DT	SP 049321
	Winchcombe Carriage Works Yard	GL54 5LB	SP 026297
	Winchcombe Station Yard	GL54 5LB	SP 025297

Gauge 4ft 8½in   www.gwsr.com

2807		2-8-0	OC	Sdn	2102	1905
2874		2-8-0	OC	Sdn	2780	1918
3850		2-8-0	OC	Sdn		1942
7820	DINMORE MANOR	4-6-0	OC	Sdn		1950
7903	FOREMARKE HALL	4-6-0	OC	Sdn		1949
35006	PENINSULAR & ORIENTAL S.N.CO					
		4-6-2	3C	Elh		1941
		reb		Elh		1959
–		0-4-0ST	OC	P	1976	1939
(D135)	45149	1Co-Co1DE		Crewe		1961
(D)1693	(47105)	Co-CoDE		BT	455	1963

(D1895)	47376 FREIGHTLINER 1995	Co-CoDE	BT	657	1965
D2182		0-6-0DM	Sdn		1962
(D2280)		0-6-0DM	_(RSHD	8098	1960
			(DC	2657	1960
(D)5081	(24081)	Bo-BoDE	Crewe		1960
D5343	(26043)	Bo-BoDE	BRCW	DEL88	1959
(D6915)	37215	Co-CoDE	_(EE	3393	1963
			(EEV	D859	1963
(D6948)	37248 LOCH ARKAIG	Co-CoDE	_(EE	3505	1964
			(EEV	D936	1964
(D8128)	20228 (2004)	Bo-BoDE	_(EE	3599	1965
			(EEV	D998	1965
(D8137)	20137	Bo-BoDE	_(EE	3608	1965
			(EEV	D1007	1965
21	(MAVIS)	0-4-0DM	JF	4210130	1957
11230		0-6-0DM	_(RSHN	7860	1956
			(DC	2574	1956
W51360		2-2w-2w-2DMR	PSteel		1960
W51363		2-2w-2w-2DMR	PSteel		1960
W51372		2-2w-2w-2DMR	PSteel		1960
W51405		2-2w-2w-2DMR	PSteel		1960
W55003		2-2w-2w-2DMR	GRC&W		1958
(9127)		4wDHR	BD	3743	1976

GOTHERINGTON STATION, GRETTON ROAD, GOTHERINGTON
Gauge 4ft 8½in   Private site with occasional open days

GL52 9QY
SP 974298

(TR2)	PWM 2779	2w-2PMR	Wkm	6878	1954

GREAT WESTERN RAILWAY MUSEUM,
THE OLD RAILWAY STATION, COLEFORD, Forest of Dean
Gauge 4ft 8½in   www.gwrmuseumcoleford.co.uk

GL16 8RH
SO 576105

2	182	0-4-0ST	OC	P	1893	1936

HOPEWELL COLLIERY MUSEUM, CANNOP HILL,
SPEECH HOUSE ROAD, near COLEFORD, Forest of Dean
Gauge 2ft 0in   www.hopewellcolliery.com

GL16 7EL
SO 603114

EILEEN	4wDM	RH	432648	1959

LEA BAILEY LIGHT RAILWAY, NEWTOWN, near ROSS-ON-WYE
Gauge 2ft 0in   www.lblr.fod.uk

SO 645196

	WHISTLING PIG	4wCA	G	EIMCO	401-216	1968
21282		4wDM		MR	21282	1957
LM 4	3	4wBE		WR	N7605	1973
3		4wBE		WR	7888R	1977
–		0-4-0BE		WR	L1009	1979

Gauge 1ft 6in

(JMLM6)	0-4-0BE	WR	7617	1973

NORTH GLOUCESTERSHIRE RAILWAY CO LTD, TODDINGTON NARROW
GAUGE RAILWAY, TODDINGTON GOODS YARD, TODDINGTON GL54 5DT
Gauge 2ft 0in   www.toddington-narrow-gauge.co.uk   SP 048318

1966	TOURSKA	0-6-0T	OC	Chrz	3512	1957	a
	CHAKA'S KRAAL No.6	0-4-2T	OC	HE	2075	1940	
1091		0-8-0T	OC	Hen	15968	1918	
7	JUSTINE	0-4-0WT	OC	Jung	939	1906	
	"IVAN"	4wPM		FH	3317	1948	
	YARD No. A497 ND 3824	4wDM		HE	6647	1967	
2	DFK 538	4wDM		L	34523	1949	
	"SPITFIRE"	4wPM		MR	7053	1937	
6		4wDM		RH	166010	1932	
5		4wDM		RH	354028	1953	
	BRYNEGLWYS	4wDH		SMH	101T023	1985	
			reb	YEC	L145	1995	
P.W.No.1	DB 965082	2w-2DMR		Wkm	7597	1957	

a   carries plate Fabrika Kotlow Toron 1966 1957

J. RAINBOW, PRIVATE SITE, GLOUCESTER
Gauge 2ft 0in   Private Site

	FOXHANGER	4wDM	House	c1974

ROYAL FOREST OF DEAN'S MINING MUSEUM,
CLEARWELL CAVES, CLEARWELL, near COLEFORD, Forest of Dean GL16 8JR
Gauge 2ft 6in   www.clearwellcaves.com   SO 576082

–	0-6-0DMF	_(HC	DM1435	1977	
		(HE	8583	1977	

Gauge 2ft 0in

–	0-4-0DMF	HC	DM739	1950	Dsm	
T42	0-6-0DMF	HC	DM841	1954		
68	0-4-0DMF	HC	DM924	1955	Dsm	
R3	0-6-0DMF	_(HC	DM1442	1980		
		(HE	8842	1980	OOU	
–	4wDHF	HE	7386	1976	a	
7	4wDHF	HE	7446	1975		
–	4wDHF	HE	8985	1981		
–	4wDHF	HE	8986	1981	a	b
1	4wBE	WR	7964	1978		

a   located at Hawthorn Tunnel, near Drybrook
b   uses parts and carries worksplate from HE 9053

P. SADDINGTON
Gauge 2ft 0in   Private Site

–	4wPM	Bg	2095	1936

TREASURE TRAIN LTD, PERRYGROVE RAILWAY, PERRYGROVE FARM, PERRYGROVE ROAD, MILKWALL, near COLEFORD, Forest of Dean   GL16 8QB
Gauge 2ft 0in   www.perrygrove.co.uk   SO 579095

HLH 003D		4wDH	HE	9352	1994

Gauge 1ft 3in

	LYDIA	2-6-2T	OC	AK	77	2008	a
	SPIRIT OF ADVENTURE	0-6-0T	OC	ESR	295	1993	
3	ANNE	0-6-2T	OC	ESR	323	2004	
	MR HALLWORTH	0-6-0T	OC	NBRES		2022	b
27	SOONY	0-4-0	OC	NemethJ		2012	c
–		4-6wDM		Guest		1960	
4	JUBILEE	4wDH		HE	9337	1994	
No.2	WORKHORSE	4wDM		MR	26014	1967	

a   carries worksplate dated 2007
b   carries worksplate dated 2023
c   constructed from parts supplied by Hillcrest Locomotives, USA;
       construction started by J. Page; carries w/n BLW 15912/1901.

WINCHCOMBE RAILWAY MUSEUM TRUST, 3 GLOUCESTER STREET, WINCHCOMBE, near CHELTENHAM   GL54 5LX
Gauge 2ft 0in   (Closed)   SP 022282

	AMOS	2w-2BE		FoxA	c1972	Dsm

VALE OF BERKELEY RAILWAY, THE ENGINE SHED, SHARPNESS DOCKS, DOCK ROAD, SHARPNESS   GL13 9YA
Gauge 4ft 8½in   www.valeofberkeleyrailway.co.uk   SO 668023

44027		0-6-0	IC	Derby		1924	
44901		4-6-0	OC	Crewe		1945	Dsm
–		0-4-0F	OC	AB	2126	1942	
No.15	EARL DAVID	0-6-0ST	IC	AB	2183	1945	
3947		4wDM		FH	3947	1960	
(7069)		0-6-0DE		HL	3841	1935	
–		4wDM		RH	210479	1942	
	99709 909139-6	2w-2PMR		Geismar ST/04/08		2004	
A156W	(PWM 2188 TR12)	2w-2PMR		Wkm	4165	1948	
(68066	DB 965564) 23 PWM 4306	2w-2PMR		Wkm	7509	1956	

HAMPSHIRE

INDUSTRIAL SITES

ARLINGTON FLEET GROUP LTD,
EASTLEIGH WORKS, CAMPBELL ROAD, EASTLEIGH　　　　　　　SO50 5AD
Vehicles for repair/overhaul/restoration/storage usually present
Gauge 4ft 8½in　　　www.arlington-fleet.com　　　　　　　**SU 457185**

(D2991)	07007		0-6-0DE		RH	480692	1962
(D3734)	08567	JOHN ARLINGTON STEPHENS - 20TH MAY 1925 TO 19TH JULY 1984					
			0-6-0DE		Crewe		1959
01508	(428)		0-6-0DH		RH	466617	1961
323·539·7	CHEVIOT		4wDH		Gmd	4861	1955	a

Preserved vehicles based here :

35005	CANADIAN PACIFIC		4-6-2	3C	Elh		1941	
				reb	Elh		1959	
(D421)	50021	RODNEY	Co-CoDE		_(EE	3791	1968	
					(EEV	D1162	1968	
(D426)	50026 (89426)		Co-CoDE		_(EE	3796	1968	
	INDOMITABLE				(EEV	D1167	1968	
(D1946)	47503) 47771		Co-CoDE		BT	708	1966	
(51371)	977987 960301		2-2w-2w-2DMR		PSteel		1960	
(51413)	977988 960301		2-2w-2w-2DMR		PSteel		1960	
55508	141 108		4wDMR		_(BRE(D)		1984	
					(Leyland R4.016		1984	
				reb	AB	759	1989	
55528	141 108		4wDMR		_(BRE(D)		1984	
					(Leyland R4.033		1984	
				reb	AB	758	1989	
62529	313201		4w-4wRE/WER		York(BRE)		1976	
62593	313201		4w-4wRE/WER		York(BRE)		1976	
68505	(61299)		4w-4wRER		Afd/Elh		1959	

　　a　　property of Northumbria Rail Ltd, Bedlington, Northumberland

BLUE FUNNEL FERRIES LTD,
HYTHE PIER RAILWAY, PROSPECT PLACE, HYTHE　　　　　　SO45 6AU
Gauge 2ft 0in　　　www.hytheferry.co.uk　　　　　　　　**SU 423081**

−		4wRE	BE	16302	1917
	GERALD YORKE	4wRE	BE	16307	1917

BRYAN HIRST LTD, BULLINGTON CROSS, SUTTON SCOTNEY　　SO21 3FN
Locomotives for scrap or resale occasionally present.　www.bryanhirst.com　　**SU 463420**

FREIGHTLINER GROUP LTD, SOUTHAMPTON MARITIME MAINTENANCE DEPOT,
off WESTERN AVENUE, SOUTHAMPTON DOCKS, SOUTHAMPTON   **SO15 0GN**
(part of the America Genesee & Wyoming Railway Co)
Gauge 4ft 8½in   www.freightliner.co.uk   **SU 380129**

(D3953)	08785	0-6-0DE	Derby	1960	**M**

GB RAILFREIGHT LTD, (part of the Hector Rail Group),
EASTLEIGH DEPOT, DUTTON LANE, EASTLEIGH   **SO50 6AA**
Gauge 4ft 8½in   www.gbrailfreight.com   **SU 458196**

(D3673)	08511	0-6-0DE	Dar	1958	a
(D3850)	08683	0-6-0DE	Hor	1959	a

    a   property of Railway Support Services Ltd, Wishaw, Warwickshire

LONDON & NORTH WESTERN RAILWAY COMPANY LTD, t/a ARRIVA TRAINCARE
EASTLEIGH TRACTION & ROLLING STOCK DEPOT,
CAMPBELL ROAD, EASTLEIGH (Arriva - a DB Company)   **SO50 5AD**
Gauge 4ft 8½in   www.arrivatc.com   **SU 458179**

(D3903)	08735		0-6-0DE	Crewe	1960
(D3978)	08810	RICHARD J. WENHAM EASTLEIGH DEPOT DECEMBER 1969 – JULY 1999			
		0-6-0DE	Derby	1960	

SIEMENS AG, SIEMENS RAIL SYSTEMS, NORTHAM TRAINCARE FACILITY,
RADCLIFFE ROAD, NORTHAM, SOUTHAMPTON   **SO14 0PS**
Gauge 4ft 8½in   www.siemens.com   **SU 429126**

–		4wBE	Niteq	B193	2002

SOLENT GATEWAY LTD, (part of the Solent Freeport Group)
SEA MOUNTING CENTRE, MARCHWOOD MILITARY PORT,
CRACKNORE HARD, MARCHWOOD   **SO40 4ZG**
(owned by Associated British Ports)
Gauge 4ft 8½in   www.solentgateway.com   **SU 395103**

01523	(259)	4wDH	TH	299V	1981
01541	(260)	4wDH	TH	300V	1982
01542	(262)	4wDH	TH	302V	1982

PRESERVATION SITES

BEAULIEU – NATIONAL MOTOR MUSEUM, BEAULIEU, BROCKENHURST SO42 7ZN
Gauge 4ft 8½in   www.beaulieu.co.uk   **SU 384024**

XJ12 596	2w-2PMR	Jaguar	1990	a

    a   road vehicle, converted to rail operation

Gauge monorail   **SU 386026**

    –   4wDM   Watson&Haig   c1974

COLIN BILLINGHURST, PRIVATE LOCATION, FAREHAM
Gauge 2ft 0in   **Private Site**

| S128 | 263 001 | 4wBE | CE | 5882A | 1971 | a |

    a   frame used as a replica; private location with no public vantage points

EXBURY GARDENS LTD, EXBURY GARDENS RAILWAY,
EXBURY GARDENS & STEAM RAILWAY, EXBURY, near SOUTHAMPTON SO45 1AZ
Gauge 1ft 0¼in   www.exbury.co.uk   **SU 423006**

	ROSEMARY	0-6-2T	OC	ESR	315	2001
	NAOMI	0-6-2T	OC	ESR	316	2002
	MARILOO	2-6-2	OC	ESR	326	2008
	EDDY	4wDH		HE	9336	1994

HAYLING LIGHT RAILWAY TRUST,
HAYLING LIGHT RAILWAY, EASTOKE, SEA FRONT, HAYLING ISLAND   **PO11 9HL**
Gauge 2ft 0in   www.haylinglightrailway.wixsite.com   **SZ 729985**

	DAVE SCALES	4wDH	s/o	AK	11	1984
No.3	JACK	0-4-0DH	s/o	AK	23	1988
No.1	ALAN B	4wDM		MR	7199	1937
No.5	EDWIN	4wDM		RH	7002-0967-5	1967

R. GAMBRILL, PRIVATE LOCATION
Locomotives not available for viewing. Locomotives may occasionally visit galas and public events.
Gauge 2ft 0in   **Private Site**

| | – | 4wPM | FH | 1747 | 1931 |
| | – | 4wDM | RH | 444193 | 1960 |

HAMPSHIRE BUILDINGS PRESERVATION TRUST LTD,
CENTRE FOR THE CONSERVATION OF THE BUILT ENVIRONMENT,
BURSLEDON BRICKWORKS, SWANWICK LANE, LOWER SWANWICK   **SO31 7HB**
Gauge 2ft 0in   www.thebrickworksmuseum.org   **SU 499098**

| | SIMPLEX ASHBY | 4wDM | MR | 8694 | 1943 |

HORSEBRIDGE STATION,
HORSEBRIDGE, KINGS SOMBOURNE, STOCKBRIDGE   **SO20 6PU**
Gauge 4ft 8½in   **SU 343303**

| | (9119) | 4wDHR | BD | 3708 | 1975 |

MARWELL'S WONDERFUL RAILWAY, MARWELL WILDLIFE, COLDEN COMMON, near WINCHESTER

SO21 1JH

Gauge **1ft 3in** R.T.C. www.marwell.org.uk **SU 508216**

PRINCESS ANNE	2-6-0DH	s/o	SL	75.3.87	1987	OOU

PHIL MASON, PRIVATE SITE

Gauge **2ft 0in** **Private Site**

T203	0-4-0DM	Dtz	11898	1934	
–	0-4-0DM	Dtz	16392	1938	Dsm
–	4wDM	RH	339209	1952	

MID HANTS RAILWAY plc, "THE WATERCRESS LINE"

Locomotives are kept at :-

New Alresford Station SO24 9JG SU 588325
Ropley Station SO24 0BL SU 629324
Medstead & Four Marks Station GU34 5EN SU 668353

Gauge **4ft 8½in** www.watercressline.co.uk

30499		4-6-0	OC	Elh		1920	
(30506)	S.R. 506	4-6-0	OC	Elh		1920	
30828	(E828) HARRY E.FRITH	4-6-0	OC	Elh		1928	
(30850)	E850 LORD NELSON	4-6-0	4C	Elh		1926	
30925	(925) CHELTENHAM	4-4-0	3C	Elh		1934	
34105	SWANAGE	4-6-2	3C	Bton		1950	
41312		2-6-2T	OC	Crewe		1952	
45379	(5379)	4-6-0	OC	AW	1434	1937	
53808		2-8-0	OC	RS	3894	1925	
73096		4-6-0	OC	Derby		1955	
75079		4-6-0	OC	Sdn		1956	
80150		2-6-4T	OC	Bton		1956	
3781	1	0-6-0ST	IC	HE	3781	1952	
	rebuilt as	0-6-0T	IC	MHR Ropley		1994	
6	KILMERSDON	0-4-0ST	OC	P	1788	1929	
(D427)	50027 LION	Co-CoDE		_(EE	3797	1968	
				(EEV	D1168	1968	
(D3044)	08032	0-6-0DE		Derby		1954	
D3358	(08288) PHOENIX	0-6-0DE		Derby		1957	
D3462	(08377)	0-6-0DE		Dar		1957	
D8188	(20188)	Bo-BoDE		_(EE	3669	1966	
				(EEV	D1064	1966	
12049	(12082 01553)	0-6-0DE		Derby		1950	
205025	S60124	4-4wDER		Afd/Elh		1959	
TRS 805	99709 901031-3	2w-2PMR		Geismar	ST03/35	2003	
(68075)	DB965991	2w-2DMR		Wkm	10707	1974	
(68082)	DB966031	2w-2DMR		Wkm	10839	1975	

MILESTONES – HAMPSHIRE'S LIVING HISTORY MUSEUM, LEISURE PARK, CHURCHILL WAY WEST, BASINGSTOKE

RG22 6PG

Gauge 4ft 8½in   www.milestonesmuseum.org.uk   SU 612524

WOOLMER	0-6-0ST	OC	AE	1572	1910

PAULTONS PARK LTD, PAULTONS RAILWAY, OWER, near ROMSEY

SO51 6AL

Gauge 1ft 3in   www.paultonspark.co.uk   SU 316167

–	2-8-0DH	s/o	SL	RG.11.86	1987

TINKERS CROSS RAILWAY, FORDINGBRIDGE

Gauge 1ft 0¼in   Private site

–	4wPM	LewAJ		
–	4wBE	Bickton	2010	

TWYFORD WATERWORKS TRUST, TWYFORD WATERWORKS, HAZELEY ROAD, TWYFORD

SO21 1QA

Gauge 2ft 0in   www.twyfordwaterworks.co.uk   SU 493248

–		4wDM	FH	1731	1931	
	a rebuild of	4wPM?	MR			Dsm
(DOE 3983)		4wDM	FH	3983	1962	
–		4wDM	L	3916	1931	
–		4wDM	L	42494	1956	
–		4wDM	LB	52886	1962	
SUZANNE		4wDM	LB	55730	1968	
–		4wPM	MR	5355	1932	
No.29	AYALA	4wDM	MR	7374	1939	
–		4wBE	Red(F)		1979	
–		4wDMF	RH	209429	1943	
JMLM2		0-4-0BE	WR	M7550	1972	
–		4wBE	WR		1973	a

a   one of WR N7606, WR N7620 or WR N7621

WELLINGTON COUNTRY PARK RAILWAY, WELLINGTON COUNTRY PARK, ODIHAM ROAD, RISELEY

RG7 1SP

Gauge 1ft 0¼in   www.wellingtoncountrypark.co.uk   SU 730627

No.100	ARTHUR	0-6-0DH	AK	100	2017

HEREFORDSHIRE

INDUSTRIAL SITES

ALAN KEEF LTD, LIGHT RAILWAY ENGINEERS & LOCOMOTIVE BUILDERS, LEA LINE, ROSS-ON-WYE · HR9 7LQ

Locomotives under construction and repair are usually present. The following is understood to be a complete FLEET LIST of locomotives currently owned by (or in the care of) Messrs Keef. Some may be hired out from time to time, as shown by footnotes.

Gauge 3ft 0in · www.alankeef.co.uk · **SO 665214**

(No.10 G.H. WOOD)	2-4-0T	OC	BP	4662	1905	

Gauge 2ft 6in

DALMUNZIE	4wPM		MR	2014	1920	

Gauge 2ft 0in

TAFFY	0-4-0VBT	VC	AK	30	1994	a
WOTO	0-4-0ST	OC	WB	2133	1924	
SKIPPY	4wDM		AK	2	1976	
NELLIE	4wDM		Coferna	3821	1950	
ADAM	4wDM		MR	9978	1954	
PLANT No.34 SUE	4wDM		MR	40SD502	1975	
–	4wDM		OK	7595	1937	
–	4wDM		RH	213834	1942	
C.C.R.89 No.1	4-4-0DH	s/o	SL	139/1.2.89	1989	
C.C.R.89 No.2	4-4-0DH	s/o	SL	139/2.1.89	1989	
C.C.R.94 No.3	4-4-0DH	s/o	SL	606.3.94	1994	
–	4wBE		BEV	323	1921	
SP 8 "BATTY"	4wBE		WR	1393	1939	
		reb	AK		2006	
JMLM19 "6502"	4wBE		WR	6502	1962	
		reb	WR	10102	1983	
JMLM22 "MURPHY"	4wBE		WR	6504	1962	
		reb	WR	10106	1983	

a   worksplate is dated 1990

GREATWEST 2003 LTD, MORETON BUSINESS PARK, MORETON-ON-LUGG, HEREFORD · HR4 8DS

Gauge 4ft 8½in

01583 422 VALIANT	0-6-0DH		RH	459517	1961	Pvd

PAINTER BROTHERS LTD, HEREFORD STEELWORKS, MORTIMER ROAD, HEREFORD · HR4 9SW

(part of the Balfour Beatty Group Ltd)

Gauge 2ft 0in · www.painterbrothers.com · **SO 508413**

–	4wBE	CE	B0142B	1973	
–	4wBE	CE	5806	1970	

WYE VALLEY GROUP – EASTSIDE 2000 LTD, FORDSHILL ROAD, ROTHERWAS INDUSTRIAL ESTATE, HEREFORD

Gauge 4ft 8½in   www.wyevalleygroup.co.uk   SO 538377

| 220 | | 0-4-0DM | AB | 359 | 1941 | Pvd |

PRESERVATION SITES

D2578 LOCOMOTIVE GROUP, PRIVATE SITE, MORETON ON LUGG

The locomotives are under restoration inside a building on a secure industrial estate and visits are not normally possible. A contact for enquiries is at :   hfdned@hotmail.com

Gauge 4ft 8½in   Private Site

(D2145)	03145	0-6-0DM		Sdn		1961	
D2302		0-6-0DM		_(RSHD	8161	1960	
				(DC	2683	1960	
D2578		0-6-0DM		HE	5460	1958	
			reb	HE	6999	1968	

STOKE EDITH STATION, STATION HOUSE, TARRINGTON   HR1 4EY

Gauge 4ft 8½in   www.stokeedithstation.co.uk   SO 614414

| – | | 4wDM | RH | 463150 | 1961 |
| A162 | PWM 2194 | 2w-2PMR | Wkm | 4171 | 1948 |

M. DEEM, LAMARO, ECCLES GREEN, NORTON CANON, HEREFORD

Gauge 1ft 3in   Private Site   SO 374488

| 101 | 2-4wPM | TaylorJ | c1964 |

TITLEY JUNCTION STATION, near KINGTON

Private collection; cannot be viewed from any public place. No visitors without prior appointment.

Gauge 4ft 8½in   Private Site   SO 328581

–		0-4-0ST	OC	P	1738	1928	
–		4wDM		FH	3906	1959	
10	BRESSINGHAM	4wDH		TH	163V	1966	a

a   carries worksplate 173V 1966 in error

K. MATTHEWS, FENCOTE OLD STATION, HATFIELD, near LEOMINSTER

Gauge 4ft 8½in   Private Site   SO 601589

| W40 | (TR40 PWM 4314) | 2w-2PMR | Wkm | 7517 | 1956 |

OWEN BROS MOTORS, c/o K. JONES, 13 NORBURY PLACE, TUPSLEY, HEREFORD
Gauge 1ft 3in   SO 542373

No.303	0-6-0PM	s/o	TaylorJ	1967	

BOB PALMER, BROMYARD & LINTON LIGHT RAILWAY ASSOCIATION LTD,
BROMYARD & LINTON LIGHT RAILWAY, BROADBRIDGE HOUSE, BROMYARD
This railway does not operate public trains.
Locomotives are kept at :-   Bromyard   SO 657548, 661546
                                                           Linton   SO 668542

Gauge 2ft 6in

–	4wDM		Bg	3406	1953	Dsm

Gauge 2ft 0in

	MESOZOIC	0-6-0ST	OC	P	1327	1913	Dsm
–		4wPM		MR	6031	1936	
–		4wDM		MR	9382	1948	
1		4wDM		MR	9676	1952	
2		4wDM		MR	9677	1952	
No.7		4wDM		MR	20082	1953	
–		4wDM		MR	102G038	1972	
–		4wDM		RH	187101	1937	
–		4wDM		RH	195849	1939	a
L 10		4wDM		RH	198241	1939	
No.3	"NELL GWYNNE"	4wDM		RH	229648	1944	
No.6	"PRINCESS"	4wDM		RH	229655	1944	
–		4wDH		RH	437367	1959	
–		4wDM		RH	444200	1960	
–		2w-2PM		Wkm	3034	1941	Dsm

  a   converted for use as a generating unit

THE WOOLHOPE LIGHT RAILWAY,
P.J. FORTEY, THE HORNETS NEST, CHECKLEY, MORDIFORD, near HEREFORD
Gauge 1ft 3in   SO 608378

202 TREVOR	0-6-0PM	s/o	TaylorJ	c1974	

HERTFORDSHIRE

INDUSTRIAL SITES

RUSH GREEN MOTORS, LONDON ROAD, LANGLEY, HITCHIN   SG4 7PQ
Gauge 4ft 8½in   www.rushgreenmotors.com   TL 210236

(C955 YOR?)	4wDM	R/R	Bruff	(514	1986?)	OOU
C966 YOR	4wDM	R/R	Bruff	524	1986	OOU
(C969 YOR?)	4wDM	R/R	Bruff	(527	1986?)	OOU

PRESERVATION SITES

B.LAWSON, BRY RAILWAY, TRING
Locomotives stored at various private locations.

Gauge 4ft 8½in   Private Site

No.25	GR5091		4wDM		RH	294269	1951	
–			0-4-0DH		RH	418793	1957	
No.28	L5 (50) HERBERT TURNER		0-4-0DH		RH	513139	1967	
No.27			0-4-0DH		RH	518190	1965	
No.26			4wDH		RH	544996	1968	

Gauge 2ft 6in

No.2	6		4wBE		BV	694	1974	
No.3	9		4wBE		BV	696	1974	
No.17	3135B SP 03 426		4wBEF		CE	B3135B	1984	
				reb	CE	B4066RF	1994	
No.16	SP 03.425		4wBEF		CE	B3204A	1985	
				reb	CE	B4139	1995	
No.24	8		4wBE		GB	2920	1958	
No.14	ND 3307 YARD No. B49		4wBE		GB	3546	1948	
	ND 3308 YARD No.B50		2w-2BE		GB	3547	1948	Dsm
No.15			4wBE		GB	3825	c1949	
No.03			4wDHF		HE	7384	1976	
–			4wDM		LB	55870	1968	Dsm
No.16			4wDM		RH	170200	1934	Dsm
No.24			4wDM		RH	247182	1947	
–			4wDMF		RH	353491	1954	
No.14			4wDM		RH	441945	1959	
No.12			0-4-0DH		RH	476133	1964	
No.17			4wDMF		RH	480680	1963	
No.13			4wDM		RH	506415	1964	
No.19			4wDM		RH	7002/0767/6	1967	
No.15			4wDM		RH	7002/0867/3	1967	

Gauge 2ft 3in

No.31	"ROBERT (BOB) MACBETH"	4wDH		RH	476108	1964

Gauge 2ft 0in

No.29			4wDM		RH	213848	1942
(LM 20)			4wDM		RH	243387	1946
No.30			4wDM		RH	246793	1947
LM 26			4wDM		RH	248458	1946
–			4wDM		RH	280866	1949
No.21	LM 264		4wDM		RH	371535	1954
No.22	LM 265		4wDM		RH	375696	1954
No.23	LM 112		4wDM		RH	375699	1954
No.20			4wDMF		RH	381704	1955
	NEATH ABBEY		4wDH		RH	476106	1964
No.20	1511		4wBEF		CE	5382	1966
No.25			4wBE		CE	5688/2	1969
No.18			4wBE		CE	B0107A	1973

No.21	1517	4wBEF		CE	B0122	1973	
	JMLM25	4wBE		CE	B0145A	1973	
			reb	CE	B3786B	1991	
"No.22"	S232 263025	4wBE		CE	B0459A	1975	
"No.23"	JMLM17 (JM93)	4wBEF		CE	B1547B	1977	
			reb	CE	B3672	1990	
No.27	LM18 (JM95)	4wBE		CE	B1534B	1977	
			reb	CE	B3672	1990	
No.28	LM16 (JM94)	4wBE		CE	B1534A	1977	
			reb	CE	B3672	1990	
SP 204		4wBE		CE	B1808	1978	
			reb	CE	B3214A	1985	
			reb	CE	B3825	1992	
No.26	LM10	4wBE		CE	B3329B	1986	
			reb	CE	B3804	1991	
			reb	CE	B4181	1996	
JMLM20		4wBE		WR	6505	1962	
			reb	WR	10105	1983	
JMLM21		4wBE		WR	6503	1962	
			reb	WR	10104	1983	
No.6	F	0-4-0BE		WR	M7544	1972	
No.1	MANDI MIS 47	0-4-0BE		WR	N7639	1973	

Gauge 1ft 6in

No.8	L 12	4wBE		CE	5965A	1973	
No.7	L 15	4wBE		CE	B0109A	1973	
–		2w-2BE		Iso	T6	1972	
–		2w-2BE		Iso	T9	1972	Dsm
–		2w-2BE		Iso	T15	1972	
–		2w-2BE		Iso	T40	1973	
–		2w-2BE		Iso	T53	1974	Dsm
–		2w-2BE		Iso	T71	1975	
–		2w-2BE		Iso	T79	1975	
–		2w-2BE		Iso	T81	1975	
–		2w-2BE		Iso			a
No.19		4wBH		Tunn		1980	
			reb	Tunn		1996	b
No.9	SP 100 35	2w-2BE		WR	L800	1983	
No.10	SP 101 35T005	2w-2BE		WR	L801	1983	
No.11	SP 102	2w-2BE		WR	544901	1984	
No.12	JM 103	2w-2BE		WR	546001	1987	
No.13	SP 104	2w-2BE		WR	546601	1987	
No.4	N7652	0-4-0BE		WR	F7117	1966	
			reb	WR	10142	1985	
No.5	1580	0-4-0BE		WR			

a   one of Iso T21 to T38
b   one of Tunn TQ121 to TQ126

Gauge monorail

-		2wPH		RM	8111	1959

C.& D. LAWSON, DORCLIFF RAILWAY, TRING

(Custodians : B. & G. Lawson) Locomotives are currently stored elsewhere.

Gauge 2ft 6in **Private Site**

No.04			4wDM		Diema	3543	1974	
No.05			4wDM		Fisons		1976	
		rebuilt as	4wDH		Lawson		1999	
No.02			4wDM		HE	7366	1974	
No.01			4wDM		HE	7367	1974	
–			4wPM		L	34652	1949	
–			2w-2PH		Lawson	2	1998	
–			2w-2PH		Lawson	3	2000	
–			4wDH		Lawson	4	2001	
–			2w-2DMR		Lawson	5	2004	
–			2w-2DH		Lawson	6	2005	
–			4wDM		LB	53976	1964	
–			4wDM		LB	53977	1964	
No.1			4wDM		RH	166045	1933	
No.7	ELLEN		4wDM		RH	200069	1939	
No.6			4wDM		RH	224315	1944	
No.9			4wDM		RH	229657	1945	
No.8			4wDM		RH	244559	1946	
No.2	CUCKOO BUSH		4wDM		RH	247178	1947	
No.3			4wDM		RH	297066	1950	
No.4			4wDM		RH	402439	1957	
No.5			4wDM		RH	432654	1959	
No.18	SIMBA		4wDH		RH	432661	1959	
				reb	Swanhaven		c1985	
No.10	TANIA		4wDM		RH	432665	1959	
No.11	SHEEBA		4wDM		RH	466594	1962	
–			2w-2PM		Wkm	3175	1942	
–			2w-2PM		Wkm	3431	1943	
		rebuilt as	2w-2DH		Lawson		1997	
–			2w-2PM		Wkm	3578	1944	
		rebuilt as	2w-2DH		Lawson		2002	

Gauge 2ft 0in

–	4wDM		RH	441944	1960	Dsm

N.V. SILL, PRIVATE SITE, near HEMEL HEMPSTEAD
Gauge 5ft 0in **Private Site**

1151	1819	2-8-0	OC	Frichs	397	1949

WARNER BROTHERS STUDIO TOUR LONDON, STUDIO TOUR DRIVE, LEAVESDEN, WATFORD
Gauge 4ft 8½in www.wbstudiotour.co.uk

WD25 7LR
TL 092005

5972	HOGWARTS CASTLE	4-6-0	OC	Sdn		1937

**ZOOLOGICAL SOCIETY OF HERTFORDSHIRE, HERTFORDSHIRE ZOO,
WHITE STUBBS LANE, BROXBOURNE** **EN10 7QA**
Gauge 4ft 8½in   www.hertfordshirezoo.com **TL 338067**

No.1	671	0-4-0DH	s/o	RH	512463	1965	

ISLE OF WIGHT

INDUSTRIAL SITES

**FIRST MTR SOUTH WESTERN TRAINS LTD, t/a ISLAND LINE,
RYDE TRAIN CARE DEPOT, ST. JOHNS HILL, RYDE** **PO33 1HS**
Gauge 4ft 8½in   www.southwesternrailway.com **SZ 596919**

–	2-2w-2wBE R/R	Harmill		c2020	
–	2w-2PMR	Bance	067	1998	a

a   operated by Network Rail

PRESERVATION SITES

**HOUSE OF CHILLI LTD,
HOLLIERS FARM, BRANSTONE, SANDOWN** **PO36 0LT**
Gauge 4ft 8½in **SZ 554836**

(10205)	124	483004	2-2w-2w-2RER	MetCam	1938
(11205)	224	483004	2-2w-2w-2RER	MetCam	1938

**ISLE OF WIGHT RAILWAY CO LTD,
ISLE OF WIGHT STEAM RAILWAY, HAVENSTREET STATION** **PO33 4DS**
Gauge 4ft 8½in   www.iwsteamrailway.co.uk **SZ 556898**

(32110)	110) No.2 YARMOUTH	0-6-0T	IC	Bton		1877	
(32640)	W11 (NEWPORT)	0-6-0T	IC	Bton		1878	
(32646)	No.8 FRESHWATER	0-6-0T	IC	Bton		1876	
41298		2-6-2T	OC	Crewe		1951	
41313		2-6-2T	OC	Crewe		1952	
No.24	CALBOURNE	0-4-4T	IC	9E		1891	
W37	INVINCIBLE	0-4-0ST	OC	HL	3135	1915	
(38)	AJAX	0-6-0T	OC	AB	1605	1918	
ARMY 92	WAGGONER	0-6-0ST	IC	HE	3792	1953	
ARMY 198	ROYAL ENGINEER	0-6-0ST	IC	HE	3798	1953	
	HAYDOCK	0-6-0T	IC	RS	2309	1876	a
D2059	(03059)	0-6-0DM		Don		1959	
D2554	(05001 97803) NUCLEAR FRED						
		0-6-0DM		HE	4870	1956	
ARMY 235	(233)	0-4-0DM		AB	369	1945	

No.2		4wDMR	_(AK	98R	2021	
			(Lowther		2021	b
(10291)	127 483007	2-2w-2w-2RER	MetCam		1938	
(11291)	227 483007	2-2w-2w-2RER	MetCam		1938	
68800		4wDHR	Perm	001	1985	
D68809	(DX 68809) "LA-LA"	4wDMR	Perm	010	1986	
(DS 3320	PWM 3766) "66 532"	2w-2PMR	Wkm	6645	1953	Dsm
6944	PWM 3959	2w-2PMR	Wkm	6944	1955	

a   works plate reads 1879
b   incorporates parts from Bg/DC 1647/1927

KENT

INDUSTRIAL SITES

AVONDALE ENVIRONMENTAL SERVICES LTD,
FORT HORSTED, PRIMROSE CLOSE, CHATHAM
ME4 6HZ
Gauge 4ft 8½in   www.avondaleuk.com
TQ 751651

C959 YOR	4wDMR	R/R	Bruff	517	1986

Other road/rail vehicles usually present

DB CARGO UK LTD,
HOO JUNCTION YARD, QUEENS FARM ROAD, SHORNE, GRAVESEND
DA12 3HU
Gauge 4ft 8½in   www.uk.dbcargo.com
TQ 698735

99709 979116-9	4wDH	R/R	Zephir	2929	2021

DEFENCE INFRASTRUCTURE ORGANISATION, DEFENCE TRAINING ESTATES -
SOUTH, CINQUE PORTS TRAINING AREA, LYDD, ROMNEY MARSH
Gauge 600mm
TR 033198

1	5210	2w-2PM	Wkm	11684	1990
3	5208	2w-2PM	Wkm	11685	1990
4	5211	2w-2PM	Wkm	11679	1990
5	1378	2w-2PM	Wkm	11681	1990
6	5209	2w-2PM	Wkm	11678	1990
7	5207	2w-2PM	Wkm	11677	1990
8	1379	2w-2PM	Wkm	11680	1990

GETLINK GROUP, EUROTUNNEL LESHUTTLE,
CHERITON TERMINAL, ASHFORD ROAD, FOLKESTONE
CT18 8XX
These locomotives, used for shunting and maintenance, are also employed in the Channel Tunnel and at
Coquelles Terminal, France, as required.
Gauge 4ft 8½in   www.eurotunnel.com
TR 185375

0001	(21901)	Bo-BoDE	MaK	1000.867	1993
0002	(21902)	Bo-BoDE	MaK	1000.868	1993

0003	(21903)		Bo-BoDE	MaK	1000.869	1993
0004	(21904)		Bo-BoDE	MaK	1000.870	1993
0005	(21905)		Bo-BoDE	MaK	1000.871	1993
0006	(6456 21906)		Bo-BoDE	MaK	1200.056	1991
0007	(6457 21907)		Bo-BoDE	MaK	1200.057	1991
0008	(6450 21908)		Bo-BoDE	MaK	1200.050	1991
0009	(6451 21909)		Bo-BoDE	MaK	1200.051	1991
0010	(6447 21910)		Bo-BoDE	MaK	1200.047	1991
0031	FRANCES		4wDH	Schöma	5366	1993
	Incorporates parts of		HE			
0032	ELISABETH		4wDH	Schöma	5367	1993
	Incorporates parts of		HE			
0033	SILKE		4wDH	Schöma	5263	1994
	Incorporates parts of		HE			
0034	AMANDA		4wDH	Schöma	5262	1994
	Incorporates parts of		HE			
0035	MARY		4wDH	Schöma	5269	1994
0036	LAURENCE		4wDH	Schöma	5268	1994
0037	LYDIE		4wDH	Schöma	5264	1994
0038	JENNY		4wDH	Schöma	5266	1994
0039	PACITA		4wDH	Schöma	5401	1994
0040	JILL		4wDH	Schöma	5402	1994
0041	KIM		4wDH	Schöma	5464	1995
0042	NICOLE		4wDH	Schöma	5465	1995
520			4wDH	Cockerill		1996
521			4wDH	Cockerill		1996
522			4wDH	Cockerill		1996
(TSV 001)			2w-2BER	Bance	017	1994
(TSV 002)			2w-2BER	Bance	018	1994

HITACHI RAIL EUROPE LTD, ASHFORD TRAIN MAINTENANCE CENTRE, ASHFORD DEPOT, ASHFORD
TN23 1EZ

Gauge **4ft 8½in**   www.hitachirail.com   **TR 016420**

CHIAKI UEDA	0-4-0DH	S	10089	1962
–	4wBE	Scul		2007

LONDON & SOUTH EASTERN RAILWAY LTD t/a SOUTH EASTERN
(part of the Govia Group) and **HITACHI RAIL EUROPE LTD,**
RAMSGATE DEPOT, NEWINGTON ROAD, RAMSGATE   **CT12 6EA**

Gauge **4ft 8½in**   www.southeasternrailway.co.uk // www.hitachirail.com   **TR 370658**

–	4wBE	R/R	Zephir	2111	2007

METROPOLITAN POLICE SPECIALIST TRAINING CENTRE, off MARK LANE, DENTON, GRAVESEND
DA12 2HN

London underground railcar used for instructional purposes.

Gauge **4ft 8½in**   www.met-police.uk   **TQ 672742**

1306	2-2w-2w-2RER	MetCam		1961

NETWORK RAIL HIGH SPEED LTD,
SINGLEWELL INFRASTRUCTURE MAINTENANCE DEPOT,
HENHURST ROAD, SINGLEWELL, GRAVESEND   DA12 3AN
Gauge 4ft 8½in   www.networkrail.co.uk   TQ 659702

DR 97001	DU 94 001 URS	4w-4wDHR	Eiv de Brive 001URS	2003	
DR 97011		4w-4wDHR	Windhoff	2625	2004
DR 97012	GEOFF BELL	4w-4wDHR	Windhoff	2626	2004
DR 97013		4w-4wDHR	Windhoff	2627	2004
DR 97014		4w-4wDHR	Windhoff	2628	2004

RIDHAM SEA TERMINALS LTD, RIDHAM DOCK, SITTINGBOURNE   ME9 8SR
Gauge 4ft 8½in R.T.C.   www.ridhamseaterminals.co.uk   TQ 918684

–	0-6-0DH	EEV	D1227	1967	OOU

SEACON TERMINALS LTD, TOWER WHARF, NORTHFLEET   DA11 9BD
Gauge 4ft 8½in R.T.C.   www.seacongroup.co.uk   TQ 612752

–	2w-2BER	PWR	BO.598W.01	1993	OOU

PRESERVATION SITES

DAVID BREAKER, STALISFIELD
Locomotive currently stored elsewhere
Gauge 4ft 8½in   Private Site

–	0-6-0ST	IC	MW	1317	1895

BREDGAR & WORMSHILL LIGHT RAILWAY,
"THE WARREN", SWANTON STREET, BREDGAR, near SITTINGBOURNE   ME9 8AT
Gauge 2ft 0in   www.bwlr.co.uk   TQ 873585

7	VICTORY	0-4-2T	OC	Dec	246	1897
3	LADY JOAN	0-4-0ST	OC	HE	1429	1922
No.10	ZAMBEZI	0-4-2T	OC	JF	13573	1912
9	LIMPOPO	0-6-0WT	OC	JF	18800	1930
2	KATIE	0-6-0WT	OC	Jung	3872	1931
6	EIGIAU	0-4-0WT	OC	OK	5668	1912
8	HELGA	0-4-0WT	OC	OK	12722	1936
No.4	ARMISTICE	0-4-0ST	OC	WB	2088	1919
14	(NG 54) MILSTEAD	4wDH		AB	765	1988
		reb		HAB		1996
5	BREDGAR	4wDH		BD	3775	1983
13	LYNE	4wDM		MR	7037	1936
	ESK	4wDM		MR	7498	1940
12	BICKNOR	4wDM		MR	9869	1953
	JENNY	4wDH		Schöma	5239	1991

BRETT GROUP, MILTON MANOR FARM, ASHFORD ROAD, CHARTHAM, near CANTERBURY

Gauge 2ft 0in   www.brett.co.uk

CT4 7PP
TR 120558

–		4wDM		MR	8730	1941

CHATHAM HISTORIC DOCKYARD TRUST, THE HISTORIC DOCKYARD, CHATHAM

Gauge 4ft 8½in   www.thedockyard.co.uk

ME4 4TE
TQ 758689

No.8 INVICTA	0-4-0ST	OC	AB	2220	1946
YARD No.361 AJAX	0-4-0ST	OC	RSHN	7042	1941
WD 42 OVERLORD	0-4-0DM		AB	357	1941
YARD No.562 ROCHESTER CASTLE	4wDM		FH	3738	1955
THALIA	0-4-0DM		_(RSHN	7816	1954
			(DC	2503	1954

Gauge 600mm

LOD 758148	4wDM		RH	226276	1944

DOVER TRANSPORT MUSEUM, WILLINGDON ROAD, OLD PARK, DOVER CT16 2HQ

Gauge 4ft 8½in   www.dovertransportmuseum.org.uk

TR 301444

"ST THOMAS"	0-6-0ST	OC	AE	1971	1927

Gauge 2ft 0in

–	4wPM	FH	3116	1946
–	4wDM	MR	8606	1941
–	4wDM	RH	349061	1953

EAST KENT RAILWAY TRUST, EAST KENT RAILWAY, STATION ROAD, SHEPHERDSWELL

Gauge 4ft 8½in   www.eastkentrailway.co.uk

CT15 7PD
TR 258483

	ST DUNSTAN	0-6-0ST	OC	AE	2004	1927
2087	ACHILLES	0-4-0ST	OC	P	2087	1948
	(SWANSEA VALE No.1)	4wVBT	VCG	S	9622	1958
(D3843)	08676 DAVE 2	0-6-0DE		Hor		1959 a
(E6037)	73130	Bo-BoDE/RE		_(EE	3709	1966
				(EEV	E369	1966
	SNOWDOWN	0-4-0DM		JF	4160002	1952

THE BUFFS, ROYAL EAST KENT REGIMENT, 1572-1961

	ARMY 427 C4 SA	0-6-0DH		RH	466616	1961
01530	(269)	4wDH		TH	311V	1984 a
331		4wDE		Werk	868	1950
42	(KEARSLEY No.1)	Bo-BoWE		HL	3682	1927
S11161S	3142	4w-4wRER		Elh		1937
S11187S	3142	2-2w-2w-2RER		EE/Elh		1937
55558	142017	2-2wDMR		_(BRE(D)		1985
				(Leyland R5.40		1985
	rebuilt as	2-2wDHR		RFSD		1988

55608	142017		2w-2DMR	_(BRE(D)		1985
				(Leyland R5.39		1985
		rebuilt as	2w-2DHR	RFSD		1988
55577	142036		2-2wDMR	_(BRE(D)		1986
				(Leyland R5.70		1986
		rebuilt as	2-2wDHR	RFSD		1989
55627	142036		2w-2DMR	_(BRE(D)		1986
				(Leyland R5.69		1986
		rebuilt as	2w-2DHR	RFSD		1989
(60154	1101 205001)		4-4wDER	Afd/Elh		1957
62385	(1399)		4w-4wRER	York(BRE)		1971
65917	365524		4w-4wRER	York(BRE)		1994
65974	365540		4w-4wRER	York(BRE)		1995
67300	7001 (316999)		4w-4wWER	York(BRE)		1989
			a rebuild of	DerbyC&W		1981	b
68509	(61280) 9110		4w-4RER	Afd/Elh		1959
	99709 901014-9		2w-2PMR	Lesmac LMS009		2006

a   property of Harry Needle Railroad Co Ltd, Barrow Hill, Derbyshire
b   rebuild of non-motorised Driving trailer

ELHAM VALLEY LINE TRUST,
COUNTRYSIDE CENTRE & RAILWAY MUSEUM, PEENE, FOLKESTONE   CT18 8AZ
Gauge 4ft 8½in   www.elhamvalleylinetrust.org   **TR 185377, 185378**

–		0-6-0T	OC	EVM		a

a   non-working replica locomotive

Gauge 900mm

RR10		4wDH	RACK	HE	9283	1988
TU 20	56404	4wDH		Moës		

KEN JACKSON, EYNSFORD LIGHT RAILWAY, EYNSFORD
Gauge 2ft 0in   **Private Site**

–		4wDM	MR	9711	1952	
–		2w-2PM	Westwood			a

a   converted ride-on lawnmower with fixed rail wheels for guidance

MICHAEL LIST-BRAIN, PRESTON SERVICES, THE STEAM MUSEUM,
COURT LANE, PRESTON, near CANTERBURY   CT3 1DH
Gauge 4ft 8½in   www.prestonservices.co.uk   **TR 244604**

No.1	0-4-0T	OC	N	4444	1892

Gauge 600mm

SMT T912	0-10-0T	OC	OK	10957	1925
SMT T908	0-10-0T	OC	OK	12470	1934

Gauge 1ft 6in

–	4-2-2	OC	WB	1425	1893

THE ONE : ONE COLLECTION,
c/o HORNBY HOBBIES LTD, THE HORNBY VISITOR CENTRE,
WESTWOOD INDUSTRIAL ESTATE, MARGATE

Gauge 4ft 8½in www.theonetoonecollection.co.uk

4270		2-8-0T	OC	Sdn	2850	1919	
6960	RAVENINGHAM HALL	4-6-0	OC	Sdn		1944	
(31178	1178) 178	0-6-0T	IC	Afd		1910	
45379	(5379)	4-6-0	OC	AW	1434	1937	
(60019)	4464 BITTERN	4-6-2	3C	Don	1866	1937	
(D6890)	37190 (37314)	Co-CoDE		_(EE	3368	1963	
				(EES	8411	1963	
D9016	(55016)	Co-CoDE		_(EE	2921	1961	
	GORDON HIGHLANDER			(VF	D573	1961	
60081	ISAMBARD KINGDOM BRUNEL	Co-CoDE		BT		983	1991
12795	4732	4w-4wRER		Lancing/Elh		1951	
(12796)	4732	4w-4wRER		Lancing/Elh		1951	
3304	373304	Bo-BoWE/RE		GEC-Alsthom		1995	
B30W	PWM 3956	2w-2PMR		Wkm	6941	1955	

Gauge 2ft 6in

	LEADER	0-4-2ST	OC	KS	926	1905

Gauge 1ft 3in

70000	BRITANNIA	4-6-2	OC	TMA	8733	1987	a

a   construction begun by Longfleet Motor & Engineering Works Ltd, 46 Fernside Road, Poole,
Dorset, in 1968; completed by TMA in 1988

J. MARTIN, THE RICHMOND LIGHT RAILWAY, near HEADCORN

Gauge 2ft 0in Private site – visits by invitation or open days only

	CHUQUITANTA	0-4-0T	OC	_(Couillet	810	1885	
				(Dec	36	1885	
	LILY	0-4-0T	OC	Dec	648	1912	
	ELIN	0-4-0ST	OC	HE	705	1899	
	SYBIL	0-4-0ST	OC	HE	827	1903	
No.8	"LEARY"	0-4-0VBT	VC	Foulds/Collins		2010	
			reb	RLR		2018	
No.3	JENNY	0-4-0WT	OC	Jung	3175	1921	
–		0-4-0WT	OC	OK	5745	1912	
–		0-6-0T	OC	KS	2442	1915	a
	PIXIE	0-4-0ST	OC	WB	2090	1919	b
No.1	(NG 49)	4wDH		BD	3701	1973	
–		4wPM		Campagne	903	1925	
"49"	"SAMPSON"	4wDM		FH	1887	1934	
–		4wDM		HE	3621	1947	
(4)		4wDM		MR	7403	1939	
N.230		4wDM		MR	8828	1943	
–		4wDM		MR	20073	1950	
5		4wDM		RH	235654	1946	
			reb	ALR	5	2002	

Gauge 1ft 3in

1904	"CAGNEY"	4-4-0	OC	McGarigle		1904
	PYLON	4wDM		L	40407	1954

a   stored off site
b   currently at the Ffestiniog Railway Company, Portmadog, North Wales

ROMNEY HYTHE & DYMCHURCH RAILWAY, NEW ROMNEY STATION, NEW ROMNEY

TN28 8PL

Gauge 1ft 3in   www.rhdr.org.uk   **TR 074249**

1	GREEN GODDESS	4-6-2	OC	DP	21499	1925	
2	NORTHERN CHIEF	4-6-2	OC	DP	21500	1925	
3	SOUTHERN MAID	4-6-2	OC	DP	22070	1926	
4	THE BUG	0-4-0TT	OC	KraussS	8378	1926	
5	HERCULES	4-8-2	OC	DP	22071	1926	
6	SAMSON	4-8-2	OC	DP	22072	1926	
7	TYPHOON	4-6-2	OC	DP	22073	1926	a
8	HURRICANE	4-6-2	OC	DP	22074	1926	
9	WINSTON CHURCHILL	4-6-2	OC	YE	2294	1931	
No.10	DOCTOR SYN	4-6-2	OC	YE	2295	1931	
11	BLACK PRINCE	4-6-2	OC	Krupp	1664	1937	
No.12	J. B. SNELL	4w-4wDH		TMA	6143	1983	
14	CAPTAIN HOWEY	4w-4wDH		TMA	2336	1989	
	SHELAGH OF ESKDALE	4-6-4DH		SL		1969	
PW3	REDGAUNTLET	4wPM		AK	[1977]	1977	
7		4wDM	s/o	L	37658	1952	
–		4wDM		MR	7059	1938	
	rebuilt as	4wDH		TMA		1988	
(PW2)	SCOOTER	2w-2PM		RHDR		1964	

a   currently at Vale of Rheidol Railway, Aberystwyth, Ceredigion

SITTINGBOURNE & KEMSLEY LIGHT RAILWAY LTD, SITTINGBOURNE and KEMSLEY

Locomotives are kept at Kemsley, access to which is by train from Sittingbourne.

Gauge 4ft 8½in   www.sklr.net   **TQ 904642, 920661**

–		0-4-0F	OC	AB	1876	1925	Pvd

Gauge 2ft 6in

	PREMIER	0-4-2ST	OC	KS	886	1905	
	MELIOR	0-4-2ST	OC	KS	4219	1924	
	UNIQUE	2-4-0F	OC	WB	2216	1923	OOU
	ALPHA	0-6-2T	OC	WB	2472	1932	OOU
	TRIUMPH	0-6-2T	OC	WB	2511	1934	OOU
	SUPERB	0-6-2T	OC	WB	2624	1940	
	VICTOR	4wDM		HE	4182	1953	
P6495	BARTON HALL	4wDH		HE	6651	1965	
	EDWARD LLOYD	0-4-0DM		RH	435403	1961	

SOUTHERN LOCOMOTIVES LTD, HOPE FARM, SELLINDGE, near ASHFORD

Gauge 5ft 0in   www.southern-locomotives.co.uk   TR 119388

799		0-6-0T	OC	TK	355	1925
1134		2-8-0	OC	TK	531	1946
(1157)		2-8-0	OC	Frichs	403	1949

Gauge 4ft 8½in

34010	SIDMOUTH	4-6-2	3C	Bton		1945
			reb	Elh		1959
34058	SIR FREDERICK PILE	4-6-2	3C	Bton		1947
			reb	Elh		1960
35025	BROCKLEBANK LINE	4-6-2	3C	Elh		1948
			reb	Elh		1956
68078		0-6-0ST	IC	AB	2212	1946
No.3180	ANTWERP	0-6-0ST	IC	HE	3180	1944
3142	11201	4w-4wRER		Elh		1937
4902	(4002) S13003S	4w-4wRER		Elh		1949
4002	S 13004 S	4w-4wRER		Elh		1949
5176	S14352	4w-4wRER		Lancing/Elh		1955
(14573)	6307	4w-4wRER		Lancing/Elh		1959

TENTERDEN RAILWAY CO LTD, (KENT & EAST SUSSEX RAILWAY)

Locomotives are kept at :-

Bodiam Station	TN32 5UD	TQ 783249
Rolvenden Station	TN17 4JP	TQ 865328
Tenterden Station	TN30 6HE	TQ 882336
Wittersham Road Station	TN17 4QA	TQ 866288

Gauge 4ft 8½in   www.kesr.org.uk

4253		2-8-0T	OC	Sdn	2640	1917
5668		0-6-2T	IC	Sdn		1926
6619		0-6-2T	IC	Sdn		1928
30065		0-6-0T	OC	VIW	4441	1943
(30070	DS 238) WD 300	0-6-0T	OC	VIW	4433	1943
(31556)	753	0-6-0T	IC	Afd		1909
(32670)	70 POPLAR	0-6-0T	IC	Bton		1872
32678		0-6-0T	IC	Bton		1880
76017		2-6-0	OC	Hor		1953
No.15	HASTINGS	0-6-0ST	IC	HE	469	1888
75008	SWIFTSURE	0-6-0ST	IC	HE	2857	1943
No.23	HOLMAN F.STEPHENS	0-6-0ST	IC	HE	3791	1952
No.25	NORTHIAM	0-6-0ST	IC	HE	3797	1953
14	CHARWELTON	0-6-0ST	IC	MW	1955	1917
376	NORWEGIAN	2-6-0	OC	Nohab	1163	1919
No.12	MARCIA	0-4-0T	OC	P	1631	1923
(W20W)		4w-4wDMR		Sdn		1940
D2023		0-6-0DM		Sdn		1958
D2024		0-6-0DM		Sdn		1958
(D)3174	(08108) DOVER CASTLE	0-6-0DE		Derby		1955
D7594	(25244)	Bo-BoDE		Dar		1964
(D8087)	20087	Bo-BoDE		_(EE	2993	1961
				(RSHD	8245	1961
D9504	(01566)	0-6-0DH		Sdn		1964

M50971		2-2w-2w-2DMR		DerbyC&W	1959	
M51571		2-2w-2w-2DMR		DerbyC&W	1960	
–		Bo-BoDE		MV	1932	
–		0-4-0DE		RH	423661 1958	
DR 98211A		4wDMR		Plasser	52766A 1985	
L111 EED		4wDM	R/R	Isuzu	2005	
RTU 7109	99709 901034-7	2w-2PMR		Geismar ST/02/19 2002		
(900393)		2w-2PMR		Wkm	673 1932	DsmT
–		2w-2PMR		Wkm	6603 1953	
7438		2w-2PMR		Wkm	7438 1956	

Colonel Stephens Railway Museum,
Tenterden Town Station, Station Road, Tenterden   TN30 6HE
Gauge 4ft 8½in   www.kesr.org.uk/museum   TQ 882336

GAZELLE		0-4-2WT	IC	Dodman	1893	
EW&BR No.1		2-2wPMR		ShuttC	c2005	

TUNBRIDGE WELLS & ERIDGE RAILWAY PRESERVATION SOCIETY LTD,
SPA VALLEY RAILWAY, WEST STATION, NEVILL TERRACE,
TUNBRIDGE WELLS   TN2 5QY
Gauge 4ft 8½in   www.spavalleyrailway.co.uk   TQ 579385

32650	SUTTON		0-6-0T	IC	Bton		1876
47493			0-6-0T	IC	VF	4195	1927
(57566)	828		0-6-0	IC	StRollox		1899
68077			0-6-0ST	IC	AB	2215	1947
80078			2-6-4T	OC	Bton		1954
2315	LADY INGRID		0-4-0ST	OC	AB	2315	1951
	"NEWSTEAD"		0-6-0ST	IC	HE	1589	1929
No.57			0-6-0ST	IC	RSHN	7668	1950
62	UGLY		0-6-0ST	IC	RSHN	7673	1950
No.10	TOPHAM		0-6-0ST	OC	WB	2193	1922
(D4114)	09026	CEDRIC WARES	0-6-0DE		Hor		1962
(D4152)	08922		0-6-0DE		Hor		1962
(D5695	31265	31530) 31430 SISTER DORA					
			A1A-A1ADE		BT	296	1961
(D)6583	(33063)	R.J. MITCHELL	Bo-BoDE		BRCW	DEL187	1962
(D6585)	33065	SEALION	Bo-BoDE		BRCW	DEL189	1962
15224			0-6-0DE		Afd		1949
(E6047)	73140		Bo-BoRE/DE		_(EE	3719	1966
					(EEV	E379	1966
2591	SOUTHERHAM		0-4-0DM		_(RSHN	7924	1959
					(DC	2591	1959
–			4wDH		TH	189C	1967
		a rebuild of	4wVBT	VCG	S	9536	1952
S60142	(207017) 1317		4-4wDER		Afd/Elh		1962
62402	(1497)		4w-4wRER		York(BRE)		1971
	99709 901125-3		2w-2PMR	R/R	Geismar M44/075		2010
	99709 901126-1		2w-2PMR	R/R	Geismar M44/076		2010
	99709 901012-3		2w-2PMR		Lesmac LMS007/2		2006
	99709 901020-6		2w-2PMR		Lesmac LMS003/3		2005	Dsm

WHITSTABLE COMMUNITY MUSEUM and GALLERY,
FORESTER'S HALL, 5A OXFORD STREET, WHITSTABLE
Gauge 4ft 8½in   www.whitstablemuseum.org

(INVICTA)	0-4-0	OC	RS	24	1830

CT5 1DB
TR 106663

WOODLANDS RAILWAY
Gauge 1ft 3in   Closed   Private Site

PAM	4wPM		Mace	1980	OOU
SIMON	6wPM	s/o	Mace	1985	OOU

LANCASHIRE

INDUSTRIAL SITES

BLACKPOOL TRANSPORT SERVICES LTD, BLACKPOOL
Rigby Road Depot & Works, Blackpool
Gauge 4ft 8½in   www.blackpooltransport.com/blackpool-tramway

FY1 5DD
SD 307350

754		4w-4wDE/WER	ELC		1992	
938	Q204 HFR	4wDM	R/R	Unimog	029065	1981
939	J271 TEC	4wDM	R/R	Unimog	172544	1992
	MX14 LRJ	4wDM	R/R	Hako		2015

Starr Gate Depot, Blackpool
Gauge 4ft 8½in

FY4 1SN
SD 304317

–	4wBE	Zephir	2372	2011

EDF ENERGY plc, HEYSHAM POWER STATION,
PRINCESS ALEXANDRA WAY, HEYSHAM
Gauge 4ft 8½in   www.edfenergy.com

LA3 2XH
SD 401599

H 055	VINCENT DE RIVAZ	4wDH	S	10037	1960	a
H 003	BRIAN LARK CME	4wDH	S	10070	1961	a

    a property of Rail Management Services Ltd, Chesterfield, Derbyshire

HEIDELBERG MATERIALS CEMENT (HANSON CEMENT),
RIBBLESDALE WORKS, WEST BRADFORD ROAD, CLITHEROE
(part of the Hanson Heidelberg Cement Group)
Gauge 4ft 8½in   www.heidelbergmaterials.com

BB7 4QF
SD 749434

10	WINSTON	0-6-0DH	GECT	5396	1975		
9	CHUG CHUG	0-6-0DH	GECT	5401	1975		
(DB 965051	DE 320477 9036T)	2w-2PMR	Wkm	7574	1956	DsmT	Pvd

HELICAL TECHNOLOGY LTD, PRECISION SPRING MANUFACTURERS, (LYTHAM MOTIVE POWER MUSEUM), DOCK ROAD, LYTHAM ST ANNES FY8 5AQ

Preserved locomotives in store, not on public display.   Private Site
Gauge 4ft 8½in   www.helical-technology.com   SD 381276

	RIBBLESDALE No.3	0-4-0ST	OC	HC	1661	1936
SNIPEY	HODBARROW No.6	0-4-0CT	IC	N	4004	1890
	GARTSHERRIE No.20	0-4-0ST	OC	NBH	18386	1908

Gauge 2ft 0in

		4wDM	HE	2198	1940
-		4w-4DMR	Helical		2017

AUSTIN MOSS, RAILWAY ENGINEER,
c/o WINDMILL ANIMAL FARM, FISH LANE, HOLMESWOOD   L40 1UQ

Gauge 1ft 3in   (operates independently from Windmill Animal Farm) SD 427156

No.4468	DUKE OF EDINBURGH	4-6-2DE	s/o	Barlow		1948	
	PRINCESS ANNE	4-6w-2DE	s/o	Barlow		1948	a
	DUKE OF EDINBURGH	4-6-2DE	s/o	Barlow		1950	
No.2510	PRINCE CHARLES	4-6-2DE	s/o	Barlow		1954	OOU
	GOLDEN JUBILEE 1911-1961	4-6wDE	s/o	Barlow		1963	
	–	4-4-2	OC	MossAJ		(2019)	u/c
	–	4-4-2	OC	MossAJ			u/c
	–	0-6-2T	OC	MossAJ		2023	
	a rebuild of	0-6-0	OC			2017	
	–	0-6-0	OC	MossAJ			u/c
	–	0-6-2D	s/o	MossAJ			
	a rebuild of	0-6-4ST		MasseyD			u/c
	PRINCESS ANNE	6-6wDH		SL		1971	
7 278		2-8-0DH	s/o	SL	17/6/79	1979	
14		2w-2PM		WalkerG		1985	

a   carries works plate dated 1962 (an overhaul date ?)

D.W. MOSS, ENGINEERS, APPLEY BRIDGE

Gauge 2ft 0in   Private Site

1863	166 C.P.HUNTINGTON	4w-2-4wPH	s/o	Chance		
				79 50166 24		1979
–		4wBE		CE	B0495	1975
7		4wBE		CE	B1854	1979
–		0-4-0BE		WR	P7731	1975
	rebuilt as	2w-2BE		Ayle		1994

Gauge 1ft 3in

3		0-4-0+0-4-0 4C	LJ		1990	
	"THE CUB"	4-4wDH	Minirail		1954	
		2w-2-2-2w+4-4DMR	MossDW		2017	a
	"FRANKENCUB"	4w-4wDE	MossDW		2022	
KD1		4-4w-4-4w-4DER	RRS		1983	Dsm

KöNIGSWINTER		2-8-2DH	s/o	SL	7217	1972
	rebuilt			CCLR	1492	1992
	rebuilt as	2-8-0DH	s/o	MossDW		2017

a   constructed using part of RRS/1983 railcar

RIBBLE RAIL LTD, off CHAIN CAUL ROAD, RIVERSWAY, PRESTON
Gauge 4ft 8½in

PR2 2PD
SD 505294

Operates freight trains between Total UK Ltd, Chain Caul Way, Preston and Strand Road exchange sidings, Preston - see **Ribble Steam Railway and Museum** entry for locomotives

WEST COAST RAILWAY COMPANY LTD, JESSON WAY, CRAG BANK, CARNFORTH
Other locomotives may be occasionally present for repairs.

LA5 9UR

Gauge 4ft 8½in   www.westcoastrailways.co.uk

SD 496708

34016	BODMIN	4-6-2	3C	Bton		1945	
		reb		Elh		1958	
34067	TANGMERE	4-6-2	3C	Bton		1947	
34073	(249 SQUADRON)	4-6-2	3C	Bton		1948	Dsm
35018	BRITISH INDIA LINE	4-6-2	3C	Elh		1945	
		reb		Elh		1956	
44767		4-6-0	OC	Crewe		1947	
44932		4-6-0	OC	Hor		1945	
45110		4-6-0	OC	VF	4653	1935	
(45690	5690 LEANDER)	4-6-0	3C	Crewe	288	1936	
45562	ALBERTA		reb	Derby		1973	
45699	GALATEA	4-6-0	3C	Crewe	297	1936	
46115	(6115) SCOTS GUARDSMAN	4-6-0	3C	NBQ	23610	1927	
		reb		Crewe		1947	
(46201)	6201 PRINCESS ELIZABETH	4-6-2	4C	Crewe	107	1933	
48151		2-8-0	OC	Crewe		1942	
62005		2-6-0	OC	NBQ	26609	1949	
No.1		0-6-0F	OC	AB	1572	1917	OOU
	W.T.T.	0-4-0ST	OC	AB	2134	1942	
1		0-4-0ST	OC	AB	2230	1947	
	GLAXO	0-4-0F	OC	AB	2268	1949	OOU
(D2084	03084)	0-6-0DM		Don		1959	OOU
D2381	03381	0-6-0DM		Sdn		1961	OOU
(D3533)	08418	0-6-0DE		Derby		1958	OOU
(D3600)	08485	0-6-0DE		Hor		1958	
(D3845	08678) 555	0-6-0DE		Hor		1959	
(625	690) H 043 16	0-6-0DE		_(EE	2122	1956	
				(VF	D312	1956	
555	PZB-90 98800 170012-5	6wDH		OK	26880	1979	
		reb		Newag	275	2016	

PRESERVATION SITES

BLACKPOOL MINIATURE RAILWAY CO LTD,
RIO GRANDE EXPRESS, c/o BLACKPOOL ZOO, ZOOLOGICAL GARDENS,
EAST PARK DRIVE, BLACKPOOL   FY3 8PP
Gauge 1ft 3in   www.blackpoolzoo.org.uk   SD 335362

–	2-8-0DH	s/o	SL	7219	1972

FOLDHOUSE PARK LTD, HOLIDAY HOME PARK,
FOLD HOUSE, HEAD DYKE LANE, PILLING, near KNOTT-END-ON-SEA   PR3 6SJ
Gauge 4ft 8½in   www.foldhouse.co.uk   SD 409477

11302 THE PILLING PIG	0-6-0ST	OC	HC	1885	1955

MERSEYSIDE TRANSPORT TRUST, OSPREY PLACE, off TOLLGATE ROAD,
BURSCOUGH INDUSTRIAL ESTATE, BURSCOUGH   L40 8TG
Gauge 4ft 8½in   www.class502.org.uk   SD 429108

(M)28361(M) JOHN M ECCLES	4w-4wRER		DerbyC&W	c1939

PLEASURE BEACH RAILWAY,
PLEASURE BEACH RESORT, SOUTH SHORE, BLACKPOOL   FY4 1EZ
Gauge 1ft 9in   www.blackpoolpleasurebeach.com   SD 305332

	BARBIE	4wDM		AK	7	1982
4472	MARY LOUISE	4-6-2DH	s/o	HC	D578	1933
4473	CAROL JEAN	4-6-2DH	s/o	HC	D579	1933
		reb	4-6-4DH	s/o	Ravenglass	1988
6200	GEOFFREY THOMPSON OBE DL	4-6-2DH	s/o	HC	D586	1935
		reb		PBR		2004

POULTON & WYRE RAILWAY PRESERVATION SOCIETY, THORNTON STATION
Gauge 4ft 8½in   www.pwrs.org   Private Site

–	0-4-0DM	JF	4210108	1955	a
(51937 977806 LO 905)	2-2w-2w-2DMR	DerbyC&W		1960	a
99709 909196-6	2w-2PMR	Geismar	ST/04/11	2004	a

a   currently stored at Hillhouse Business Park, Thornton (private site)

RIBBLE STEAM RAILWAY AND MUSEUM,
off CHAIN CAUL ROAD, RIVERSWAY, PRESTON   PR2 2PD
Gauge 4ft 8½in   www.ribblesteam.org.uk   SD 504295, 500293

19	(11243)	0-4-0ST	OC	Hor	1097	1910
1439		0-4-0ST	IC	Crewe	842	1862
	JOHN HOWE	0-4-0ST	OC	AB	1147	1908
	EFFICIENT	0-4-0ST	OC	AB	1598	1918
(No.20)	"NIDDRIE"	0-6-0ST	OC	AB	1833	1924

	Name	Type		Maker	No.	Date	
	ALEXANDER	0-4-0ST	OC	AB	1865	1926	
	"HEYSHAM"	0-4-0F	OC	AB	1950	1928	
	J.N.DERBYSHIRE	0-4-0ST	OC	AB	1969	1929	
No.6		0-4-0ST	OC	AB	2261	1949	
No.4	BRITISH GYPSUM	0-4-0ST	OC	AB	2343	1953	
	LUCY	0-6-0ST	OC	AE	1568	1909	
	"MDHB No 26"	0-6-0ST	OC	AE	1810	1918	
	"JOAN"	0-6-0ST	OC	AE	1883	1922	
	EARL FITZWILLIAM	0-6-0ST	OC	AE	1917	1923	
	WINDLE	0-4-0WT	OC	EB	53	1909	
–		0-4-0ST	OC	GR	272	1894	
–		0-6-0ST	OC	HC	1450	1922	
	rebuilt as	0-6-0T	OC	Byworth		2000	
	"KINSLEY"	0-6-0ST	IC	HE	1954	1939	
WD 75105	WALKDEN	0-6-0ST	IC	HE	3155	1944	
1R	"RESPITE"	0-6-0ST	IC	HE	3696	1950	
	"SHROPSHIRE"	0-6-0ST	IC	HE	3793	1953	
No.4	GLASSHOUGHTON No.4	0-6-0ST	IC	HE	3855	1954	
No.13		0-4-0ST	OC	HL	3732	1928	
21	"LINDA"	0-6-0ST	OC	HL	3931	1938	
	"DAPHNE"	0-4-0ST	OC	P	737	1899	
	FONMON	0-6-0ST	OC	P	1636	1924	
	CALIBAN	0-4-0ST	OC	P	1925	1937	
	HORNET	0-4-0ST	OC	P	1935	1937	
"NORTH WESTERN GAS BOARD"		0-4-0ST	OC	P	1999	1941	
	JOHN BLENKINSOP	0-4-0ST	OC	P	2003	1941	
–		0-4-0ST	OC	P	2103	1950	
	AGECROFT No.2	0-4-0ST	OC	RSHN	7485	1948	
	"GASBAG"	4wVBT	VCG	S	8024	1929	
	"ST MONANS"	4wVBT	VCG	S	9373	1947	
	COURAGEOUS	0-6-0ST	OC	WB	2680	1942	
D2148		0-6-0DM		Sdn		1960	
(D2189 03189)		0-6-0DM		Sdn		1961	
D2595		0-6-0DM		HE	7179	1969	
	a rebuild of			HE	5644	1959	
D9539		0-6-0DH		Sdn		1965	
–		4wWE		BTH		1908	Dsm
–		4wBE		EEDK	788	1930	
(663)		0-6-0DE		_(EE	2160	1956	
				(VF	D350	1956	a
–		4wBE		GB	2000	1945	
	"HOTTO"	4wPM		H	965	1930	
	MIGHTY ATOM	0-4-0DM		HC	D628	1943	
D629		0-4-0DM		HC	D629	1945	
	"MARGARET"	0-4-0DM		HC	D1031	1956	
	PERSIL	0-4-0DM		JF	4160001	1952	
–		0-4-0DM		JF	4200003	1946	
	BICC	0-4-0DH		NBQ	27653	1956	
	ENERGY	4wDH		RR	10226	1965	a
	ENTERPRISE	4wDH		RR	10282	1968	a
	PROGRESS	4wDH		RR	10283	1968	a
	STANLOW No.4	0-4-0DH		TH	160V	1966	

D2870		0-4-0DH	YE	2677	1960	
E79960		2w-2DMR	WMD	1265	1958	
(98404)		4wDHR	Perm	MTU 001	1991	a

a   locomotives owned and operated by Ribble Rail Ltd

Furness Railway Trust
Gauge 4ft 8½in   www.furnessrailwaytrust.org.uk

4979	(WOOTTON HALL)	4-6-0	OC	Sdn		1930
5643		0-6-2T	IC	Sdn		1925
No.20		0-4-0	IC	SS	1448	1863
	rebuilt as	0-4-0ST	IC	BHSC		1870
	rebuilt as	0-4-0	IC	FRT		1998
–		0-4-0	IC	SS	1585	1865
	rebuilt as	0-4-0ST	IC	BHSC		1873
	CUMBRIA	0-6-0ST	IC	HE	3794	1953
No.2	FLUFF	0-4-0DM		JF	21999	1937

WEST LANCASHIRE LIGHT RAILWAY TRUST, STATION ROAD, HESKETH BANK, near PRESTON   PR4 6SP
Gauge 2ft 0in   www.westlancsrailway.org   SD 448229

(45)		0-6-0T	OC	Chrz	3506	1957	
"No.3"	IRISH MAIL	0-4-0ST	OC	HE	823	1903	
47		0-8-0T	OC	Hen	14676	1917	a
VII		0-4-2T	OC	JF	15513	1920	a
–	(CHEETAL)	0-6-0WT	OC	JF	15991	1923	
	JOFFRE	0-6-0WTT	OC	KS	2405	1915	
"No.34"	No.22 MONTALBAN	0-4-0WT	OC	OK	6641	1913	
	(SYBIL)	0-4-0ST	OC	WB	1760	1906	
1	"CLWYD"	4wDM		RH	264251	1951	
2	TAWD	4wDM		RH	222074	1943	
"No.4"	"BRADFIELD"	4wPM		FH	1777	1931	
"No.5"		4wDM		RH	200478	1940	
"No.7"		4wDM		MR	8992	1946	
10		4wDM		FH	2555	1942	
11		4wDM		MR	5906	1934	
12		4wDM		MR	11258	1964	
"No.12"	No.2	4wDM		MR	7955	1945	
			reb	WLLR	No.2	1987	b Dsm
"No.16"		4wDM		RH	202036	1941	Dsm
19		4wBE		BV	613	1972	a c
"No.19"		4wPM		L	10805	1939	
"No.20"		4wPM		Bg	3002	1937	
21		4wDM		HE	1963	1939	
25		4wBE		BV	692	1974	a c
"No.25"		4wDM		RH	297054	1950	
"No.26"	8	4wDM		MR	11223	1963	
No.31	"No.27" MILL REEF	4wDM		MR	7371	1939	
36		4wDM		RH	339105	1953	
"No.38"	750	0-4-0DMF		HC	DM750	1949	
"No.39"	P37829 "BLACK PIG"	4wDM		FH	3916	1959	
40	DAME VERA DUCKWORTH	4wDM		RH	381705	1956	

8	"PATHFINDER"	4wDM	HE	4478	1953	d
–		0-6-0DMF	HC	DM1393	1967	Dsm
640	"No.44" WELSH PONY	4wWE	BEV	640	1926	
–		4wBE	GB	1840	1942	
–		4wDM	L	29890	1946	
–		4wDM	MR	8995	1946	Dsm
52		4wBE	WR	C6765	1963	
	"BREDBURY"	2w-2PM	Bredbury		c1954	
(A155W	TR11 PWM 2187) "81"	4wDMR	CravenJ		1987	e

a	currently stored elsewhere
b	converted into a brake van
c	to be regauged from 2ft 6in
d	plate reads 4480/1953
e	built by J. Craven, Walesby, Nottinghamshire. Incorporates parts from Wkm 4164. Convertible to all gauges from 2ft 0in to 4ft 8½in.

WINDMILL ANIMAL FARM MINIATURE RAILWAY,
WINDMILL ANIMAL FARM LTD, FISH LANE, HOLMESWOOD   L40 1UQ
Gauge 1ft 3in   www.windmillanimalfarm.co.uk   **SD 427156**

	DUDLEY	4w-4wDH	Guest	1957

LEICESTERSHIRE

INDUSTRIAL SITES

AGGREGATE INDUSTRIES UK LTD
BARDON HILL QUARRY, BARDON HILL, COALVILLE   LE67 1TD
Gauge 4ft 8½in   www.aggregate.com   **SK 446129**

No.59	DUKE OF EDINBURGH	6wDH	RR	10273	1968	
No.159	"DUCHESS"	6wDH	TH	297V	1981	
(D3587)	08472	0-6-0DE	Crewe		1958	a

a	property of Hunslet Ltd, Barton-under-Needwood, Staffordshire

CLAYTON EQUIPMENT LTD,
FALCON WORKS, NOTTINGHAM ROAD, LOUGHBOROUGH   LE11 1HL
Gauge 4ft 8½in   www.claytonequipment.co.uk   **SK 542210**

(18002)	9770 0018002-5	4w-4wBE	CE	B4660.2	2021

Clayton locomotives for overhaul/testing/repair occasionally present

Site also occupied by **UK Rail Leasing Ltd** - see entry below

KING RAIL, KING VEHICLE ENGINEERING LTD, RIVERSIDE, MARKET HARBOROUGH

LE16 7PX

SP 743876

UK agent for Zagro.   www.king.uk.com

Road/Rail vehicles under conversion/repair occasionally present

LOCOMOTIVE MAINTENANCE SERVICES LTD, 14 BAKEWELL ROAD, LOUGHBOROUGH

LE11 5QY

SK 526212

Gauge 4ft 8½in   www.locomotivemaintenanceservices.com

60163	TORNADO	4-6-2	3C	Darlington	2195	2008
62712	(MORAYSHIRE)	4-4-0	3C	Dar	(1391?)	1928
65033		0-6-0	IC	Ghd		1889
76077		2-6-0	OC	Hor		1956
(17)		0-6-0ST	IC	HE	2880	1943
S.121	PRIMROSE No.2	0-6-0ST	IC	HE	3715	1952
	LOCOMOTION	0-4-0	VC	LocoEnt	No.1	1975
"No.11"		0-4-0ST	OC	N	2937	1882
	LINDSAY	0-6-0ST	IC	WCI		1887
No.3		0-4-0ST	OC	YE	2474	1949

Locomotives for overhaul or repairs usually present

NETWORK RAIL INFRASTRUCTURE LTD, RAIL INNOVATION and DEVELOPMENT CENTRE

Locomotives are kept at :-

Unit F and Sidings, Melton Commercial Park,		
formerly known as Asfordby Business Park, Melton Mowbray	LE14 3JL	SK 727207
Old Dalby Test Centre, Station Road, Old Dalby	LE14 3NQ	SK 680239

Gauge 4ft 8½in   www.networkrail.co.uk

CD40	B4618/3	GARY	4wDH	CE	B4618.3	2016	a
CD40	B4618/4	PUSHY-PULLY	4wDH	CE	B4618.4	2016	a

a   property of Railway Support Services Ltd, Wishaw, Warwickshire

TARMAC plc - A CRH Company, BARROW RAILHEAD, SILEBY ROAD, BARROW-UPON-SOAR

LE12 8LX

SK 587168

Gauge 4ft 8½in   www.tarmac.com

ERNIE	4wDH	R/R	Zephir	1928	2005
TED	4wDH	R/R	Zephir	2136	2008
TRIGGER	4wDH	R/R	Zephir	2779	2018

UK RAIL LEASING LTD,
Beal Street, Leicester   **LE2 0AA**
Gauge 4ft 8½in   www.ukrl.co.uk   **SK 597044**

(D5410 27059)		Bo-BoDE	BRCW	DEL253	1962
(D6703) 37003		Co-CoDE	_(EE	2866	1960
			(VF	D582	1960
58016 050		Co-CoDE	Don		1984
58023		Co-CoDE	Don		1984

Other mainline and preserved locomotives usually present

25 Shop, Falcon Works, Nottingham Road, Loughborough   **LE11 1HL**
Gauge 4ft 8½in   www.ukrl.co.uk   **SK 542210**

–		4wDH	TH	184V	1967	a
89001	(AVOCET)	Co-CoWE	_(Crewe		1986	
			(BT	875	1986	

a   property of Railway Support Services Ltd, Wishaw, Warwickshire

PRESERVATION SITES

ARMOURGEDDON MILITARY MUSEUM,
SOUTHFIELDS FARM, HUSBANDS BOSWORTH, near LUTTERWORTH   **LE17 6NW**
Gauge 600mm   www.militarymuseum.uk   **SP 638861**

(RTT/767149)	PRESIDENT	2w-2PM	Wkm	3151	1943	
RTT/767163		2w-2PM	Wkm	3236	1943	Dsm

THE BATTLEFIELD LINE, (THE SHACKERSTONE RAILWAY SOCIETY LTD),
SHACKERSTONE STATION, MARKET BOSWORTH
Locomotives are kept at :-   Shackerstone   CV13 0BS   SK 378067, 379064

Market Bosworth   CV13 0PF   SK 392030

Gauge 4ft 8½in   www.battlefieldline.co.uk

5199			2-6-2T	OC	Sdn		1934	
	SIR GOMER		0-6-0ST	OC	P	1859	1932	
	TEDDY		0-4-0ST	OC	P	2012	1941	
(D14)	45015		1Co-Co1DE		Derby		1960	OOU
(D318	97408)	40118	1Co-Co1DE		_(EE	2853	1961	
					(RSHD	8148	1961	
(D2310	04110)		0-6-0DM		_(RSHD	8169	1960	
					(DC	2691	1960	
(D3820)	08653	VERNON	0-6-0DE		Hor		1959	a
(D3868)			0-6-0DE		Hor		1960	a
(D3849	08682)	LIONHEART	0-6-0DE		Hor		1959	a
(D3972)	08804		0-6-0DE		Derby		1960	a
(D4135)	08905		0-6-0DE		Hor		1962	a
D6508	(33008)	EASTLEIGH	Bo-BoDE		BRCW	DEL100	1960	
(D6586)	33201		Bo-Bo DE		BRCW	DEL157	1962	
D6593	(33208)		Bo-BoDE		BRCW	DEL164	1962	
D7523	(25173)		Bo-BoDE		Derby		1965	

(D8063	20063) 2002 AT3 DJ 054	Bo-BoDE		_(EE	2969	1961	
				(RSHD	8221	1961	b
58012		Co-CoDE		Don		1984	
58048		Co-CoDE		Don		1986	
12083	(201276 M413)	0-6-0DE		Derby		1950	
(19)		0-6-0DH		AB	594	1974	
–		4wBE/WE		EEDK	905	1935	
"DAVY"		0-6-0DE		EEDK	1901	1951	
"MAZDA"		0-4-0DE		RH	268881	1949	
–		0-4-0DM		RH	281271	1950	a
01547	(266)	4wDH		TH	308V	1983	a
M51131		2-2w-2w-2DMR		DerbyC&W		1958	
W51321	(977753) P464	2-2w-2w-2DMR		BRCW		1960	
M55005		2-2w-2w-2DMR		GRC&W		1958	
(S65321	5791 977505 932053)	4w-4RER		Elh		1954	
	C958 YOR	4wDMR	R/R	Bruff	516	1986	
	C951 YOR	4wDMR	R/R	Bruff	519	1986	c
–		2w-2PMR		Geismar ST/00/16		2000	

a   property of Harry Needle Railroad Co Ltd, Derbyshire
b   property of Railway Support Services Ltd, Wishaw, Warwickshire
c   incorporates chassis from Bruff 519 and parts from 518 & 520

"CONKERS", RAWDON ROAD, MOIRA, near ASHBY-DE-LA-ZOUCH   DE12 6GA
(operated by Planning Solutions Ltd)
Gauge 2ft 0in   www.visitconkers.com   **SK 312161**

CONKACHOO	4-4-0DH	s/o	SL	2121	2001
"CONKACHOO TWO"	4w-4wDH	s/o	SL	8204	2011

EASTWELL HISTORY GROUP, CROSSROADS FARM, SCALFORD ROAD, EASTWELL, near MELTON MOWBRAY   LE14 4EF
Gauge 3ft 0in   **SK 772279**

LORD GRANBY	0-4-0ST	OC	HC	633	1902

GREAT CENTRAL RAILWAY plc, GREAT CENTRAL STATION, LOUGHBOROUGH
Locomotives are kept at :-

Loughborough Loco Shed	LE11 1RW	SK 543194
Rothley Carriage & Wagon Works	LE7 7LD	SK 569122
Swithland Sidings	LE7 7SL	SK 563132
Quorn and Woodhouse	LE12 8AW	SK 549161

Gauge 4ft 8½in   www.gcrailway.co.uk

6990	WITHERSLACK HALL	4-6-0	OC	Sdn		1948
(30777)	777 SIR LAMIEL	4-6-0	OC	NBH	23223	1925
34039	BOSCASTLE	4-6-2	3C	Bton		1946
			reb	Elh		1959
45305	ALDERMAN A.E. DRAPER	4-6-0	OC	AW	1360	1937
45491		4-6-0	OC	Derby		1943
48305		2-8-0	OC	Crewe		1943
48624	(8624)	2-8-0	OC	Afd		1943

63601	(102)		2-8-0	OC	Gorton		1912	
70013	OLIVER CROMWELL		4-6-2	OC	Crewe		1951	
73156			4-6-0	OC	Don		1956	
78018			2-6-0	OC	Dar		1954	
78019			2-6-0	OC	Dar		1954	
92214			2-10-0	OC	Sdn		1959	
–			0-6-0ST	IC	HE	3809	1954	
4	"MEAFORD"		0-6-0T	OC	RSHN	7684	1951	
–			4wVBT	VCG	S	9370	1947	
D123	(45125 89423) LEICESTERSHIRE AND DERBYSHIRE YEOMANRY							
			1Co-Co1 DE		Crewe		1961	
D1705	(47117) SPARROWHAWK		Co-CoDE		BT	467	1965	
(D2989	07005)		0-6-0DE		RH	480690	1960	
				reb	Resco	L106	1978	
(D3101)	13101		0-6-0DE		Derby		1955	
D4067	No.1802/B4 MARGARET ETHEL - THOMAS ALFRED NAYLOR							
			0-6-0DE		Dar		1961	
D4137	(08907)		0-6-0DE		Hor		1962	
D5185	(25735 25035) CASTELL DINAS BRAN							
			Bo-BoDE		Dar		1963	
(D)5401	(27056)		Bo-BoDE		BRCW	DEL244	1962	
D5830	(31563)		A1A-A1A DE		BT	366	1962	
(D6535	33116)		Bo-BoDE		BRCW	DEL127	1960	
D6700	(37350)		Co-CoDE		_(EE	2863	1960	
					(VF	D579	1960	
(D6724	37024) 37714		Co-CoDE		_(EE	2887	1961	
	CARDIFF CANTON				(VF	D603	1961	
(D6907)	37207		Co-CoDE		_(EE	3385	1963	
					(EEV	D851	1963	a
D8098	(20098)		Bo-BoDE		_(EE	3003	1961	
					(RSHD	8255	1961	
	ARTHUR WRIGHT		0-4-0DM		JF	4210079	1952	
(50193)	960992 977898		2-2w-2w-2DMR		MetCam		1957	
50203	(960992 977897)		2-2w-2w-2DMR		MetCam		1957	
E50266	(53266)		2-2w-2w-2DMR		MetCam		1957	
M50321	(960993 977900)		2-2w-2w-2DMR		MetCam		1958	
51396	L720		2-2w-2w-2DMR		PSteel		1960	
E51427	(960 993 977899)		2-2w-2w-2DMR		MetCam		1959	
52308	153308		4w-4DHR		Leyland	R6.016	1988	
				reb	AB		1991	
55009	109		2-2w-2w-2DMR		GRC&W		1958	
57371	153371		4w-4DHR		Leyland	R6.041	1988	
				reb	AB		1992	
M79900	(975010) IRIS		2-2w-2w-2DMR		DerbyC&W		1956	
2			2w-2PMR		Bance	002	1995	
3			2w-2PMR		Bance	037	1995	
–			2w-2BER		Bance	104	2002	Dsm
	99709 901205-3		2w-2PMR	R/R	Gelsmar M44/155		2012	
	99709 901214-5		2w-2PMR	R/R	Geismar M44/161		2012	

a   property of Meteor Power Ltd, Silverstone, Northamptonshire

Mountsorrel Railway

Line runs from Swithland Sidings to Mountsorrel, with trains operated by the Great Central Railway. Listed seperately as **Mountsorrel & Rothley Community Heritage Centre** - see entry below)

LEICESTER CITY COUNCIL, LEISURE & CULTURE, ABBEY PUMPING STATION - LEICESTER CITY'S MUSEUM OF SCIENCE & TECHNOLOGY, CORPORATION ROAD, LEICESTER

LE4 5PX

Gauge 2ft 0in   www.abbeypumpingstation.org   **SK 589067**

LEONARD	0-4-0ST	OC	WB	2087	1919	
NEW STAR	4wPM		L	4088	1931	
–	4wPM		MR	5260	1931	
–	4wDM		RH	223700	1943	
PETER	4wDM		SMH	40SD515	1979	

LEICESTERSHIRE COUNTY COUNCIL, LEICESTERSHIRE COUNTRY PARKS, SNIBSTON MINE, SNIBSTON COLLIERY PARK, ASHBY ROAD, COALVILLE

LE67 3LF

Gauge 4ft 8½in   www.visitleicester.info   **SK 420144**

No.2	0-4-0F	OC	AB	1815	1924	
CADLEY HILL No.1	0-6-0ST	IC	HE	3851	1962	
(MARS II)	0-4-0ST	OC	RSHN	7493	1948	
–	0-6-0DH		HE	6289	1966	

Gauge 2ft 6in

(T42.1994) T15 (2416)	4wBEF		_(EE	2416	1957	
			(RSHN	7935	1957	
–	4wBEF		_(EE	2086	1955	
			(Bg	3434	1955	Dsm
–	4wBEF		GECT	5424	1976	
–	0-6-0DMF		HC	DM1238	1960	
–	4wDH		HE	8973	1979	

Guided tours on advertised open days

MOUNTSORREL & ROTHLEY COMMUNITY HERITAGE CENTRE, MOUNTSORREL RAILWAY, NUNCKLEY HILL QUARRY, SWITHLAND LANE, ROTHLEY

LE7 7SJ

Gauge 4ft 8½in   www.heritage-centre.co.uk   **SK 569142**

No.3 (COLIN McANDREW)	0-4-0ST	OC	AB	1223	1911	
–	0-4-0ST	OC	BE	314	1906	
No.72 2235/72	0-6-0ST	IC	VF	5309	1945	
–	4wDM		RH	393304	1956	
(DB 965080)	2w-2PMR		Wkm	7595	1957	
	rebuilt as 2w-2DMR				1999	

Nunckley Narrow Gauge Railway
Gauge 2ft 0in

–	4wDM		MR	22070	1960	
85049 MALCOLM	4wDM		RH	393325	1952	

OAKBERRY TREES, OAKBERRY FARM,
LUTTERWORTH ROAD, DUNTON BASSETT, LUTTERWORTH   **LE17 5JU**
Gauge 1ft 0¼in   www.oakberrytrees.co.uk   **SP 548899**

D1102	TENACITY	6wPM		Perriton	1992	
37309	ENDURANCE	6w-6wDH		Perriton	2009	
–		4wPM		?		

PETER THOMAS, BLABY "WAKES" SHOWGROUND, LEICESTER ROAD, BLABY
Gauge 4ft 8½in   **SK 567983**

–		0-6-0F	OC	WB	2370	1929	OOU

TWINLAKES PARK, MELTON SPINNEY ROAD, MELTON MOWBRAY   **LE14 4FF**
Gauge 1ft 3in   www.twinlakespark.co.uk   **SK 774213**

–		2-6-0DH	s/o	SL	RG11-86	1986	
–		2-6-0DH	s/o	SL	76.3.88	1988	

WELLAND VALLEY VINTAGE TRACTION CLUB,
GLEBE ROAD, MARKET HARBOROUGH
Gauge 3ft 0in   **Private Site**   **SP 742868**

KETTERING FURNACES No.8	0-6-0ST	OC	MW	1675	1906	a

a   currently under renovation at a private location

DAVID WHITE, GREENLEA LIGHT RAILWAY, WORKHOUSE LANE, BURBAGE
Gauge 2ft 0in   **SP 447915**

SIR GEORGE	4wDH	s/o	AK	12	1984	
C.P.HUNTINGTON	4w-4wDH	s/o	Chance			
			78-50157-24	1978		
GOLIATH	4wDM		MR	5881	1935	

LINCOLNSHIRE

INDUSTRIAL SITES

BRITISH STEEL LTD,
HEAD OFFICE, BRIGG ROAD, SCUNTHORPE   **DN16 1XA**
(part of the Jingye Group)   www.britishsteel.co.uk
Appleby Coke Ovens
Gauge 4ft 8½in   (Closed)   **SE 917108**

5	4wRE	Schalke	10-310-0054	1973		OOU
6	4wRE	Schalke	10-310-0055	1973		OOU
7	4wRE	Schalke	10-310-8070	1979	a	OOU

a   built under licence by Starco Engineering, Winterton Road, Scunthorpe

Appleby-Frodingham Works, Scunthorpe

DN16 1BP

Gauge 4ft 8½in

SE 910110, 915110, 916105

4	0448-73-04	0-4-0DE	_(BD	3737	1977		
			(GECT	5437	1977		OOU
29		0-6-0DE	YE	2938	1964		Dsm
30	FUSION	Bo-BoDE	CNES		2013	a	OOU
44		0-6-0DE	YE	2768	1960		OOU
51	5	0-6-0DE	YE	2709	1959		
58		0-6-0DH	HE	7409	1976		
61		6wDH	RR	10277	1968	b	OOU
63		6wDH	TH	V317	1987	b	OOU
70	BIG KEITH	Bo-BoDE	HE	7281	1972		
71		Bo-BoDE	HE	7282	1972		
72		Bo-BoDE	HE	7283	1972		
		reb	BS(S)		2019		
73		Bo-BoDE	HE	7284	1972		Dsm
74		Bo-BoDE	HE	7285	1972		
75	GENERAL	Bo-BoDE	HE	7286	1972		
76		Bo-BoDE	HE	7287	1973		Dsm
79		Bo-BoDE	HE	7290	1973		Dsm
80		Bo-BoDE	HE	7474	1977		Dsm
(D8056	20056) 81	Bo-BoDE	_(EE	2962	1961		
			(RSHD	8214	1961	b	OOU
90		0-6-0DE	YE	2943	1965		OOU
91		0-6-0DE	YE	2944	1965		OOU
92		0-6-0DE	YE	2788	1960		OOU
93	3	0-6-0DE	YE	2902	1963		
94		0-6-0DE	RR	10238	1967		OOU
95	(29)	0-6-0DE	YE	2690	1959		OOU
8.701		Bo-BoDE	MaK	1600.001	1996	c	
8.702	92 76 0308702-8	Bo-BoDE	MaK	1600.002	1996	c	
8.703		Bo-BoDE	MaK	1600.003	1996	c	
8.704	PAT	Bo-BoDE	MaK	1600.004	1996	c	
8.708	92 76 0308708-5	Bo-BoDE	MaK	1600.008	1996	c	
8.712		Bo-BoDE	MaK	1600.012	1996	c	OOU
8.716		Bo-BoDE	MaK	1600.016	1996	c	
(8.717)	817	Bo-BoDE	MaK	1600.017	1996	c	
8.718		Bo-BoDE	MaK	1600.018	1996	c	Dsm
8.719		Bo-BoDE	MaK	1600.019	1996	c	
(8.720)	820 POPPY	Bo-BoDE	MaK	1600.020	1996	c	
(18003)	9770 0018003-3	4w-4wBE	CE	B4660.3	2022		

a constructed using parts from HE 7290
b property of Harry Needle Railroad Co, Derbyshire
c operated by GBRF Ltd and maintained by Electro-Motive Diesel Inc

Blast Furnace Highline, Appleby-Frodingham Works, Scunthorpe

Gauge 4ft 8½in

SE 917104

HL 1	0449-73-01	0-4-0DE	_(BD	3734	1977
			(GECT	5434	1977
HL 2	0448-73-02	0-4-0DE	_(BD	3735	1977
			(GECT	5435	1977

No.3	0448/73/03	0-4-0DE	_(BD	3736	1977	
			(GECT	5436	1977	
No.5	0448-73-05	0-4-0DE	_(BD	3738	1977	
			(GECT	5438	1977	
6	0448-73-06	0-4-0DE	_(BD	3739	1977	
			(GECT	5439	1977	
HL 7		0-4-0DE	_(BD	3740	1977	
			(GECT	5440	1977	

Rail Mill, Appleby-Frodingham Works, Scunthorpe
Gauge 4ft 8½in   SE 910113

DAISY	4wCE	Vollert		2008

DEPOT RAIL LTD, UNIT 5-6 SANDARS ROAD,
HEAPHAM ROAD INDUSTRIAL ESTATE, GAINSBOROUGH   DN21 1RZ
Gauge 4ft 8½in   www.depotrail.co.uk   SK 835890

–	4wDH	R/R	Zephir	1470	1997	a
–	4wDH	R/R	Zephir	1496	1997	
–	4wBE	R/R	Zephir	1890	2004	

     a   currently at Alstom Transport, Crewe Works, Cheshire

    UK agents for Zephir

EXOLUM IMMINGHAM LTD, ABP PORT OF IMMINGHAM,
IMMINGHAM WEST TERMINAL, IMMINGHAM DOCK, IMMINGHAM   DN40 2QU
(part of the Exolum Group)
Gauge 4ft 8½in   www.exolum.com   TA 195167

45	TMC	4wDH	NNM	83506	1984	a

     a   permanently coupled to a 4w wagon for brake assistance

H.M. DETENTION CENTRE, NORTH SEA CAMP, FREISTON, nr BOSTON   PE22 0QX
Gauge 2ft 0in   www.justiceinspectorates.gov.uk   TF 385405

–	4wDM	LB	55413	1967	Pvd

PHILLIPS 66 LTD, HUMBER REFINERY, SOUTH KILLINGHOLME   DN40 3DW
Gauge 4ft 8½in   www.phillips66.co.uk   TA 163168

CD40	B4618/5	4wDH	CE	B4618.5	2016	a
CD40	B4618/7	4wDH	CE	B4618.7	2016	a
(DH60)		6wDH	HE	9372	2010	
	EARL OF YARBOROUGH	6wDH	RFSK	V336	1991	
		refurbished	HE	9383	2013	

     a   property of Railway Support Services Ltd, Wishaw, Warwickshire

RAMCO UK LTD, CHURCH ROAD SOUTH, SKEGNESS   PE25 3RS
Gauge 4ft 8½in   www.ramco.co.uk   TF 512607

Locomotives for auction or resale occasionally present

RMS TRENT PORTS, FLIXBOROUGH WHARF,
FLIXBOROUGH STATHER, FLIXBOROUGH (part of the RMS Group)   DN15 8RS
Gauge 4ft 8½in   www.rms-humber.co.uk   SE 859147

TNS 107	2w-2DMR	Robel		
		56.27-10-AG38	1982	a

a   property of Rail Management Services Ltd, Chesterfield, Derbyshire

TOTAL UK LTD, TOTAL LINDSEY OIL REFINERY,
EASTFIELD ROAD, NORTH KILLINGHOLME, IMMINGHAM   DN40 3LW
Gauge 4ft 8½in   www.totalenergies.co.uk   TA 160176

BEAVER	0-6-0DH		AB	630	1978
		reb	YEC	L168	1999
BADGER	0-6-0DH		AB	658	1980
DUNDERS	0-6-0DH		HE	6971	1968
TIGGA	0-6-0DH		TH	285V	1979
		reb	HAB		1992

VICTORIA GROUP HOLDINGS LTD, t/a VICTORIA GROUP,
PORT OF BOSTON, BOSTON   PE21 6BN
Gauge 4ft 8½in   www.victoriagroup.co.uk   TF 329431

(D3460)	08375	PB357	0-6-0DE	Dar	'	1957
(D4110)	09022	PB144	0-6-0DE	Hor		1961

PRESERVATION SITES

JASON ALLEN, GRIMOLDBY
Gauge 2ft 0in   Private Site   TF 388877

–	2w-2PM	AllenJ		c2005
	a rebuild of	Wkm	4092	1946

Gauge 1ft 3in

E1	NUCLEAR ELECTRIC	4wBE	HardyK	E1	1992

APPLEBY-FRODINGHAM RAILWAY PRESERVATION SOCIETY,
c/o BRITISH STEEL, APPLEBY-FRODINGHAM WORKS, SCUNTHORPE   (DN16 1XA)
Gauge 4ft 8½in   www.afrps.co.uk   SE 913109

No.54	No.22	0-4-0ST	OC	AB	2320	1952
(No.8)		0-4-0ST	OC	AB	2369	1955

	CRANFORD	0-6-0ST	OC	AE	1919	1924
3138	HUTNIK A11	0-6-0T	OC	Chrz	3138	1954	a
22		0-6-0ST	IC	HE	3846	1956	b
–		0-4-0ST	OC	P	1438	1916
2	"HORSA"	0-4-0DH		_(RSHD	8368	1962
				(WB	3213	1962
	ARNOLD MACHIN	0-6-0DE		YE	2661	1958
1		0-6-0DE		YE	2877	1963
55817	144017	2-2wDMR		_(BRE(D)		1986
				(Alex	2785/33	1986
	rebuilt as	2-2wDHR		AB		1991
55840	144017	2w-2DMR		_(BRE(D)		1986
				(Alex	2785/34	1986
	rebuilt as	2w-2DHR		AB		1991
55853	(144017)	2-2wDMR		_(BRE(D)		1988
				(Alex	1187/4	1988
	rebuilt as	2-2wDHR		AB		1991

a   carries plate Chrz 3140 in error
b   rebuilt using parts from HE 3844 and WB 2758; carries worksplate HE 3844; currently at North Norfolk Railway, Sheringham, Norfolk

P. CLARK, FULSTOW STEAM CENTRE, CARPENTERS ROW, MAIN STREET, FULSTOW, near LOUTH

Private Site

Gauge 4ft 8½in   Private site with occasional public open days   **TF 331972**

No.1		0-4-0ST	OC	RSHN	7680	1950

CLEETHORPES COAST LIGHT RAILWAY LTD, THE MERIDIAN LINE, LAKESIDE STATION, KINGS ROAD, CLEETHORPES

DN35 0AG

Gauge 1ft 3in   www.cclr.co.uk   **TA 319070, 321072**

	"SEA BREEZE"	4-6-2	OC	CCLR		(1998)	a Dsm
15	RACHEL	0-6-0DM		MossDW		2018	
	a rebuild of	0-6-0PE		MossAJ		2017	
	a rebuild of	0-6-0PE/BE		MossDW		2013	
	a rebuild of	0-6-0DM		G&S		1961	b
	EFFIE	0-4-0T+T	OC	GNS	11	1999	
	BONNIE DUNDEE	0-4-0WT	OC	KS	720	1901	
	rebuilt as	0-4-2T	OC	Ravenglass		1981	
	rebuilt as	0-4-2	OC	Ravenglass		1996	
5	CEAWLIN	2-8-2DH	s/o	Longleat		1989	
	a rebuild of	0-6-2DH	s/o	Longleat			
	a rebuild of	2-8-0DH	s/o	SL	75 356	1975	
	THE FLOWER OF THE FOREST	2w-2VBT	VCG	Ravenglass	5	1985	
DA1	6	4wDM		Bush Mill Rly		1986	
			reb	Bush Mill Rly		c1995	
–		0-6-0DH		RVM	27	2021	

a   not yet completed
b   carries worksplate G&S 15/1959

COLIN COPCUTT, SILVERLEAF POPLAR RAILWAY, SIBSEY ROAD, OLD LEAKE

Gauge 2ft 0in / 600mm Visitors by appointment only

PE22 9QS
Private Site

05/580		4wDH	AK	19	1985	
–		4wPM	FH	1767	1931	Dsm
–		4wPM	FH	3424	1949	
–		4wDM	L	41803	1955	
–		4wPM	MH	110	1925	
R14	JANE	4wDM	MR	8565	1940	
"37"		4wDM	RH	172901	1935	
5		4wPM	VIW	4196	1936	
SP 202		4wBE	CE	B0182C	1974	
–		4wBE	WR	918	1936	

CROWLE PEATLAND RAILWAY SOCIETY, CROWLE PEATLAND RAILWAY, THE OLD PEATWORKS, off DOLE ROAD, CROWLE

Gauge 3ft 0in www.peatland.co.uk

DN17 4BL
SE 757141

H 11023	LITTLE PEAT	4wDM		MR	40S302	1967	
H23		4wDH		Schöma	5129	1990	
			reb	AK		1998	
S20		4wDH		Schöma	5130	1990	
			reb	AK		1998	
–		4wDH		Schöma	5131	1990	
			reb	AK		1998	a
–		4wDH		Schöma	5132	1990	
			reb	AK		1998	a
S21	THE THOMAS BUCK	4wDH		Schöma	5220	1990	
–		4wDH		Schöma	5221	1990	a
–		2w-2PMR		Wkm	4091	1946	
–		2w-2PMR		Wkm			DsmT

a   slave unit for use with a master unit (5129, 5130 or 5220)

Gauge 2ft 0in - Crowle Brickworks Locomotive Group

No.1881	4wPM	FH	1881	1934	

J. HUGHES, UNKNOWN LOCATION

Gauge 1ft 0¼in

Private Site

	PRINCE EDWARD	4-4-2	OC	FlooksG	1935
1005	(GORDON)	4-6-4	OC	Thurston	c1948
1010	(HENRY)	4-6-4	OC	Thurston	1947
No.1	(PRINCESS MARGARET)	4-4-0PM	s/o	Barnard	1937

HUMBER BRIDGE GARDEN CENTRE, OLD TILE WORKS, FAR INGS TILERIES, FAR INGS ROAD, BARTON-UPON-HUMBER

Gauge 2ft 0in www.humberbridgegardencentre.co.uk

DN18 5RF
TA 023233

–	4wDM	MR	8678	1941	OOU

Loco stored on former William Blyth, Far Ings Tile Works site.

LINCOLNSHIRE COAST LIGHT RAILWAY, SKEGNESS WATER LEISURE PARK, WALLS LANE, INGOLDMELLS, SKEGNESS

Gauge 2ft 0in   www.lclr.co.uk   **PE25 1JF**

     TF 561671

No.2	JURASSIC	0-6-0ST	OC	P	1008	1903	
(7)	NOCTON	4wDM		MR	1935	1920	
1	PAUL	4wDM		MR	3995	1934	
4	WILTON	4wDM		MR	7481	1940	
No.6	GRICER	4wDM		MR	8622	1941	
T3		4wDM		MR	8738	1942	
	rebuilt as	4wDH				c1998	
9	SARK	4wDM		MR	8825	1943	
No.5	MAJOR J.E. ROBINS R.E.	4wDM		MR	8874	1944	
11140		4wDM		MR	8905	1944	Dsm
8	FRED	4wDM		MR	9264	1947	

LINCOLNSHIRE COUNTY COUNCIL, MUSEUM OF LINCOLNSHIRE LIFE, BURTON ROAD, LINCOLN

Gauge 4ft 8½in   www.lincolnshire.gov.uk   **LN1 3LY**

     SK 972723

–	4wDM		RH	463154	1961

Gauge 2ft 6in

–	4wPM		RP	52124	1918

Gauge 2ft 3in

–	4wDM		RH	192888	1938

Gauge 2ft 0in

–	4wDM		RH	421432	1959

LINCOLNSHIRE WOLDS RAILWAY plc, LUDBOROUGH STATION, STATION ROAD, LUDBOROUGH

Gauge 4ft 8½in   www.lincolnshirewoldsrailway.co.uk   **DN36 5SQ**

     TF 309960

1313		4-6-0	OC	Motala	586	1917
	SPITFIRE	0-4-0ST	OC	AB	1964	1929
	LION	0-4-0ST	OC	P	1351	1914
	FULSTOW	0-4-0ST	OC	P	1749	1928
	ZEBEDEE	0-6-0T	OC	RSHN	7597	1949
D3167	(08102)	0-6-0DE		Derby		1955
–		4wDM		HE	5308	1960
			reb	Resco	L107	1981
–		0-4-0DM		JF	4210131	1957
–		0-4-0DM		JF	4210145	1958
4		0-4-0DM		RH	375713	1954
6		0-4-0DM		RH	414303	1957
7		4wDM		RH	421418	1958
–	"LITTLE DEBBIE"	0-4-0DE		RH	423657	1958
	DEBBIE	0-6-0DM		WB	3151	1962

M. MAITLAND, THE OLD STATION, FEN ROAD, RIPPINGALE

Gauge 4ft 8½in   TF 115283

ELIZABETH	0-4-0ST	OC	AE	1865	1922	
WD 8313	0-4-0DH		TH	132C	1963	
a rebuild of	0-4-0DM		JF	22982	1942	

A. NEALE, PRIVATE SITE, near GAINSBOROUGH

Gauge 1ft 3in   Private Site

ANNIE	0-4-0T	OC	FMB	001	1992	a
–	4wPM		L	35811	1950	
–	4wPM		StanhopeT			Dsm

a   carries plate DB/1896

NORTH INGS FARM MUSEUM,
FEN ROAD, DORRINGTON, near RUSKINGTON   LN4 3QB

Gauge 2ft 0in   www.northingsfarmmuseum.co.uk   TF 098527

No.9 SWIFT	4wVBT	G	HallT	1859401	1994	
rebuild of [ODDSON]	4wVBT	G	MarshallJ		1970	
–	4wDM		ClayCross		1961	a
1	4wDM		HE	6013	1961	OOU
BULLFINCH	4wDM		HE	7120	1969	
–	4wDM		MR	7493	1940	
LOD/758022 PENELOPE	4wDM		MR	8826	1943	
–	4wDM		OK		c1932	
–	4wDM		RH	183773	1937	OOU
INDIAN RUNNER	4wDM		RH	200744	1940	
No.1	4wDM		RH	371937	1954	
–	4wDM		RH	375701	1954	Dsm
–	4wDM		RH	421433	1959	

a   constructed from parts supplied by Lister

M.C. PALMER, THE GARDEN RAILWAY, METHERINGHAM

Gauge 1ft 3in   Private Site

No.1 EFFIE	0-4-0T	OC	Waterfield	2010	
No.3 URSULA	0-6-0T	OC	Waterfield	1999	a
No.4	4wPM		L		

a   carries worksplate DB 1916, incorporating parts from original locomotive

KELVIN SCULLY, LOUTH

Gauge 1ft 3in   Private Site

2870 CITY OF LONDON	4-6-0DM	s/o	JMR		1978

TONY SINCLAIR, PRIVATE LOCATION, near BROTHERTOFT, BOSTON

Gauge 4ft 8½in   **Private Site**

7514 (PWM 4311)	2w-2PMR	Wkm	7514	1956

STATION ROAD STEAM LTD,
UNIT 16, MOORLANDS INDUSTRIAL ESTATE, METHERINGHAM   **LN14 3HX**

Miniature locomotives under construction, repair or for resale, usually present.
www.stationroadsteam.com   **TF 078614**

TYSDALE FARM, TYDD ST. MARY

Gauge 2ft 0in   **Private Site**

–	4wDM	OK	6931	1937
–	4wDM	OK	7734	1938

JAMES WATERFIELD, BOSTON

Gauge 1ft 3in   **Private Site**

AO-TE-AROA	4-4-0	OC	Herschell

H. WILLIAMS,
STAMFORD HOUSE, THORNTON STREET, BARROW-ON-HUMBER   **DN19 7DG**

Gauge 2ft 0in (nominal)   **Private Site**   **TA 069208**

-	2w-2BE	_(WilliamsH	2022
		(PL Services	2022

GREATER LONDON

INDUSTRIAL SITES

ALSTOM TRANSPORT (part of Alstom Holdings S.A.)
Ilford Depot, Ley Street, Ilford   **IG1 4BP**

Gauge 4ft 8½in   www.alstom.com   **TQ 445869**

(D3755)	08588	H 047	0-6-0DE	Crewe		1959	a
(D3930)	08762	H 067	0-6-0DE	Hor		1961	a
–			4wBE	Niteq	B238	2005	
1			4wBE	R/R	Zwiehoff	2017	
2			4wBE	R/R	Zwiehoff	2018	
3			4wBE	R/R	Zwiehoff	2018	

    a   property of Rail Management Services Ltd, Chesterfield, Derbyshire

Old Oak Common Depot, Old Oak Common Lane, London · **NW10 6DW**
Gauge 4ft 8½in   www.alstom.com   TQ 217822

–		4wBE	R/R	Zephir	2566	2015
–		4wBE		Zephir	2672	2017

Stonebridge Park Carriage Maintenance and Wheel Lathe Depot,
Argenta Way, Wembley   **NW10 0RW**
Gauge 4ft 8½in   www.alstom.com   TQ 191845

(D3778)	08611	0-6-0DE		Derby		1959
–		4wDH	R/R	Unilok	4010	2002

Wembley Traincare Centre, Argenta Way, Wembley   **NW10 0RW**
Gauge 4ft 8½in   www.alstom.com   TQ 192843

(D3863)	08696	0-6-0DE	Hor	1959

ARRIVA RAIL LONDON LTD
(operating a concession agreement for Transport for London)
New Cross Gate Depot, Juno Way, New Cross   **SE14 5RW**
Gauge 4ft 8½in   www.arrivaraillondon.co.uk   TQ 359775, 359778

–	4wBE	Niteq	B244	2007
–	4wBE	Niteq	B281	2009
–	4wBE	Niteq		2014

Willesden Traction & Rolling Stock Maintenance Depot, Station Approach,
Willesden Junction, Willesden, Brent   **NW10 4UY**
Gauge 4ft 8½in   www.arrivaraillondon.co.uk   TQ 221828

D3671	09007	0-6-0DE	Dar	1959

BAM NUTTALL LTD, CIVIL ENGINEERS, PLANT DEPOT,
RAYLAMB WAY, off MANOR ROAD, SLADE GREEN, ERITH   **DA8 2LD**
Locomotives may be present at this depot between use on contracts.
Gauge 2ft 0in   www.bamnuttall.co.uk   TQ 537777

TO.11	4wBE	CE	B4071.4	1995

BAZALGETTE TUNNEL LTD, t/a TIDEWAY, THAMES TIDEWAY TUNNEL
(25km sewerage storage and stormwater tunnel for Thames Water /2016 to /2024)
Tideway East,
Cvjv - Costain Ltd, Vinci Construction Grands Projets
and Bachy Soletanche Joint Venture
(Bermondsley to Stratford contract)
Gauge 750mm   www.tideway.london

2	4wBE	CE	B4539.1	2012

-		4wBE		CE	B4539.2	2012
			reb	CE	B4573	2012
3		4wBE		CE	B4539.4	2012
-		4wBE		CE	B4539.6	2012
	DUKE OF YORK	4wBE		CE	B4540	2012

Tideway Central,
Ferrovial Agroman UK Ltd and Laing O'Rourke Construction Joint Venture
(Fulham To Blackfriars Contract)

Kirtling Street, Nine Elms, Wandsworth SW8 5BP   TQ 293775

Gauge 900mm   www.tideway.london

7	4wDH	Schöma	6917	2017
-	4wDH	Schöma	7050	2018
-	4wDH	Schöma	7051	2018
-	4wDH	Schöma	7052	2018
-	4wDH	Schöma	7053	2018
-	4wDH	Schöma	7054	2018
-	4wDH	Schöma	7055	2019
-	4wDH	Schöma	7056	2019
-	4wDH	Schöma	7057	2019

Tideway West,
Bam Nuttall Ltd, Morgan Sindall Plc and Balfour Beatty Ltd Joint Venture
(Acton to Fulham Contract)

Gauge 900mm   www.tideway.london

–	4wDH	Schöma	6987	2018
–	4wDH	Schöma	6988	2018
–	4wDH	Schöma	6989	2018
–	4wDH	Schöma	6990	2018

BBVS (Balfour Beatty Vinci Systia) JV, HS2 LTD,
CHANNEL GATE ROAD, WILLESDEN JUNCTION   NW10 6UA
Gauge 4ft 8½in   www.balfourbeattyvinci.co.uk   **TQ 214829**

(D3772	08605)	0-6-0DE	Derby	1959	a b

a	property of Railway Support Services Ltd, Wishaw, Warwickshire
b	operated by GB Railfreight Ltd

EUROSTAR GROUP LTD, EUROSTAR ENGINEERING CENTRE,
TEMPLE MILLS DEPOT, 2 ORIENT WAY, LONDON   E10 5YA
Gauge 4ft 8½in   www.eurostar.com   **TQ 372864, 374862**

(D4178)	08948		0-6-0DE		Dar		1962	
NP B2 WP C24			4wBE		BEMO	12162	2003	OOU
TM-LT-02			4wBE	R/R	Niteq	B334	2012	
TM-LT-03			4wBE	R/R	Niteq	B335	2012	
TM-LT-01			6wBE		Scul		2007	OOU
3314	373314	ENTENTE CORDIALE	Bo-BoWE/RE		GEC-Alsthom		1992	Pvd

GOVIA THAMESLINK RAILWAY LTD
(subsidiary of Govia Ltd, a Go-Ahead Group and Keolis Joint Venture)
Gatwick Express, Battersea Stewarts Lane Train Maintenance Depot, Dickens Street, Nine Elms, Battersea   SW8 3EP
Gauge 4ft 8½in   www.gatwickexpress.com   TQ 289763

–	4wBE	R/R	Zephir	2484	2013

Southern, Selhurst Traincare Depot, Selhurst Road, Croydon   SE25 6LJ
Gauge 4ft 8½in   www.southernrailway.com   TQ 332675

–	4wBE	R/R	Zephir	2254	2009
–	4wBE	R/R	Zephir	2979	2021
–	4wBE	R/R	Zephir	2980	2021

Thameslink, Hornsey Train Servicing Centre, Hampden Road, Hornsey   N8 0HF
Gauge 4ft 8½in   www.thameslinkrailway.com   TQ 310892

–	4wBE		Niteq	B155	2000
–	4wBE		Niteq	B248	2007
–	4wBE		SET		
–	4wBE	R/R	Zagro		2019
–	4wBE		Zephir	2276	2009

FIRST MTR SOUTH WESTERN TRAINS LTD, t/a SOUTH WESTERN RAILWAY, WIMBLEDON TRAINCARE DEPOT, DURNSFORD ROAD, WIMBLEDON   SW19 8EG
(a First Group and MTR Joint Venture)
Gauge 4ft 8½in   www.southwesternrailway.com   TQ 256723

–	4wBE	R/R	Zephir	2455	2013
–	4wBE		Zephir	2591	2015

FORD MOTOR CO LTD, DAGENHAM ENGINE PLANT, THAMES AVENUE, DAGENHAM   RM9 6SA
Gauge 4ft 8½in   www.ford.co.uk   TQ 496825, 499827

1		0-4-0DH	EEV	D1124	1966
2 "HUDSWELL"		0-6-0DH	HC	D1396	1967
		reb	HAB	6385	1996
3 MALCOLM		0-4-0DH	S	10127	1963
		reb	Wilmott		2003
–		4wDH	CE	B4637	2018
GT/PL/1P 260 C		4wDM	Robel 21.11.RK1		1966

HITACHI RAIL EUROPE LTD, NORTH POLE TRAIN MAINTENANCE CENTRE, MITRE WAY, LONDON   W3 7DX & W10 6AU
Gauge 4ft 8½in   www.hitachirail.com   TQ 219820, 228822

–	4wBE		Zephir	2508	2014

HOTCHIEF MURPHY JV, LONDON POWER TUNNELS

Wimbledon (TQ 259718) to New Cross (TQ 347777)
New Cross (TQ 347777) to Hurst (TQ 498727)
Hurst (TQ 498727) to Crayford
Gauge 610mm

–	4wBE	Schöma		2021
–	4wBE	Schöma		2021

LOCOMOTIVE SERVICES LTD and WEST COAST RAILWAY COMPANY LTD, SOUTHALL M.P.D., SOUTHALL   UB2 4SE
Gauge 4ft 8½in   www.locoservicesgroup.co.uk / www.westcoastrailways.co.uk   **TQ 133798**

(30064)	WD 1959	0-6-0T	OC	VIW	4432	1943
(D3905)	08737	0-6-0DE		Crewe		1960
D2447	LORD LEVERHULME	0-4-0DM		AB	388	1953
A.E.C. No.1		4wDM		AEC		1938
S61229	7105 (99229)	4w-4RER		Afd/Elh		1958
S61230	7105 (99230)	4w-4RER		Afd/Elh		1958
S68001	931091	4w-4RE/BER		Afd/Elh		1959
68002	(9002) 931092	4w-4RE/BER		Afd/Elh		1959
(S68008)	(9008) 931098 093	4w-4RE/BER		Afd/Elh		1961
S68009	9009	4w-4RE/BER		Afd/Elh		1961

Locomotives for maintenance, overhaul and repair usually present

LONDON NORTH EASTERN RAILWAY LTD, t/a LNER, BOUNDS GREEN MAINTENANCE DEPOT, BRIDGE ROAD, ALEXANDRA PALACE   N22 7SN
Gauge 4ft 8½in   www.lner.co.uk   **TQ 301907**

–	4wBE	Niteq	B245	2006

LONDON & SOUTH EASTERN RAILWAY LTD, t/a SOUTH EASTERN, SLADE GREEN MAINTENANCE DEPOT, MOAT LANE, ERITH   DA8 2JN
Gauge 4ft 8½in   www.southeasternrailway.co.uk   **TQ 527759**

–	4wBE	R/R	Zephir	2602	2016
–	4wBE	R/R	Zephir	2840	2018

J. MURPHY & SONS LTD, PLANT DEPOT, HIGHVIEW HOUSE, HIGHGATE ROAD, KENTISH TOWN   NW5 1TN

Locomotives are present at this depot between use on contracts.
Gauge 750mm   www.murphygroup.co.uk   **TQ 287855**

–	4wBE	Schöma	7105	2019	a
–	4wBE	Schöma	7106	2019	a
–	4wBE	Schöma	7107	2019	a
–	4wBE	Schöma	7109	2019	a

Gauge 2ft 0in

PLM11		4wBE		CE	B3329A	1986	
			reb	CE	B3791	1991	
			reb	CE	B4181	1996	
			reb	CE	B4512/3	2010	
LM12	JM	4wBE		CE	B3070C	1983	
			reb	CE	B3804	1991	
			reb	CE	B4512/2	2010	
PLM13		4wBE		CE	B3070D	1983	
			reb	CE	B3782	1991	
			reb	CE	B4181	1996	
LM14		4wBE		CE	B3070B	1983	
			reb	CE	B3804	1991	
			reb	CE	B4181	1996	
			reb	CE	B4512/1	2010	
PLM15		4wBE		CE	B3070A	1983	
			reb	CE	B3782	1991	
(LM24)	PLM13	4wBE		CE	B0167	1974	
			reb	CE	B3786A	1991	
			reb	CE	B4512/4	2010	
LM26JM		4wBE		CE	B0145C	1973	
			reb	CE	B3799	1991	
PLM 27		4wDH		Schöma	5694	2001	
PLM 28		4wDH		Schöma	5695	2001	b
PLM 29		4wDH		Schöma	5696	2001	c
PLM 31		4wDH		Schöma	5698	2001	
PLM 32		4wDH		Schöma	5699	2001	
PLM 33		4wDH		Schöma	5700	2001	
PLM 34		4wDH		Schöma	5701	2001	

a   currently at North Bristol Sewer Relief contract, Bristol
b   carries worksplate Schöma 5696
c   carries worksplate Schöma 5695

PLASSER UK LTD, MANOR ROAD, WEST EALING

W13 0PP

Track maintenance vehicles under construction, overhaul or repair usually present

Gauge 4ft 8½in   www.plasser.co.uk   **TQ 161809**

HERBERT	0-4-0DM	Bg/DC	2724	1963

PROCAT (PROSPECTS COLLEGE of ADVANCED TECHNOLOGY), TUNNELLING AND UNDERGROUND CONSTRUCTION ACADEMY, ALDERSBROOK SIDINGS, LUGG APPROACH, ILFORD

E12 5LN

Gauge 2ft 0in   www.tfl.gov.uk   **TQ 430862**

2	S238	4wBE	CE	B0471B	1975
–		4wDH	Schöma	5572	1998

QUATTRO PLANT LTD, RAIL DIVISION, STRATFORD DEPOT and HEAD OFFICE, GREENWAY COURT, CANNING ROAD, STRATFORD

E15 3ND

Gauge 4ft 8½in   www.quattroplant.co.uk   **TQ 390832**

1340	99709 977079-1	4wDM	R/R	_(Unimog	215665	2008
				(Zagro	3810	2008

1341	99709 977056-9	4wDM	R/R	_(Unimog 240772	2015	
				(Zagro/King	2015	
1343	99709 977058-5	4wDM	R/R	_(Unimog 240933	2016	
				(King/Zagro 4346	2016	
1342	99709 977057-7	4wDM	R/R	_(Unimog 240938	2016	
				(King/Zagro 4345	2016	

ROYAL MAIL LETTERS LTD, THE POST OFFICE UNDERGROUND RAILWAY
(part of London Region of Post Office Letters Ltd)
System closed, on care and maintenance.
Locomotives are kept at various disused sidings and stations throughout the system, including:-

King Edward Building, St Pauls   TQ 321816
Mount Pleasant Parcels Office, Clerkenwell   TQ 311823
New Western District Office, Rathbone Place   TQ 296814

Gauge 2ft 0in

1			4wBE	EEDK		702	1926	
2			4wBE	EEDK		703	1926	
66	[101	104]	2w-2-2-2wRE	_(EE		3335	1962	a
				(EES		8314	1962	OOU
01	[169	170]	2w-2-2-2wRE	HE		9134	1982	OOU b
02	[103	104]	2w-2-2-2wRE	GB		420461/2	1980	OOU
03	[105	106	2w-2-2-2wRE	GB		420461/3	1980	OOU
04	[107	108]	2w-2-2-2wRE	_(GB		420461/4	1980	
				(HE		9103	1980	OOU
05	[109	110]	2w-2-2-2wRE	_(GB		420461/5	1980	
				(HE		9104	1980	OOU
06	[111	112]	2w-2-2-2wRE	_(GB		420461/6	1980	
				(HE		9105	1980	OOU
07	[113	114]	2w-2-2-2wRE	_(GB		420461/7	1980	
				(HE		9106	1980	OOU
08	[115	116]	2w-2-2-2wRE	_(GB		420461/8	1980	
				(HE		9107	1980	OOU
09	[117	118]	2w-2-2-2wRE	_(GB		420461/9	1981	
				(HE		9108	1981	OOU
10	[119	120]	2w-2-2-2wRE	_(GB	420461/10		1981	
				(HE		9109	1981	OOU
11	[121	122]	2w-2-2-2wRE	_(GB	420461/11		1981	
				(HE		9110	1981	OOU
12	[123	124]	2w-2-2-2wRE	_(GB	420461/12		1981	
				(HE		9111	1981	OOU
13	[125	126]	2w-2-2-2wRE	_(GB	420461/13		1981	
				(HE		9112	1981	OOU
14	[127	128]	2w-2-2-2wRE	_(GB	420461/14		1981	
				(HE		9113	1981	OOU
15	[129	130]	2w-2-2-2wRE	_(GB	420461/15		1981	
				(HE		9114	1981	OOU
16	[131	132]	2w-2-2-2wRE	_(GB	420461/16		1981	
				(HE		9115	1981	OOU
17	[133	134]	2w-2-2-2wRE	_(GB	420461/17		1981	
				(HE		9116	1981	OOU
18	[135	136]	2w-2-2-2wRE	_(GB	420461/18		1981	
				(HE		9117	1981	OOU
19	[137	138]	2w-2-2-2wRE	_(GB	420461/19		1981	
				(HE		9118	1981	OOU

20	[139 140]	2w-2-2-2wRE	_(GB 420461/20	1981		
			(HE 9119	1981	OOU	
23	[145 146]	2w-2-2-2wRE	_(GB 420461/23	1981		
			(HE 9122	1981	OOU	
24	[147 148]	2w-2-2-2wRE	_(GB 420461/24	1981		
			(HE 9123	1981	OOU	
25	[149 150]	2w-2-2-2wRE	_(GB 420461/25	1981		
			(HE 9124	1981	OOU	
26	[151 152]	2w-2-2-2wRE	_(GB 420461/26	1981		
			(HE 9125	1981	OOU	
27	[153 154]	2w-2-2-2wRE	_(GB 420461/27	1982		
			(HE 9126	1982	OOU	
28	[155 156]	2w-2-2-2wRE	_(GB 420461/28	1982		
			(HE 9127	1982	OOU	
29	[157 158]	2w-2-2-2wRE	_(GB 420461/29	1982		
			(HE 9128	1982	OOU	
30	[159 160]	2w-2-2-2wRE	_(GB 420461/30	1982		
			(HE 9129	1982	OOU	
31	[161 162]	2w-2-2-2wRE	_(GB 420461/31	1982		
			(HE 9130	1982	OOU	
32	[163 164]	2w-2-2-2wRE	_(GB 420461/32	1982		
			(HE 9131	1982	OOU	
33	GREAT EAST EXPRESS [165 166]	2w-2-2-2wRE	_(GB 420461/33	1982		
			(HE 9132	1982	OOU	
34	[167 168]	2w-2-2-2wRE	_(GB 420461/34	1982		
			(HE 9133	1982	OOU	
752	[101 102]	2w-2-2-2wRE	EEDK	752	1930	OOU	c
–		2w-2-2-2wRE	EEDK	753	1930		d
35	[107 108]	2w-2-2-2wRE	EEDK	755	1930	OOU	
36	[105 106]	2w-2-2-2wRE	EEDK	756	1930	OOU	
759	[110 115]	2w-2-2-2wRE	EEDK	759	1930	OOU	c
39	[121 122]	2w-2-2-2wRE	EEDK	762	1930	OOU	
763	[123 124]	2w-2-2-2wRE	EEDK	763	1930	OOU	c
793	[183 184]	2w-2-2-2wRE	EEDK	793	1930	OOU	c
795	[187 188]	2w-2-2-2wRE	EEDK	795	1930	OOU	c
797	[191 192]	2w-2-2-2wRE	EEDK	797	1930	OOU	
799	[195 196]	2w-2-2-2wRE	EEDK	799	1930	OOU	c
802	[201 202]	2w-2-2-2wRE	EEDK	802	1930	OOU	c
804	[205 206]	2w-2-2-2wRE	EEDK	804	1930	OOU	c
41	[207 208]	2w-2-2-2wRE	EEDK	805	1930	OOU	
810	[217 218]	2w-2-2-2wRE	EEDK	810	1930	OOU	
43	[219 220]	2w-2-2-2wRE	EEDK	811	1930	OOU	
813	[223 224]	2w-2-2-2wRE	EEDK	813	1930	OOU	c
45	[226]	2-2wRE	EEDK	814	1930	OOU	e
46	[227 228]	2w-2-2-2wRE	EEDK	815	1930	OOU	
816	[229 230]	2w-2-2-2wRE	EEDK	816	1930	OOU	c
817	[231 232]	2w-2-2-2wRE	EEDK	817	1930	OOU	c
818	[233 234]	2w-2-2-2wRE	EEDK	818	1930	OOU	c
820	[237 238]	2w-2-2-2wRE	EEDK	820	1931	OOU	c
–	[239 240]	2w-2-2-2wRE	EEDK	821	1931	OOU	d
822	[241 242]	2w-2-2-2wRE	EEDK	822	1931	OOU	c
826	[249 250]	2w-2-2-2wRE	EEDK	826	1931	OOU	c
49	[251 252]	2w-2-2-2wRE	EEDK	827	1931	OOU	
830	[257 258]	2w-2-2-2wRE	EEDK	830	1931	OOU	c

925	[445	446]	2w-2-2-2wRE	EEDK	925	1936	OOU	c
51	[457	458]	2w-2-2-2wRE	EEDK	931	1936	OOU	
932	[459	460]	2w-2-2-2wRE	EEDK	932	1936	OOU	c
50	[235	451]	2w-2-2-2wRE	EE			OOU	f
55	[236	452]	2w-2-2-2wRE	EE			OOU	f

a	also contains parts of EE 3334-EES 8313 / 1962
b	supplied as spare power units for 01 to 34 (orig 501 to 534) but now forming 01
c	stored in a disused tunnel at Rathbone Place
d	converted to a passenger car
e	one power unit only; other preserved at Mail Rail exhibition
f	contains a power unit from EEDK 819/1930 and EEDK 928/1936

see also entry for **Postal Heritage Trust** in the Preservation Section

TRANSPORT for LONDON, DOCKLANDS LIGHT RAILWAY LTD

(Operated by Keolis Amey Docklands Ltd)

Locomotives are kept at :		
Castor Lane, Poplar	E14 0DS	TQ 376806
Armada Way, Beckton	E6 7FB	TQ 442812

Gauge 4ft 8½in   www.tfl.gov.uk

992	DANNY		4wDH	HE	9388	2019	
		a rebuild of	4wDM	Wkm	11622	1986	a
993	KYLIE		4wBE/RE	HE	9385	2014	
		a rebuild of		RFSK	V339	1991	
994	KEVIN KEARNEY		0-4-0DH	GECT	5577	1979	
88			4w-4-4wRER	Bombardier(BN)		1995	b

a	currently at EP Industries Ltd, Alfreton, Derbyshire
b	used as a shunter at Beckton Depot

TRANSPORT for LONDON, LONDON TRAMLINK,
THERAPIA LANE DEPOT, COOMBER WAY, CROYDON   CR0 4TQ

Gauge 4ft 8½in   www.tfl.gov.uk   TQ 300669

-		4wBE	R/R	Zephir	2331	2011

TRANSPORT for LONDON, LONDON UNDERGROUND LTD

London underground railways maintenance vehicles.

Locomotives are kept at :-

AW	Acton Works, Bollo Lane	W3 8BZ	TQ 196791
EC	Ealing Common Depot, Uxbridge Road	W5 3PA	TQ 189802
GG	Golders Green Depot, Finchley Road	NW11 7NU	TQ 253875
LB	Lillie Bridge Depot	SW6 1TP	TQ 250782
N	Neasden Depot	NW10 1PH	TQ 206858
NF	Northfields Depot, Northfields Avenue	W5 4UB	TQ 167789
NP	Northumberland Park Depot	N17 0XE	TQ 349907
U	Upminster Depot	RM14 1XL	TQ 570871
WR	West Ruislip Depot	HA4 6NS	TQ 094862

Locomotives may also be found between duties at :-	Cockfosters Depot	N14 4UT	TQ 288962
	Hainault Depot	IG6 2UU	TQ 450918
	Hammersmith Depot	W6 7PY	TQ 234787
	Highgate Depot	(N10 3JN)	TQ 279886
	London Road Depot	SE1 6LW	TQ 315793
	Morden Depot	SM4 5PT	TQ 255680

Gauge 4ft 8½in   www.tfl.gov.uk

1 (L1)	BRITTA LOTTA		4wDH	Schöma	5403	1996	WR OOU
2 (L2)	NIKKI		4wBE/RE	CE	B4607.02	2015	
		a rebuild of	4wDH	Schöma	5404	1996	WR OOU
3 (L3)	CLAIRE		4wDH	Schöma	5405	1996	WR OOU
4 (L4)	PAM		4wBE/RE	CE	B4607.04	2015	
		a rebuild of	4wDH	Schöma	5406	1996	WR OOU
5 (L5)	SOPHIE		4wBE/RE	CE	B4607.05	2015	
		a rebuild of	4wDH	Schöma	5407	1996	WR OOU
6 (L6)	DENISE		4wBE/RE	CE	B4607.06	2015	
		a rebuild of	4wDH	Schöma	5408	1996	WR OOU
7 (L7)	ANNEMARIE		4wBE/RE	CE	B4607.07	2015	
		a rebuild of	4wDH	Schöma	5409	1996	WR OOU
8 (L8)	EMMA		4wBE/RE	CE	B4607.08	2015	
		a rebuild of	4wDH	Schöma	5410	1996	WR OOU
9 (L9)	DEBORA		4wDH	Schöma	5411	1996	WR OOU
10 (L10)	CLEMENTINE		4wBE/RE	CE	B4607.10	2015	
		a rebuild of	4wDH	Schöma	5412	1996	WR OOU
11 (L11)	JOAN		4wBE/RE	CE	B4607.11	2015	
		a rebuild of	4wDH	Schöma	5413	1996	WR OOU
12 (L12)	MELANIE		4wDH	Schöma	5414	1996	WR OOU
13 (L13)	MICHELE		4wBE/RE	CE	B4607.13	2015	
		a rebuild of	4wDH	Schöma	5415	1996	WR OOU
14 (L14)	CAROL		4wBE/RE	CE	B4607.14	2015	
		a rebuild of	4wDH	Schöma	5416	1996	WR OOU
L 15	69015 97715		4w-4wBE/RE	MetCam		1970	
		reb		Acton		2018	
L 16	69016 (97716)		4w-4wBE/RE	MetCam		1970	
		reb		Acton		2017	
L 17	(97717)		4w-4wBE/RE	MetCam		1970	
L 18	69018 (97718)		4w-4wBE/RE	MetCam		1970	
		reb		Acton		2013	
L 19	69019 (97719)		4w-4wBE/RE	MetCam		1970	
		reb		Acton		2016	
L 20	64020 97720		4w-4wBE/RE	MetCam		1964	
		reb		Acton		2017	
L 21	64021 (97721)		4w-4wBE/RE	MetCam		1964	
		reb		Acton		2018	
L 22	(97722)		4w-4wBE/RE	MetCam		1965	
		reb		Acton		2014	
L 23	(97723)		4w-4wBE/RE	MetCam		1965	
		reb		Acton		2012	
L 24	64024 (97724)		4w-4wBE/RE	MetCam		1965	
		reb		Acton		2009	
L 25	64025 (97725)		4w-4wBE/RE	MetCam		1965	
		reb		Acton		2010	
L 26	64026 (97726)		4w-4wBE/RE	MetCam		1965	
		reb		Acton		2012	
L 27	64027 (97727)		4w-4wBE/RE	MetCam		1965	
		reb		Acton		2011	
L 28	64028 (97728)		4w-4wBE/RE	MetCam		1965	
		reb		Acton		2013	
L 29	64029 (97729)		4w-4wBE/RE	MetCam		1965	

L 30	64030 (97730)		4w-4wBE/RE	MetCam		1965		
		reb		Acton		2012		
L 31	(97731)		4w-4wBE/RE	MetCam		1965		
		reb		Acton		2012		
L 32	64032 97732		4w-4wBE/RE	MetCam		1965		
		reb		Acton		2014		
L 44	73044		4w-4wBE/RE	Don	L44	1973		
		reb		Acton		2015		
L 45	73045		4w-4wBE/RE	Don	L45	1974		
		reb		Acton		2017		
L 46	73046		4w-4wBE/RE	Don	L46	1974		
		reb		Acton		2016		
L 47	73047 97747		4w-4wBE/RE	Don	L47	1974		
		reb		Acton		2018		
L 48	73048		4w-4wBE/RE	Don	L48	1974		
		reb		Acton		2017		
L 49	73049		4w-4wBE/RE	Don	L49	1974		
		reb		Acton		2014		
L 50	73050 97750		4w-4wBE/RE	Don	L50	1974		
		reb		Acton		2016		
L 51	97751		4w-4wBE/RE	Don	L51	1974		
		reb		Acton		2014		
L 52	73052 97752		4w-4wBE/RE	Don	L52	1974		
		reb		Acton		2017		
L 53	73053 97753		4w-4wBE/RE	Don	L53	1974		
		reb		Acton		2016		
L 54			4w-4wBE/RE	Don	L54	1974		
		reb		Acton		2017		
L 132	(3901)		4w-4wRE	Cravens		1960		a
		reb		Derby		1987		
L 133	(3905)		4w-4wRE	Cravens		1960		a
		reb		Derby		1987		
PC2	6724		4wDH	Schöma	6724	2014		b
PC1	6725		4wDH	Schöma	6725	2014		b
12	SARAH SIDDONS		4w-4wRE	VL		1922		
–			4wBE	Harmill	1	2013	N	
–			4wBE R/R	Harmill	2	2013	N	
–			4wBE	Harmill	(3?)	c2014	EC	
–			4wBE	Niteq	B184	2001	NP	
–			4wBE	Niteq	B222	2005	NF	
–			4wBE	Niteq	B285	2009	N	
–			4wBE	Niteq	B379	2016	U	
–			4wBE	SET		2003	GG	
–			4wBE R/R	Zephir	2556	2015	AW	
1406			2-2w-2w-2RER	MetCam		1962		c
1407			2-2w-2w-2RER	MetCam		1962		c
1441			2-2w-2w-2RER	MetCam		1962		c
1506			2-2w-2w-2RER	MetCam		1962		d
1507			2-2w-2w-2RER	MetCam		1962		d
1570			2-2w-2w-2RER	MetCam		1963		c
1681			2-2w-2w-2RER	MetCam		1963		c
1682			2-2w-2w-2RER	MetCam		1963		c
1690			2-2w-2w-2RER	MetCam		1963		OOU
1691			2-2w-2w-2RER	MetCam		1963		
3233			2-2w-2w-2RER	MetCam		1973		e

3333	2-2w-2w-2RER	MetCam		1973	e
3451	2-2w-2w-2RER	MetCam		1972	e
3551	2-2w-2w-2RER	MetCam		1972	e
9125	2-2w-2w-2RER	MetCam		1961	c
9441	2-2w-2w-2RER	MetCam		1962	c
9459	2-2w-2w-2RER	MetCam		1962	c
9507	2-2w-2w-2RER	MetCam		1962	d
9577	2-2w-2w-2RER	MetCam		1963	c
9691	2-2w-2w-2RER	MetCam		1963	c
92442	4w-4wRER	BRE(D)		1994	f
93442	4w-4wRER	BRE(D)		1994	f
AC795709	4wBER	Bance	068/98	1998	
AC795701	2w-2BER	Bance	070/98	1998	
AC795704	2w-2BER	Bance	071/98	1998	
AC795703	2w-2BER	Bance	072/98	1998	
AC795705	2w-2BER	Bance	073/98	1998	
AC795706	2w-2BER	Bance	074/98	1998	
AC795702	2w-2BER	Bance	075/98	1998	
AC795710	2w-2BER	Bance	076/99	1998	
–	2w-2BER	Bance	077/99	1999	
AC795707	2w-2BER	Bance	078/99	1999	
AC795708	2w-2BER	Bance	079/99	1999	
AC790309	2w-2BER	Bance	107/03	2003	
AC790310	2w-2BER	Bance	108/03	2003	
–	2w-2BER	Bance	263	2013	
AC795713	2w-2BER	Consillia	05/001	2005	
MTP106002	2w-2BER	Consillia	05/002	2005	
AC795711	2w-2BER	Consillia	05/003	2005	
AC795712	2w-2BER	Consillia	05/004	2005	
MTP106003	2w-2BER	Consillia	05/005	2005	
MTP106005	2w-2BER	Consillia	05/007	2005	
MTP106001	2w-2BER	Consillia	05/008	2005	
AC795724	2w-2BER	Consillia	06/009	2006	
AC795723	2w-2BER	Consillia	06/010	2006	
AC795722	2w-2BER	Consillia	06/011	2006	
AC795721	2w-2BER	Consillia	06/013	2006	
AC795716	2w-2BER	Consillia	06/014	2006	
AC795717	2w-2BER	Consillia	06/017	2006	
AC795714	2w-2BER	Consillia	06/018	2006	
AC795720	2w-2BER	Consillia	06/019	2006	
AC795726	2w-2BER	Consillia	13/0046	2013	
AC795725	2w-2BER	Consillia	13/0047	2013	
MEC3016003	2w-2BER	Consillia	14/0049	2014	
MEC3016002	2w-2BER	Consillia	14/0050	2014	
MEC3016001	2w-2BER	Consillia	14/0051	2014	
AC304101	2w-2BER	Consillia	15/0055	2015	
AC 2828101	2w-2BER	Consillia	17/0057	2017	

a	track recording pilot motor cars
b	rail grinding train power cars
c	Sandite Car Unit
d	spares for sandite car units

e track recording unit
f Sandite / ATO Test Unit

PRESERVATION SITES

ARCHITECTURAL SUPPLIES LTD,
158 DORSET ROAD, MERTON PARK, WIMBLEDON
Gauge 2ft 0in

SW19 3EF
TQ 256693

(7)		4wFER	PPM	7	1994

COOPERS LANE PRIMARY SCHOOL, GROVE PARK, LEWISHAM
Gauge 4ft 8½in   www.cooperslane.lewisham.sch.uk

SE12 0LF
TQ 405728

7027	2-2w-2w-2RER	MetCam	1980	Dsm

CRAVENS HERITAGE TRAINS LTD,
c/o LONDON UNDERGROUND LTD, NORTHFIELDS DEPOT, LONDON
Gauge 4ft 8½in   www.cravensheritagetrains.co.uk

W5 4UB
TQ 167789

3906	4w-4wRER	Cravens	1960
3907	4w-4wRER	Cravens	1960

CROSSNESS ENGINE TRUST, THE CROSSNESS PUMPING STATION,
THE OLD WORKS, CROSSNESS SEWAGE TREATMENT WORKS,
BELVEDERE ROAD, ABBEY WOOD
Gauge 2ft 0in - Rang Railway   www.crossness.org.uk

SE2 9AQ
TQ 485810

BAZALGETTE	4wDH	s/o	SL	1986

THE FRIENDS OF THE PUMP HOUSE, WALTHAMSTOW PUMP HOUSE MUSEUM,
10 SOUTH ACCESS ROAD, LOW HALL LANE, WALTHAMSTOW
Gauge 4ft 8½in   www.walthamstowpumphouse.org.uk

E17 8AX
TQ 363882

3016		2-2w-2w-2RER	MetCam	1968	a
3186	(3122)	2-2w-2w-2RER	MetCam	1968	

a bodyshell fitted with cab from 3049

THE HAMPTON & KEMPTON WATERWORKS RAILWAY LTD,
c/o KEMPTON STEAM ENGINES TRUST, KEMPTON PARK WATER TREATMENT
WORKS, SNAKEY LANE, HANWORTH
Private site - operates on advertised public opening dates only

TW13 6XH

Gauge 2ft 0in   www.hamptonkemptonrailway.org.uk

TQ 109709

3	DARENT		0-4-0T	OC	AB	984	1903
		rebuilt as	0-4-0ST	OC	Provan Group		2003
−			4wDM		FH	3787	1956

–		4wDH	HE		9338	1994
SPELTHORNE		4wDH	HE		9357	1994
–		4wDM	MR		4023	1926
–		4wDM	RH		200748	1940
HOUNSLOW		4wPH	SPL		No.1	2008

KINGS COLLEGE LONDON, STRAND

Gauge 1ft 3in   www.kcl.ac.uk

WC2R 2LS
TQ 308808

PEARL	2-2-2WT	IC	B(C)		1860

LONDON MUSEUM OF WATER & STEAM, KEW BRIDGE WATERWORKS
RAILWAY, GREEN DRAGON LANE, BRENTFORD

Gauge 2ft 0in   www.waterandsteam.org.uk

TW8 0EN
TQ 188780

	THOMAS WICKSTEED	0-4-0ST	OC	_(Kew		1995	
				(HE	3906	2009	a
2	ALISTER	4wDM		L	44052	1958	
	OTTO	4wDM		Schöma	1676	1955	

a   locomotive part built at Kew and completed by HE

NETWORK RAIL, STEWARTS LANE DEPOT,
DICKENS STREET, NINE ELMS

Privately preserved locomotives and multiple units based here.

Gauge 4ft 8½in

SW8 3EP

TQ 288766

35028	CLAN LINE	4-6-2	3C	Elh		1948
			reb	Elh		1959
3053	CAR No.92	4w-4wRER		MetCam		1932
3053	CAR No.93	4w-4wRER		MetCam		1932

NEWHAM BOROUGH COUNCIL,
MERIDIAN SQUARE, STRATFORD STATION, STRATFORD

Gauge 4ft 8½in

E15 1XD
TQ 386843

ROBERT	0-6-0ST	OC	AE		2068	1933

PLUMCROFT PRIMARY SCHOOL - PLUM LANE CAMPUS,
PLUM LANE, PLUMSTEAD

Gauge 4ft 8½in   www.plumcroftprimary.co.uk

SE18 3HW
TQ 441774

5720	4w-4wRER	MetCam		1977	Dsm

Grounded body with bogies and underfloor equipment removed

POSTAL HERITAGE TRUST, THE POSTAL MUSEUM and MAIL RAIL, FREELING HOUSE, PHOENIX PLACE, MOUNT PLEASANT, LONDON   WC1X ODA

Gauge 2ft 0in   www.postalmuseum.org   TQ 309823

–		4w Atmospheric Car			1861
–		4wRE	EEDK	601	1926
3		4wBE	EEDK	704	1926
45	[225]	2w-2RE	EEDK	814	1930 a
824	[245 246]	2w-2-2-2wRE	EEDK	824	1931
(21)	141 142	2w-2-2-2wRE	_(GB	420461/21	1981
			(HE	9120	1981

a   single power unit only

Gauge 2ft 0in - Mail Rail experience

–	4-4w-4-4BER	SL 21625/21626	2016
–	4-4w-4-4BER	SL 21630/21631	2016

THE ROYAL GREENWICH TRUST SCHOOL, FERRANTI CLOSE, GREENWICH, LONDON   SE7 8LJ

Gauge 4ft 8½in   www.rgtrustchool.net   TQ 417790

5701	4w-4wRER	MetCam	1977

RUISLIP LIDO RAILWAY SOCIETY LTD, RUISLIP LIDO RAILWAY, RESERVOIR ROAD, RUISLIP   HA4 7TY

Gauge 1ft 0in   www.ruisliplidorailway.org   TQ 089889

6	MAD BESS	2-4-0ST+T OC	_(Winson		1996	
			(Ruislip		1998	+
5	LADY OF THE LAKE	2w-2-2-2wDH	Ravenglass No.6		1985	
	rebuilt as	4w-4wDH	SL		1991	
	rebuilt		SL	8110	2007	
3	ROBERT	4w-4wPH	SL	DDC25-73	1973	
	rebuilt as	4w-4wDH	FMB		1995	
7	GRAHAM ALEXANDER	4w-4wDM	SL	222.3.90	1990	
	rebuilt		SL	8071	2005	
	rebuilt as	4w-4wBE	AK	115R	2024	
8	BAYHURST	4w-4wDH	SL	2140-2-03	2003	
	rebuilt as	4w-4wDM	Ruislip		2012	
9	JOHN RENNIE	4w 4wDH	SL	11/04-8011	2004	
	rebuilt as	4w-4wDM	Ruislip		2012	

+   started by Winson Engineering /1986 and completed by Ruislip /1998

SCIENCE MUSEUM, EXHIBITION ROAD, SOUTH KENSINGTON   SW7 2DD

Gauge 5ft 0in   www.sciencemuseum.org.uk   TQ 268793

"PUFFING BILLY"	4w	VCG	Hedley	1827-1832	a

a   incorporates parts of locomotive of same name built c1814

Gauge 4ft 8½in

1868	2-2-2	OC	Crewe	20	1845

SOUTHERN ELECTRIC TRACTION GROUP,
c/o SIEMENS RAIL SYSTEMS, STRAWBERRY HILL DEPOT,
SHACKLEGATE LANE, TEDDINGTON, TWICKENHAM   **TW11 8SF**
Gauge 4ft 8½in   www.setg.org.uk   **TQ 154720**

3417	62236 GORDON PETTITT	4-4wRER	York	1969	

TRANSPORT for LONDON, LONDON TRANSPORT MUSEUM
Covent Garden Piazza   **WC2E 7BB**
Gauge 4ft 8½in   www.ltmuseum.co.uk

23		4-4-0T	OC	BP	710	1866
5	JOHN HAMPDEN	Bo-BoRE		VL		1922
4248		4w-4RER		GRC&W	O/5026	1924
11182		2-2w-2w-2RER		MetCam		1939
No.13		4wRE		_(BP		1890
				(M&P		1890	a

a   carried number plate from No.1 until probable identity established c1990

The Depot, Museum Depot, Gunnersbury Lane, Acton Town   **W3 9BQ**
Gauge 4ft 8½in   www.ltmuseum.co.uk   **TQ 193798**

4416	(L126)	2w-2-2-2wRE	BRCW		1939
ESL107		4w-4-4-4wRE	_(BRCW		1903
			(MetAmal		1903
		reb	Acton		1939
5034		4w-4RER	Cravens		1962
61		4w-4wRER	EEDK	1151	1940
L35		4w-4wBE/RE	GRC&W	O/7903	1938
4184	(662 274)	4w-4RER	GRC&W	O/5026	1924
4417	(L127)	2w-2-2-2wRE	GRC&W	O/8060	1939
(3370)	L134	4w-4RE	MetC&W		1927
		reb	Acton		1967
3327		4w-4RER	MetCam		1929
(3693)	L131	4w-4RE	MetCam		1934
		reb	Acton		1967
12048		4w-4wRER	MetCam		1938
10012		4w-4wRER	MetCam		1938
11012		4w-4wRER	MetCam		1938
22679		2w-2-2-2wRER	MetCam		1952
16		4w-4wRER	MetCam		1986
3052		4w-4wRER	MetCam		1968
3530		4w-4wRER	MetCam		1972
3734		4w-4wRER	MetCam		1988
5721		4w-4wRER	MetCam		1978
L85	C622 EWT	4wDM	R/R	_(Unimog 126262	1986
				(Zweiweg 1130	1986

VILLAGE UNDERGROUND,
54 HOLYWELL LANE, GREAT EASTERN STREET, SHOREDITCH   **EC2A 3PQ**
Gauge 4ft 8½in   www.villageunderground.co.uk   **TQ 334823**

3662	4w-4wRER	Metcam	1988
3733	4w-4wRER	Metcam	1988

GREATER MANCHESTER

INDUSTRIAL SITES

ALSTOM TRANSPORT (part of Alstom Holdings S.A.)
Manchester Traincare Centre, Kirkmanshulme Lane, Longsight   **M12 4HR**
Gauge 4ft 8½in   www.alstom.com   **SJ 868960**

(D3958)	08790	LONGSIGHT T.M.D.	0-6-0DE	Derby	1960
(D4117)	08887		0-6-0DE	Hor	1962

Longsight Wheel Lathe Depot, New Bank Street, Longsight   **M12 4HW**
Gauge 4ft 8½in   www.alstom.com   **SJ 865962**

–	4wDH	R/R	Unilok	4009	2002

ASHTON PACKET BOAT CO LTD,
ASHTON BOATYARD, HANOVER STREET NORTH, GUIDE BRIDGE
Gauge 2ft 0in   **SJ 920978**

–	4wDM	FH	2325	1941	a
–	4wDM	HE	2820	1943	
–	4wDM	HE	6012	1960	
–	4wDM	JF	18892	1931	a
–	4wDM	RH	200761	1941	
DORIS	4wDM	RH	418776	1958	
–	4wDM	Strüver	60327	c1955	
–	4wBE	WR	M7556	1972	
	reb	WR	10114	1984	
6	4wBE	WR	7967	1978	

a   stored on behalf of owners

V.J. DONEGHAN (PLANT) LTD, TUNNELLING & CIVIL ENGINEERING CONTRACTORS,
BIRD HALL INDUSTRIAL PARK, BIRD HALL LANE, STOCKPORT   **SK3 0WT**
Locomotives maybe present in this yard between contracts.
Gauge 1ft 6in   www.doneghan.co.uk   **SJ 879886**

PL 158	4wBE	CE	5640	1969
2	4wBE	CE		
–	4wBE	CE		

| 4 | | 4wBE | | CE | | | |

Three of the four battery boxes used by the above locomotives carry CE plates:-
5841/68, 5628/69 and 5866/71 respectively.

FREIGHTLINER GROUP LTD, TRAFFORD PARK TERMINAL,
off JOHN GILBERT WAY, TRAFFORD PARK, MANCHESTER   M17 1FA
(part of the American Genesee & Wyoming Railway Co)
Gauge 4ft 8½in   www.freightliner.co.uk   **SJ 798960**

(D3752)	08585	VICKY	0-6-0DE		Crewe		1959	**M**

FREIGHTLINER MAINTENANCE LTD, VEHICLE MAINTENANCE FACILITY,
off GUIDE LANE, GUIDE BRIDGE, ASHTON-UNDER-LYNE   M34 5HF
(part of the American Genesee & Wyoming Railway Co)
Gauge 4ft 8½in   www.freightliner.co.uk   **SJ 930977**

R314	R314 ADB		4wDH	R/R	_(Thwaites 6-93355	1997	
					(Rexquote 1160		
				reb	Shovlin	2020	a

a   converted from road/rail dumper; property of Shovlin Plant Hire Ltd, West Gorton

JOHN MARROW, BRYN ENGINEERING SERVICES, UNIT C,
SCOT LANE INDUSTRIAL ESTATE, SCOT LANE, BLACKROD, BOLTON   BL6 5SL
Other locomotives occasionally present for overhaul or repairs   **Private Site**
Gauge 4ft 8½in   **SD 624090**

6686		0-6-2T	IC	AW	974	1928
92245		2-10-0	OC	Crewe		1958
–		0-6-0ST	OC	AE	1600	1912
WD 71499		0-6-0ST	IC	HC	1776	1944
S112	REVENGE	0-6-0ST	IC	HE	2414	1941
	SIR ROBERT PEEL	0-6-0ST	IC	HE	3776	1952

NORTHERN TRAINS LTD, (owned by DfT OLR Holdings Ltd)
NEWTON HEATH TRAINCARE DEPOT, DEAN LANE, MANCHESTER   M40 3AB
Gauge 4ft 8½in   www.northernrailway.co.uk   **SD 876008**

–		4wBE		Niteq	B200	2005

RILEY & SON (E) LTD,
PREMIER LOCOMOTIVE WORKS, SEFTON STREET, HEYWOOD   OL10 2JF
Locomotives usually present for maintenance, repair and restoration.
Gauge 4ft 8½in   www.rileysuk.com   **SD 863101**

34007	WADEBRIDGE	4-6-2	3C	Bton		1945
(35009)	(SHAW SAVILL)	4-6-2	3C	Elh		1942
		reb		Elh		1957
45212		4-6-0	OC	AW	1253	1935
47298		0-6-0T	IC	HE	1463	1924

(47445)		0-6-0T	IC	HE	1529	1927
(61572) No.8572		4-6-0	IC	BP	6488	1928
			reb	MaLoWa		1994

SHOVLIN PLANT HIRE LTD,
THE BROOK BUSINESS COMPLEX, BENNETT STREET, WEST GORTON M12 5AU
Gauge 4ft 8½in   www.shovlin.co.uk   SJ 866970

P720	P720 OVR	4wDH	R/R	_(Benford ES09AH504	1997		
	99709 949063-6			(Rexquote 1078			
			reb	Shovlin		c2020	+	a
R314	R314 ADB	4wDH	R/R	_(Thwaites 6-93355	1997		
				(Rexquote 1160			
			reb	Shovlin		2020	+	b

+   converted from road/rail dumper;
a   currently at Freightliner Maintenance Ltd, Hope Cement Works, Derbyshire
b   currently at Freightliner Maintenance Ltd, Guide Bridge, Ashton-under-Lyne
Other road/rail vehicles usually present

SIEMENS AG, SIEMENS RAIL SYSTEMS, ARDWICK TRAINCARE FACILITY,
RONDIN ROAD, ARDWICK, MANCHESTER M12 6BF
Gauge 4ft 8½in   www.mobility.siemens.com   SJ 864972

(01551) LANCELOT	0-4-0DH	EEV	D1122	1966
–	4wBE	CPM 1731.02UK	2005

TRANSPORT for GREATER MANCHESTER, METROLINK
(Operated by Keolis Amey Metrolink Ltd)
Ayres Road Depot, Old Trafford, Manchester M16 0PX
Gauge 4ft 8½in   www.tfgm.com   SJ 813957

–	4wBE	Zephir	2385	2012

Queens Road Depot, Cheetham Hill, Manchester M8 0RY
Gauge 4ft 8½in   www.tfgm.com   SD 848005

1027		4wDH		_(GECT	5862	1991
				(RFSK	V337	1991	OOU
YX22 CVC		4wDM	R/R	_(Unimog	266048	2021
				(Zagro	4715	2021
–		4wBE	R/R	Zephir	2933	2020

TRAIN PLANT REALISATIONS LTD (administrators for TXM PLANT LTD)
HASLEMERE INDUSTRIAL ESTATE, PARK LANE, WIGAN ROAD, WIGAN WN4 0EL
Gauge 4ft 8½in   SD 569017

7146	99709 977048-9	4wDM	R/R	Unimog	072061	1981
7150	99709 977046-0	4wDM	R/R	Unimog	181336	1995
7147	99709 977044-5	4wDM	R/R	Unimog	184949	1999
7149	99709 977045-2	4wDM	R/R	Unimog	185657	c1999

5020 99709 979023-7 FA51 XOV 4wDM   R/R   Unimog 199725 2001

    Unimog vehicles are based here but work at other locations as required

WYKE COMMERCIAL SERVICES,
HUNTSMAN DRIVE, NORTHBANK INDUSTRIAL PARK, IRLAM   M44 5EG
Gauge 4ft 8½in   www.wykeommercials.co.uk   SJ 717932

      DORA   0-4-0ST   OC   AE   1973 1927 a

a   currently located off site for restoration

A.E. YATES LTD, CRANFIELD ROAD,
LOSTOCK INDUSTRIAL ESTATE, LOSTOCK, BOLTON   BL6 4SB
Locomotives are present in this yard between contracts.
Gauge 2ft 0in   www.aeyates.co.uk   SD 650098

  5137   4wBE   CE   5866A 1971
   –   4wBE   CE

Gauge 1ft 6in

   –   4wBE   CE   B2200B 1979
                      reb CE   B3990 1993
  2019 29   0-4-0BE   WR

PRESERVATION SITES

S. BRADLEY, SPECIAL NEEDS SPEED SHOP, TRAFFORD PARK, MANCHESTER
Gauge 4ft 8½in   **Private Site**

   –   2w-2BER   Bance   080 1999

Gauge 2ft 0in

   –   2w-2PMR   BradleyS   c2012
      "SPEEDBIRD"   2w-2PE   BradleyS   2020 a

a   powered by Kerosene / Propane

EAST LANCASHIRE LIGHT RAILWAY CO LTD
Locomotives are kept at :-   Bolton Street Station, Bury   BL9 0EY   SD 802107
                       Buckley Wells Carriage Shops, Bury   BL9 0TY   SD 800103
                       Bury Diesel Depot, Bury   BL9 0LN   SD 799101

Gauge 4ft 8½in   www.eastlancsrailway.org.uk

  3855   2-8-0   OC   Sdn   1942
  7229   2-8-2T   OC   Sdn   1935
 (30072) 72   0-6-0T   OC   VIW   4446 1943
 34092 CITY OF WELLS   4-6-2   3C   Bton   1949
 (42765) 13065   2-6-0   OC   Crewe 5757 1927
 42859   2-6-0   OC   Crewe 5981 1930   Dsm
 45337   4-6-0   OC   AW   1392 1937
 45407 THE LANCASHIRE FUSILIER 4-6-0   OC   AW   1462 1937

(46428)		2-6-0	OC	Crewe		1948	
47324		0-6-0T	IC	NBH	23403	1926	
80097		2-6-4T	OC	Bton		1954	
51456	(752 11456)	0-6-0	IC	BP	1989	1881	
	rebuilt as	0-6-0ST	IC	Hor		1896	
2890		0-6-0	IC	Gartell		2001	
	a rebuild of	0-6-0ST	IC	HE	3882	1962	
	a rebuild of			HE	2890	1943	
(32)	1 (GOTHENBURG)	0-6-0T	IC	HC	680	1903	
	MET	0-4-0ST	OC	HL	2800	1909	
35	NORMAN	0-6-0ST	IC	RSHN	7086	1943	
7164	ANN	4wVBT	VCG	S	7232	1927	
(D99)	45135 3rd CARABINIER	1Co-Co1DE		Crewe		1961	
(D120)	45108	1Co-Co1DE		Crewe		1961	
(D335)	40135	1Co-Co1DE		_(EE	3081	1961	
				(VF	D631	1961	
(D)345	(40145)	1Co-Co1DE		_(EE	3091	1961	
				(VF	D641	1961	
(D415)	50015 VALIANT	Co-CoDE		_(EE	3785	1967	
				(EEV	D1156	1967	
(D)832	ONSLAUGHT	B-BDH		Sdn		1960	
D1041	WESTERN PRINCE	C-CDH		Crewe		1962	
(D1643	47059 47631) 47765	Co-CoDE		Crewe		1964	
D2062	(03062)	0-6-0DM		Don		1959	
(D2956	01003) 11506	0-4-0DM		AB	398	1956	
(D2997)	07013	0-6-0DE		RH	480698	1962	
(D3232)	08164 PRUDENCE	0-6-0DE		Dar		1956	
(D3594	08479) 13594	0-6-0DE		Hor		1958	
(D4112)	09024	0-6-0DE		Hor		1961	
(D4174)	08944	0-6-0DE		Dar		1962	
D5054	(24054) PHIL SOUTHERN	Bo-BoDE		Crewe		1959	
D5705	(TDB 968006)	Co-BoDE		MV		1958	
(D6525)	33109 CAPTAIN BILL SMITH R.N.R.						
		Bo-BoDE		BRCW	DEL117	1960	
(D)6536	(33117) ETHEL 4	Bo-BoDE		BRCW	DEL128	1960	
(D6553)	33035	Bo-BoDE		BRCW	DEL145	1961	
(D6564	33046)	Bo-BoDE		BRCW	DEL156	1961	Dsm
(D6809)	37109	Co-CoDE		_(EE	3238	1963	
				(VF	D763	1963	
(D)7076		B-BDH		BPH	7980	1963	
D7629	(25279)	Bo-BoDE		BP	8039	1965	
D8233	(ADB 968001)	Bo-BoDE		BTH	1131	1959	a
D9502		0-6-0DH		Sdn		1964	
D9531	ERNEST	0-6-0DH		Sdn		1965	
56006		Co-CoDE		Electro		1976	
	DUKE OF EDINBURGH	0-4-0DM		AB	400	1956	b
	(BENZOLE)	4wDM		FH	3438	1950	
4002	ARUNDEL CASTLE	0-6-0DE		HC	D1076	1959	
–		4wDM		MR	9009	1948	
(M50437)	53437	2-2w-2w-2DMR		BRCW		1957	
M50455	(53455)	2-2w-2w-2DMR		BRCW		1957	
(M50494)	53494	2-2w-2w-2DMR		BRCW		1957	
M50517	(53517)	2-2w-2w-2DMR		BRCW		1957	

50531	(53531)	2-2w-2w-2DMR		BRCW		1958
Sc51485		2-2w-2w-2DMR		Cravens		1959
E 51813		2-2w-2w-2DMR		BRCW		1961
E 51842		2-2w-2w-2DMR		BRCW		1961
W55001	L101 (TDB 975023)	2-2w-2w-2DMR		GRC&W		1958
55809	144009	2-2wDMR		_(BRE(D)		1986
				(Alex	2785/17	1986
	rebuilt as	2-2wDHR		BRE(D)		1988
55832	144009	2w-2DMR		_(BRE(D)		1986
				(Alex	2785/18	1986
	rebuilt as	2w-2DHR		BRE(D)		1988
55810	144010	2-2wDMR		_(BRE(D)		1986
				(Alex	2785/19	1986
	rebuilt as	2-2wDHR		BRE(D)		1988
55833	144010	2w-2DMR		_(BRE(D)		1986
				(Alex	2785/20	1986
	rebuilt as	2w-2DHR		BRE(D)		1988
M 65451		4w-4wRER		Wolverton		1959
8099		4w-4wDER		DerbyC&W		1978
	L253 HKK	4wDM	R/R	Multicar	000406	1994

a   carries plate 1118/1959
b   currently stored off site

Bury Transport Museum, Castlecroft Goods Warehouse, Bury   BL9 0LN
Gauge 4ft 8½in   SD 801109

(42500)	2500	2-6-4T	3C	Derby		1934
No.1		0-4-0ST	OC	AB	1927	1927
	VESTA	0-6-0T	IC	HC	1223	1916
–		4wBE		_(EE	1378	1944
				(Bg	3217	1944

Gauge 1ft 6in

	WREN	0-4-0STT	OC	BP	2825	1887

FRIENDS OF IRLAM STATION, IRLAM STATION, STATION ROAD, IRLAM   M44 5ZR
Gauge 4ft 8½in   www.thestationirlam.co.uk   SJ 713931

2027	IRLAM	0-4-0ST	OC	P	2027	1942

LANCASHIRE MINING MUSEUM - AT ASTLEY GREEN,
ASTLEY GREEN COLLIERY MUSEUM,
HIGHER GREEN LANE, ASTLEY GREEN, TYLDESLEY   M29 7JB
Gauge 4ft 8½in   www.lancashireminingmuseum.org   SJ 705998

–		4wDM		RH	244580	1946

Gauge 3ft 0in

No.17		0-6-0DMF		HC	DM781	1953
No.11		0-6-0DMF		HC	DM1058	1957
No.20	BEM 401	0-6-0DMF		HC	DM1120	1958
No.18	9306/103 1	0-6-0DMF		HC	DM1270	1961
No.15	20/105/624	0-6-0DMF		_(HC	DM1439	1978
				(HE	8821	1978

–		0-6-0DMF		HE	4816	1955
No.9	20/125/101 M.T.R.	4w-4wDHF		HE	9227	1986

Gauge 2ft 6in

8566	(1-44-251)	4wDHF		_(AB	618	1981
				(HE	8566	1981
–		4wDHF		_(AB	621	1981
				(HE	8567	1981
–		4wDHF		_(AB	622	1981
				(HE	8568	1981
8802	(1-44-161A)	4wDHF		_(AB	637	1987
				(HE	8802	1987
7		4wBEF		CE	B4055	1995
		a rebuild of		CE	B3772	1991
3		4wBEF		_(EE	2417	1957
				(RSHN	7936	1957
4	1/44/17	0-6-0DMF		HC	DM1173	1959
5	1/44/18	0-6-0DMF		HC	DM1352	1967
6	PLT No.1/44/19	0-6-0DMF		HC	DM1413	1970
7	1/44/20	0-6-0DMF		HC	DM1414	1970
–		0-4-0DMF		HE	3411	1947
	1-44-170	0-6-0DMF		HE	8575	1978
	1-44-174	0-6-0DMF		HE	8577	1978

Gauge 2ft 4in

(No.6)	0-6-0DMF		HC	DM970	1957

Gauge 2ft 1in

2	KESTREL	0-6-0DMF		HC	DM647	1954

Gauge 2ft 0in

	HELEN KATHRYN	0-4-0WT	OC	Hen	28035	1948
No.21	UTRILLAS	0-4-0WT	OC	OK	2378	1907
	(STACY)	0-6-0DMF		HC	DM804	1951
T1		0-6-0DMF		HC	DM840	1954
14	GEORGE	0-6-0DMF		HC	DM929	1955
No.16	WARRIOR	0-6-0DMF		HC	DM933	1956
2305/54	TYNESIDE GEORGE	0-6-0DMF		HC	DM1119	1958
–		0-4-0DMF		HC	DM1164	1959
No.5	2201/266	0-6-0DMF		HC	DM1170	1960
R4		0-6-0DMF		_(HC	DM1443	1980
				(HE	8843	1980
3		0-4-0DMF		HE	6048	1961
75	ROGER BOWEN	0-4-0DMF		HE	7375	1973
09	CALVERTON	4wDHF		HE	7519	1977
	MOLE	4wDMF		HE	8834	1978
03	LIONHEART	4wDHF		HE	8909	1979
	NEWTON	4wDH		HE	8975	1979
	SANDY	4wDM		MR	11218	1962
	POINT OF AYR	4wDMF		RH	497547	1963
03-418	(SP 418)	4wDH		CE	B1563Q	1978
	SP 419	4wDH		CE		
SP 973	(72292/2)	4wBE		CE	B3686B	1990
A19		4wBEF		MV/BP	989	1955

THE MUSEUM OF SCIENCE & INDUSTRY IN MANCHESTER, LIVERPOOL ROAD, CASTLEFIELD, MANCHESTER

M3 4FP

Gauge **5ft 6in**   www.scienceandindustrymuseum.org.uk   **SJ 831978**

No.3157	4-4-0	IC	VF	2759	1911	a

Gauge **4ft 8½in**

PLANET	2-2-0	IC	MSI		1992	
"NOVELTY"	2-2-0VBWT	VC	ScienceMus		1929	b
AGECROFT No.1	0-4-0ST	OC	RSHN	7416	1948	
(27001) ARIADNE 1505	Co-CoWE		Gorton	1066	1954	

Gauge **3ft 6in**

2352	4-8-2+2-8-4T	4C	BP	6639	1930

Gauge **3ft 0in**

No.3 PENDER	2-4-0T	OC	BP	1255	1873	c

a	carries works plates VF 3064/1911
b	a replica of original locomotive built 1829 by Braithwaite & Ericsson
c	locomotive is sectioned to show moving parts

SALFORD CITY COUNCIL, CADISHEAD BY - PASS, CADISHEAD WAY, IRLAM

M44 5AL

Gauge **4ft 8½in**

SJ 716922

–	0-4-0F	OC	P	2155	1955

VICTORIA WAREHOUSE, TRAFFORD WHARF ROAD, OLD TRAFFORD, STRETFORD

M17 1AB

Gauge **4ft 8½in**   www.victoriawarehouse.com   **SJ 811964**

–	0-4-0DM	JF	4210074	1952

WIGAN COUNCIL, WIGAN LEISURE & CULTURE TRUST LTD, HAIGH RAILWAY, HAIGH WOODLAND PARK, COPPERAS LANE, HAIGH

WN2 1PE

Gauge **1ft 3in**   www.haighwoodlandpark.co.uk   **SD 601082**

6284	2-8-0	OC	_(TurnerT/BVR (Crome/Loxley		2009
HELEN	0-6-2DH	s/o	AK	41	1992

MERSEYSIDE

INDUSTRIAL SITES

**DIRECT RAIL SERVICES LTD, c/o FORD MOTOR CO LTD,
SPEKE ROAD, SPEKE, LIVERPOOL** **L19 2JR**
Gauge 4ft 8½in   www.directrailservices.com   **SJ 399844**

| (D4169) | 08939 | | 0-6-0DE | Dar | 1962 | a |

a   property of Railway Support Services Ltd, Wishaw, Warwickshire

**EMERGENCY SERVICES TRAINING CENTRE LTD,
EAST STREET, SEACOMBE, WIRRAL** **CH41 1BY**
Gauge 4ft 8½in   www.emergencyservices-training.com   **SJ 325905**

| 64649 | 508201 (508101) | 4w-4wRER | York(BRE) | 1980 |
| 64712 | 508209 (508121) | 4w-4wRER | York(BRE) | 1980 |

**MERSEYRAIL ELECTRICS 2002 LTD t/a MERSEYRAIL, KIRKDALE DEPOT,
off MARSH STREET, KIRKDALE** (an Abellio-SERCO joint venture) **L20 2BL**
Gauge 4ft 8½in   www.merseyrail.org   **SJ 346938**

| – | | 4wBE | Niteq | B226 | 2006 |

**MERSEYSIDE FIRE & RESCUE SERVICE, KIRKDALE COMMUNITY FIRE STATION,
STUDHOLME STREET, BANK HALL, BOOTLE** **L20 8ET**
Gauge 4ft 8½in   www.merseyfire.gov.uk   **SJ 341935**

ERV 1	99709 901016-4	2w-2wBE	Bance ERV2-186/07	2007
ERV 2	99709 901017-2	2w-2wBE	Bance ERV2-204/08	2008
ERV 3	99709 901018-0	2w-2wBE	Bance ERV2-205/08	2008

parts from each vehicle are interchangeable

NORTHERN TRAINS LTD (owned by DfT OLR Holdings Ltd)
**ALLERTON TRAINCARE DEPOT, KNAVESMIRE WAY,
off WOOLTON ROAD, ALLERTON, LIVERPOOL** **L19 5NA**
Gauge 4ft 8½in   www.northernrailway.co.uk   **SJ 414849**

| (D4002) | 08834 | 0-6-0DE | Derby | | 1960 | a |
| | – | 4wBE | Windhoff | 101005675/60 | 2008 | |

a   property of Harry Needle Railroad Co Ltd, Derbyshire

PRESERVATION SITES

G. FAIRHURST, PRIVATE LOCATION
Gauge 2ft 0in   **Private Site**

–	4wPM		LancTan	1958	Dsm

KNOWSLEY SAFARI PARK, KNOWSLEY HALL, near PRESCOT   **L34 4AN**
Gauge 1ft 3in   www.knowsleysafariexperience.co.uk   **SJ 460936**

–	2-6-0DH	s/o	SL	343.2.91	1991

A.E. LEWIS, MEADOW VIEW, 170 HALL LANE, SIMONSWOOD
Gauge 1ft 3in   **Private Site**

No.1	GORAM	2w-2-4BER s/o	Hayne	1977		
–		4w-4BE	Hayne	1977	Dsm	a

a   unpowered and in use as a wagon

MERSEYSIDE DEVELOPMENT CORPORATION,
junction of DERBY ROAD and BANKFIELD STREET,
BANK HALL, BOOTLE, LIVERPOOL   **L20 8EA**
Gauge 4ft 8½in   **SJ 339937**

THE ATLANTIC AVENUE 1998	4wDM	RH	224347	1945

NATIONAL MUSEUMS LIVERPOOL,
Museum Of Liverpool, Pier Head, Liverpool   **L3 1DG**
Gauge 4ft 8½in   www.liverpoolmuseums.org.uk/mol   **SJ 340899**

LION	0-4-2	IC	TK&L	1838
No.3	4w-4wRER		BM	1892

Vehicle Workshop, Juniper Street, Bootle   **L20 8QJ**
Gauge 4ft 8½in   **SJ 344936**

M.D.& H.B. No.1 M.D.E. 1	0-6-0ST	OC	AE	1465	1904
(CECIL RAIKES)	0-6-4T	IC	BP	2605	1885

Locomotives not on public display. Visitors by appointment only.

NEWTON-LE-WILLOWS PRIMARY SCHOOL,
SANDERLING ROAD, NEWTON-LE-WILLOWS   **WA12 9UF**
Gauge 2ft 6in   www.newton.st-helens.sch.uk   **SJ 584954**

PARKSIDE	4wDH	HE	8825	1978

SOUTHPORT PIER RAILWAY, SOUTHPORT

Gauge 3ft 6in   Closed

–	2-2w-2w-2+2-2w-2w-2BER	SL/UK LOCO	2005	OOU	a

PR8 1QX
SD 335176

a   stored off site at Sefton Council yard, Southport

SUPREME ATTRACTIONS LTD,
LAKESIDE MINIATURE RAILWAY, MARINE LAKE, SOUTHPORT

Gauge 1ft 3in   www.lakesideminiaturerailway.com

PR8 1RX
SD 331174

No.3	JENNY	2-6-2DM	s/o	MossAJ		2006
	rebuilt as	2-6-2BE	s/o	LSLL		2022
–		2-8-0DH	s/o	SL	15/2/79	1979

NORFOLK

INDUSTRIAL SITES

ABELLIO EAST ANGLIA LTD, t/a GREATER ANGLIA (part of the Transport UK Group),
CROWN POINT DEPOT, CREMORNE LANE, NORWICH   **NR1 1TZ**

Gauge 4ft 8½in   www.greateranglia.co.uk   TG 246078

(D3556)	08441		0-6-0DE	Derby		1958	a
(D3599)	08484	CAPTAIN NATHANIEL DARELL					
			0-6-0DE	Hor		1958	a
–			4wBE	R/R	Zephir	2947	2020
–			2-2w-2BE	R/R	(Harmill?)		

a   property of Railway Support Services Ltd, Wishaw, Warwickshire

**EASTERN RAIL SERVICES LTD, YARMOUTH VAUXHALL TRACTION and ROLLING
STOCK MAINTENANCE DEPOT, off ACLE ROAD, GREAT YARMOUTH**   **NR30 1TB**

Gauge 4ft 8½in   www.easternrailservices.co.uk   TG 516087

(D4038)	08870	(H 024)	0-6-0DE	Dar		1960		
(D5809	31279)	31452	A1A-A1ADE	BT	310	1961		
(653)	H 050		0-6-0DE	_(EE	2150	1956		
				⁻(VF	D340	1956		
(01573)	H 006	15	0-6-0DH	HE	6294	1965	a	OOU
H 032			0-6-0DH	HE	7541	1976	a	OOU
H 057	01582		4wDH	S	10177	1964	a	
63128	(321434)		4w-4wWER	York(BRE)		1990		

a   property of Rail Management Services Ltd, Chesterfield, Derbyshire

GREAT EASTERN TRACTION,
HARDINGHAM STATION, HARDINGHAM, near WYMONDHAM

Gauge 4ft 8½in   www.greateasterntraction.co.uk   TG 050055

No.9	HARDINGHAM	0-4-0DE	RH	512842	1965

8	DL 82		0-6-0DH	RR	10272	1967	
	99709 909246-0		2w-2PMR	Bance	091	2000	

Gauge 1ft 6in

	CLIFF		4wDH	Ashbyl	2015	
		a rebuild of	4wPM	TEE	2010	

NORTH NORFOLK RAILWAY ENGINEERING, WEYBOURNE   NR25 7HN
Locomotives under restoration usually present   TG 118419

OLYMPIC AQUATIC ENGINEERS, UNIT 10, OAKTREE BUSINESS PARK,
BASEY ROAD, RACKHEATH INDUSTRIAL ESTATE, NORWICH   NR13 6PZ
Gauge monorail   www.olympicaquaticengineers.co.uk   TG 282140

002	IVOR	2a-2DH	AK	M002	1989	
–		2a-2DH	AK	M003	1989	
–		2w-2wDH	OAE			

Monorail locomotives present in yard between contracts

PRESERVATION SITES

BRESSINGHAM STEAM PRESERVATION COMPANY,
BRESSINGHAM LIVE STEAM MUSEUM, LOW ROAD, near DISS   IP22 2AA
Gauge 5ft 0in   www.bressingham.co.uk   TM 080806

(1144)		2-8-0	OC	TK	571	1948	OOU

Gauge 4ft 8½in

(30102)	GRANVILLE	0-4-0T	OC	9E		1893
32662	(662)	0-6-0T	IC	Bton		1875
(41966)	80 THUNDERSLEY	4-4-2T	OC	RS	3367	1909
(62785)	No.490	2-4-0	IC	Str	836	1894
(68633)	No.87	0-6-0T	IC	Str	1249	1904
	(ROBERT KETT)	0-4-0F	OC	AB	1472	1916
6841	WILLIAM FRANCIS	0-4-0+0-4-0T 4C	BP	6841	1937
No.25		0-4-0ST	OC	N	5087	1896
377	7	2-6-0	OC	Nohab	1164	1919
5865	PEER GYNT	2-10-0	OC	Schichau	4216	1944
1	COUNTY SCHOOL	0-4-0DH		RH	497753	1963
11103 // 11104		0-4-0DM		_(VF	D297	1956
				(DC	2583	1956

Gauge 2ft 0in (line laid to accommodate 1ft 11½in & 1ft 10¾in gauge locomotives)

No.2	BEVAN	0-4-0PT	OC	_(Braithwaite No.2		
				(Bressingham		2009
	GWYNEDD	0-4-0STT	OC	HE	316	1883
	GEORGE SHOLTO	0-4-0ST	OC	HE	994	1909
3	FERNILEE	0-4-0VBTT	VC	_(Wbton (3)		2015
				(Brookside No.8		2015
–		4wDHF		HE	8911	1980

	BOVIS	4wBEF		HE	9155	1991	
–		4wDM		MR	21619	1957	
7	"TOBY"	4wDM	s/o	MR	22210	1964	

Gauge 1ft 3in — Waveney Valley Railway

	ST.CHRISTOPHER	2-6-2T	OC	ESR	311	2001	
No.1662	ROSENKAVALIER	4-6-2	OC	Krupp	1662	1937	
1663	MÄNNERTREU	4-6-2	OC	Krupp	1663	1937	
D6353	JOE BROWN ENGINEER	4w-4wDM		BrownJ		1998	
	IVOR	4wDH		Frenze		1979	
			reb	‡	No.1	1986	

‡ rebuilt by Gray's Engineering, Diss

Gauge 1ft 0in

No.1		4-4-0T	OC	GoodyearJ		1935	

BURE VALLEY RAILWAY (1991) LTD, THE BROADLAND LINE, AYLSHAM NR11 6BW

Gauge 1ft 3in   www.bvrw.co.uk   **TG 197265**

1	WROXHAM BROAD	2-6-2DM	s/o	Guest		1964	a
	rebuilt as	2-6-4T	OC	Winson		1992	
6	BLICKLING HALL	2-6-2	OC	Winson	12	1994	
7	SPITFIRE	2-6-2	OC	Winson	14	1994	
8	JOHN OF GAUNT	2-6-2T	OC	Winson/BVR	16	1998	
9	MARK TIMOTHY	2-6-4T	OC	Winson	20	1999	
			reb	AK	69R	2003	
3	2nd AIR DIVISION USAAF	4w-4wDH		BVR		1989	
PW 7		2-2w-4BEF		GB	6132	1966	b
4	RUSTY	4wDM		HE	4556	1954	
	rebuilt as	4wDH		EAGIT		1994	
7		4wDM		LB	51989	1960	

a   carries plates G&S 20/1964
b   converted to a non-powered wagon

CAISTER CASTLE CAR COLLECTION, CAISTER CASTLE, CASTLE LANE, CAISTER-ON-SEA, near GREAT YARMOUTH   NR30 5SN

Gauge 4ft 8½in   www.caistercastle.co.uk   **TG 504123**

42	RHONDDA	0-6-0ST	IC	MW	2010	1921	

FENGATE AGRICULTURAL LIGHT RAILWAY, FENGATE FARM, WEETING, near BRANDON   IP27 0QF

Gauge 5ft 0in   www.weetingrally.co.uk   **TL 770885**

1060		2-8-2	OC	LO	172	1954	

Gauge 4ft 8½in

	LITTLE BARFORD	0-4-0ST	OC	AB	2069	1939	

Gauge 3ft 0in

No.7	(TYNWALD)	2-4-0T	OC	BP	2038	1880	Dsm

Visiting locomotives occasionally present.

R.& A. JENKINS, FRANSHAM STATION, GREAT FRANSHAM, near SWAFFHAM

Gauge 4ft 8½in   Private Site   TF 888135

–		4wDM	RH	398611	1956

Gauge 2ft 0in

–		4wDM	RH	422573	1958

M.W. MAYES, MECHANICAL ENGINEER,
STATION YARD, STATION ROAD, YAXHAM, near DEREHAM   NR19 1RD

Gauge 4ft 8½in   Private Site   TG 003102

	MILLFIELD	0-4-0CT	OC	RSHN	7070	1942
No.2		0-4-0ST	OC	RSHN	7818	1954
	"GEORGE"	4wVBT	VCG	S	9596	1955

Gauge 1ft 11½in

No.1		0-4-0VBT	VCG	Potter		1970
No.2	RUSTY	4wDM		L	32801	1948

MID-NORFOLK RAILWAY SOCIETY, MID-NORFOLK RAILWAY,
DEREHAM STATION, STATION ROAD, DEREHAM   NR19 1DF

Gauge 4ft 8½in   www.midnorfolkrailway.co.uk   TF 995128

(D419)	50019	RAMILLIES	Co-CoDE		_(EE	3789	1967
					(EEV	D1160	1967
(D1886)	47367		Co-CoDE		BT	648	1965
D1933	(47255 47596)	ALDBURGH FESTIVAL					
			Co-CoDE		BT	695	1966
(D2063	03063)		0-6-0DM		Don		1959
(D2197)	03197		0-6-0DM		Sdn		1961
D2334			0-6-0DM		_(RSHD	8193	1961
					(DC	2715	1961
(D5683)	31255		A1A-A1A DE		BT	284	1961
(D6587)	33202	DENNIS G. ROBINSON	Bo-BoDE		BRCW	DEL158	1962
D9520	45		0-6-0DH		Sdn		1964
B.S.C.1			0-6-0DH		EEV	D1049	1965
(51226)			2-2w-2w-2DMR		MetCam		1958
W51370			2-2w-2w-2DMR		PSteel		1960
W51412			2-2w-2w-2DMR		PSteel		1960
(51434)	MATTHEW SMITH 1974-2002		2-2w-2w-2DMR		MetCam		1959
(51499)			2-2w-2w-2DMR		MetCam		1959
51503			2-2w-2w-2DMR		MetCam		1959
51942	L233		2-2w-2w-2DMR		DerbyC&W		1961
55579	142038		2-2wDMR		_(BRE(D)		1986
					(Leyland	R5.78	1986
		rebuilt as	2-2wDHR		RFSD		1988
55629	142038		2w-2DMR		_(BRE(D)		1986
					(Leyland	R5.75	1986
		rebuilt as	2w-2DHR		RFSD		1988
55711	142061		2-2wDMR		_(BRE(D)		1986
					(LeylandR5.166		1986
		rebuilt as	2-2wDHR		AB		1989

55757	142061		2w-2DMR	_(BRE(D)		1986
				(LeylandR5.167		1986
		rebuilt as	2w-2DHR	AB		1989
55818	144018		2-2wDMR	_(BRE(D)		1987
				(Alex 2785/35		1987
		rebuilt as	2-2wDHR	AB		1990
55841	144018		2-2wDMR	_(BRE(D)		1987
				(Alex 2785/36		1987
		rebuilt as	2-2wDHR	AB		1990
55854	(144018)		2w-2DMR	_(BRE(D)		1988
				(Alex 1187/5		1988
		rebuilt as	2w-2DHR	AB		1990
68004	(9004) 931 094		4w-4wRE/BER	Afd/Elh		1960
960225	GEORGE T. RASEY		2w-2PMR	Wkm	1308	1933

NORTH NORFOLK RAILWAY plc

Locomotives are kept at :-

Holt	NR25 6AJ	TG 094398
Sheringham	NR26 8RA	TG 156430
Weybourne	NR25 7HN	TG 118419

Gauge 4ft 8½in   www.nnrailway.co.uk

53809	(13809)		2-8-0	OC	RS	3895	1925	
(65462	7564) 564		0-6-0	IC	Str		1912	
76084			2-6-0	OC	Hor		1957	
90775	THE ROYAL NORFOLK REGIMENT							
			2-10-0	OC	NBH	25438	1943	
92203	BLACK PRINCE		2-10-0	OC	Sdn		1959	
	WISSINGTON		0-6-0ST	IC	HC	1700	1938	
22			0-6-0ST	IC	HE	3846	1956	a
	NICK		0-4-0ST	OC	WB	2565	1936	
D3935	(08767)		0-6-0DE		Hor		1961	
D3940	(08772)		0-6-0DE		Derby		1960	
D5631	(31207)		A1A-A1A DE		BT	231	1960	
D6732	(37032)	MIRAGE	Co-CoDE		_(EE	2895	1961	
					(VF	D611	1961	
12131			0-6-0DE		Dar		1952	
50479	(53479)		2-2w-2w-2DMR		BRCW		1958	
M51188			2-2w-2w-2DMR		MetCam		1958	
M51192			2-2w-2w-2DMR		MetCam		1958	
E51228			2-2w-2w-2DMR		MetCam		1958	
98801	DAVID PINKERTON		4wDHR		Kershaw			
	RAILWAY ENGINEER					45.121A	1992	

a   rebuilt using parts from HE 3844 and WB 2758; carries worksplate HE 3844;

ALAN POWELL, NORWICH
Gauge 1ft 3in   **Private Site**

| – | | 2-4wPM | s/o | Powell | 1997 |

FRED ROUT,
BROOMHILL FARM, LEYS LANE, BUCKENHAM, near ATTLEBOROUGH   NR17 1NS
Gauge 4ft 8½in   **Private Site**

No.21		0-6-0DM	HC	D707	1950

STOW BARDOLPH OLD STATION,
THE CAUSEWAY, STOW BRIDGE, near DOWNHAM MARKET   PE34 3PH
Gauge 4ft 8½in   TF 606070

55638 142047		2w-2DMR	_(BRE(D)		1986
			(Leyland R5.77		1986
	rebuilt as	2w-2DHR	RFSD		1989

STRUMPSHAW STEAM MUSEUM, STRUMPSHAW HALL, near ACLE   NR13 4HR
Gauge 1ft 11½in   www.strumpshawsteammuseum.co.uk   TG 345065

No.6	GINETTE MARIE	0-4-0WT	OC	Jung	7509	1937	OOU
5	JIMMY	4wDM	s/o	MR	7192	1937	

Gauge 1ft 3in

2	CAGNEY	4-4-0	OC	McGarigle		1902

THURSFORD COLLECTION,
THURSFORD GREEN, THURSFORD, near FAKENHAM   NR21 0AS
Gauge 1ft 10¾in   www.thursford.com   TF 980345

CACKLER	0-4-0ST	OC	HE	671	1898	Pvd

WHITWELL & REEPHAM STATION, WHITWELL ROAD, REEPHAM   NR10 4GA
Gauge 4ft 8½in   www.whitwellstation.com   TG 092216

	ANNIE	0-4-0ST	OC	AB	945	1904	
M.E.D. Yd. No.8 VICTORY No.4		0-4-0ST	OC	AB	2199	1945	
	AGECROFT No.3	0-4-0ST	OC	RSHN	7681	1951	
1	"GEORGIE"	4wDH		BD	3733	1977	
30	"WESTERN PRIDE"	0-6-0DM		HC	D1171	1959	
D2700		0-4-0DH		NBQ	27426	1955	
–		4wDM		RH	393303	1956	
	"TIPOCKITY"	4wDM		RH	466629	1962	
	SWANWORTH	4wDM		RH	518494	1967	
	Q454 JTT	4wDM	R/R	Unimog	027869	1999	
(DX 68051 TR 6)		2w-2PMR		Wkm	6901	1954	
	rebuilt as	2w-2DHR					a
DB 965097 68044		2w-2PMR		Wkm	7612	1957	
MNN 0005		2w-2PMR		Wkm			DsmT
–		2w-2PMR		Wkm		1932	Dsm b

a   currently off site for restoration
b   carries plate Wkm 529/1932

YAXHAM LIGHT RAILWAY, STATION ROAD, YAXHAM, near DEREHAM

Gauge 1ft 11½in

NR19 1RD

TG 003102

	KIDBROOKE		0-4-0ST	OC	WB	2043	1917
No.3	PEST		4wDM		L	40011	1954
–			4wDM		LB	54684	1965
No.4	GOOFY		4wDM		OK	(6501?)	c1936
No.6	LOD/758097	COLONEL	4wDM		RH	202967	1940
No.7			4wDM		RH	170369	1934
No.10	OUSEL		4wDM		MR	7153	1937
No.13			4wDM		MR	7474	1940
No.14	LOD 758375 ARMY 25		4wDM		RH	222100	1943
No.18			4wDM		FH	3982	1962
No.19	PENLEE		4wDM		HE	2666	1942 a
44			4wDM		Moës		c1955
–			4wDM		RH	179880	1936

a   currently off site for restoration

NORTHAMPTONSHIRE

INDUSTRIAL SITES

GEISMAR (UK) LTD, SALTHOUSE ROAD, BRACKMILLS INDUSTRIAL ESTATE, NORTHAMPTON

NN4 7EX

Railcars under construction or repair are usually present.   www.geismar.com   SP 781858

METEOR POWER LTD, UNIT 2245, SILVERSTONE TECHNOLOGY PARK, SILVERSTONE

NN12 8GX

Gauge 4ft 8½in   www.meteorpower.com   **Private Site**   SP 745612

(D3816)	08649	BRADWELL	0-6-0DE	Hor	1959

PROLOGIS R.F.I., D.I.R.F.T., DAVENTRY INTERNATIONAL RAILFREIGHT TERMINAL, RAILPORT APPROACH, CRICK

NN6 7ES

Gauge 4ft 8½in   www.prologis.co.uk   SP 566725, 573721

(D3560)	08445		0-6-0DE	Derby	1958	a
(D3796)	08629		0-6-0DE	Derby	1959	b
(425	60 460)		6wDE	GECT	5425	1977 c
(261)			6wDE	GECT	5430	1977 c
–			2-4wDMR	R/R	_(John Deere 039380 (Harsco DP270/039	

a   property of Hunslet Ltd, Barton-under-Needwood, Staffordshire
b   property of Railway Support Services Ltd, Wishaw, Warwickshire
c   property of Ed Murray & Sons Ltd, Hartlepool, Teesside

SIEMENS AG, SIEMENS RAIL SYSTEMS, KINGS HEATH TRAINCARE FACILITY, HEATHFIELD WAY, KINGS HEATH, NORTHAMPTON NN5 7QP

Gauge 4ft 8½in   www.mobility.siemens.com   SP 745612

–	4wBE		CPM	1731.01	2005

PRESERVATION SITES

BILLING AQUADROME LTD, WILLOW LAKE MINIATURE TRAIN, CROW LANE, GREATER BILLING, near NORTHAMPTON NN3 9DA

(part of the Pure Leisure Group Ltd)

Gauge 1ft 3in   www.billingaquadrome.com   SP 808615

362	2-8-0gasH	s/o	SL	15.5.78	1978
rebuilt as	2-8-0DH				

CORBY and DISTRICT MODEL RAILWAY SOCIETY, QUARRY CLOSE, off KELVIN GROVE, CORBY NN17 1AZ

Gauge 4ft 8½in   www.corbymrs.great-site.net   SP 891893

55831 144008	2w-2DMR		_(BRE(D)		1986
			(Alex	2785/16	1986
rebuilt as	2w-2DHR		AB		1991

CORBY BOROUGH COUNCIL, LEISURE & CULTURE, EAST CARLTON COUNTRY PARK, EAST CARLTON, near CORBY LE16 8YF

Gauge 4ft 8½in   www.corby.gov.uk   SP 834893

–	0-6-0ST	OC	HL	3827	1934

J.EVANS, PRIVATE LOCATION, NORTHAMPTON

Gauge 2ft 0in

–	2w-2PMR		EvansJ	2009	a

a   built using parts from Wkm 2449

HIGHAM FERRERS LOCOMOTIVES
Private Site, near Rushden
Gauge 1000mm

ND 3647	4wDM		MR	22144	1962

Private Site, near Wellingborough
Gauge 3ft 0in

18	4wBE		GB	1570	1938

IRCHESTER NARROW GAUGE RAILWAY TRUST, IRCHESTER COUNTRY PARK, GIPSY LANE, LITTLE IRCHESTER

NN29 7DL

Gauge 1000mm   www.irchesterrailwaymuseum.co.uk   **SP 906660**

(4)	CAMBRAI		0-6-0T	OC	Corpet	493	1888	
No.85	1		0-6-0ST	OC	P	1870	1934	
(No.86	2)		0-6-0ST	OC	P	1871	1934	
(No.87	3	89-12) 8315/87	0-6-0ST	OC	P	2029	1942	
	9	THE ROCK	0-4-0DM		HE	2419	1941	
ND 3645	10	MILFORD	4wDM		RH	211679	1941	
(ED 10)	11	EDWARD CHARLES HAMPTON						
			4wDM		RH	411322	1958	

Gauge 3ft 0in

LR 3084	14		4wDM		MR	3797	1926	
		a rebuild of		MR	1363	1918	
H 048	BERTIE		0-6-0DM		RH	281290	1949	a
	5		4wDMF		RH	338439	1953	

a   on loan to Rail Management Services Ltd, Chesterfield, Derbyshire

NORTHAMPTON STEAM RAILWAY

Locomotives are kept at :   Pitsford & Brampton   NN6 8BA   SP 734667, 737662
Brampton Lane   NN6 8AA   SP 737656

Gauge 4ft 8½in   www.nlr.org.uk

3862		2-8-0	OC	Sdn		1942	
5967	BICKMARSH HALL	4-6-0	OC	Sdn		1937	
(RRM 23)	"FIREFLY"	0-4-0ST	OC	AB	776	1899	
–		0-4-0ST	OC	AB	2323	1952	
3193	NORFOLK REGIMENT	0-6-0ST	IC	HE	3193	1944	
			reb	HE	3887	1964	
No.4	"SWANSCOMBE"	0-4-0ST	OC	HL	3719	1928	a
–		0-4-0ST	OC	P	2104	1950	b
(D1855)	47205 47395	Co-CoDE		Crewe		1965	
(D5821)	31289 PHOENIX	A1A-A1ADE		BT	322	1961	
(D6571)	33053	Bo-BoDE		BRCW	DEL175	1961	
No.21		0-4-0DM		JF	4210094	1954	
	rebuilt as	0-4-0DH		JF		1966	
(764	SIR GILES ISHAM)	0-4-0DM		RH	319286	1953	
No.53	SIR ALFRED WOOD	0-6-0DM		RH	319294	1953	
	(99709 901013-1)	2w-2PMR		Lesmac	LMS008	2006	

a   includes parts from HL 3718
b   actually built 1948 but plates are dated as shown

NORTHAMPTONSHIRE IRONSTONE RAILWAY TRUST LTD, HUNSBURY HILL INDUSTRIAL MUSEUM, NORTHAMPTON

NN4 9UW

Gauge 4ft 8½in   www.northantsironstonerailway.co.uk   **SP 735584**

No.7		0-4-0ST	OC	P	2130	1951
9365	89-17 BELVEDERE	4wVBT	VCG	S	9365	1946
	(MUSKETEER)	4wVBT	VCG	S	9369	1946
D2867		0-4-0DH		YE	2850	1961

–		4wDH	FH	3967	1961	
–		0-4-0DM	HC	D697	1950	
–		0-4-0DM	HE	2087	1940	Dsm
–		0-4-0DM	JF	4200022	1948	
	CHARLES WAKE	0-4-0DM	JF	4220001	1959	
	FLYING FALCON	0-4-0DH	JF	4220016	1962	
No.46	HEATHER	4wDM	RH	242868	1946	
–		4wDM	RH	321734	1952	a Dsm
5176	S14351	4w-4wRER	Lancing/Elh		1955	

a   frame in use as a wagon

PRIVATE OWNER, near NORTHAMPTON
Gauge 3ft 0in   Private Site

–	4wDM	RH	418764	1957	

PRIVATE OWNER, near WELLINGBOROUGH
Gauge 2ft 0in   Private Site

–		4wDM	OK	3685	1929
11	NEEDHAM	4wDM	OK	6504	1936
–		4wDM	OK	7728	1937 Dsm

RUSHDEN HISTORICAL TRANSPORT SOCIETY,
THE OLD STATION, STATION APPROACH, RECTORY ROAD, RUSHDEN NN10 0AW
Gauge 4ft 8½in   www.rhts.co.uk   SP 957672

	EDMUNDSONS	0-4-0ST	OC	AB	2168	1943
	"CHERWELL"	0-6-0ST	OC	WB	2654	1942
(D5630)	31206	A1A-A1A DE		BT	230	1960
(D2179)	03179 CLIVE	0-6-0DM		Sdn		1961
	WD 70048	0-4-0DM		AB	363	1942
	LES FORSTER	4wDH		S	10159	1963
	YARD No.10433	0-4-0DM		_(VF	5258	1945
				(DC	2177	1945
		reb		YEC	L121	1993
-		4wBE		(SET?)		
55734	142084	2-2wDMR		_(BRE(D)		1987
				(Leyland R5.126		1987
	rebuilt as	2-2wDHR		RFSD		1989
55780	142084	2w-2DMR		_(BRE(D)		1985
				(Leyland R5.107		1985
	rebuilt as	2w-2DHR		RFSD		1988
55741	142091	2-2wDMR		_(BRE(D)		1987
				(Leyland R5.144		1987
	rebuilt as	2-2wDHR		RFSD		1989
55787	142091	2w-2DMR		_(BRE(D)		1987
				(Leyland R5.117		1987
	rebuilt as	2w-2DHR		RFSD		1989
(55029)	977968	2-2w-2w-2DMR		PSteel		1960
059	(53.0551-1)	2w-2DMR		Robel 54.13-RW82		1975
	99709 901041-2	2w-2PMR		Geismar ST/01/34		2001

| 99709 909025-7 | | 2w-2PMR | Thomson TRE025 2009 | DsmT |

MARK SAVILLE, PRIVATE SITE, NORTHAMPTON
Gauge 4ft 8½in   **Private Site**

| PSL 193 | "PLIMSOLL" | 4wPMR | R/R | Landrover | 1957 | + |

+   road vehicle with exchangeable bolt-on rail wheels

WICKSTEED PARK LAKESIDE RAILWAY,
WICKSTEED LEISURE PARK, KETTERING   **NN15 6NJ**
Gauge 2ft 0in   www.wicksteedpark.org   **SP 883770**

	MERLIN	0-4-0DH	s/o	AK	86	2010
2042	LADY OF THE LAKE	0-4-0DM	s/o	Bg	2042	1931
2043	KING ARTHUR	0-4-0DH	s/o	Bg	2043	1931
	CHEYENNE	4wDM	s/o	MR	22224	1966

J. WOOLMER, 15 BAKERS LANE, WOODFORD, KETTERING
Gauge 2ft 0in   **SP 969769**

| 89-18 | 4wPM | L | 14006 | 1940 |
| – | 4wDM | L | 36743 | 1951 |

NORTHUMBERLAND

Our definition of "Northumberland" includes those locations which were situated within that part of the former county of Tyne & Wear which is North of the River Tyne and now consists of a number of unitary authorities.

INDUSTRIAL SITES

ALVANCE BRITISH ALUMINIUM LTD, (part of the GFG Alliance group of companies),
SHIP UNLOADING FACILITY, NORTH BLYTH   **NE24 1SE**
(operated by Port of Blyth - Blyth Habour Commissioners)
Gauge 4ft 8½in   www.alvancebritishaluminium.com   **NZ 315821**

| 1046 | 4wDH | R/R | Unilok | 4020 | 2011 |

AYLE COLLIERY CO LTD, QUARRY DRIFT, ALSTON
The site of the colliery is located in Northumberland, though the original Ayle East Drift (abandoned) and related workings were located in Cumbria.
Gauge 2ft 6in — Surface   www.lastpit.simdif.com   **NY 728498**

15/PLB6	152183	No.5261	4wBEF	CE	5074	1965	
15PLB15	15/19	No.5097	4wBEF	CE	5097	1966	
	15/27	No.B0909B	4wBEF	CE	B0909B	1976	OOU
–			4wBEF	CE	B3168	1985	
–			4wBEF	CE	B3482B	1988	

6/43		4wBE	WR	6297	1960	Dsm

Gauge 2ft 0in — Surface & Underground

EL 16	5667		4wBE	CE	5667	1969	Dsm
No.2			4wBE	CE	5667	1969	
–			4wBE	CE	5667	1969	Dsm
	263 022		4wBE	CE	5955A	1972	Dsm
–			4wBE	CE	(5955C?	1972)	
1520			4wBEF	CE	B0445	1975	
–			4wBEF	CE	B3084	1984	
	263 075		4wBE	CE	B3642A	1990	
		reb		CE	B3766A	1991	
	263 073		4wBE	CE	B3642B	1990	
		reb		CE	B3766B	1991	
	263 074		4wBE	CE	B3766D	1991	
–			4wBEF	GB	2382	1953	Dsm
–			4wDMF	HE	3496	1947	Dsm
–			4wDMF	HE	4569	1956	OOU
–			0-4-0DMF	HE	4991	1955	
		rebuilt as	4wDMF	Ayle		1977	OOU
–			0-4-0DM	HE	5222	1958	
LE/12/75	P17271		4wBE	WR	E6807	1965	
LM03			4wBE	WR	N7607	1973	
LM 5 1			4wBE	WR		1973	a
7664			0-4-0BE	WR	P7664	1975	

Gauge 1ft 7½in — Surface

–	0-4-0BE	WR	D6754	1964	Dsm

Gauge 1ft 6in — Surface

(14) 77 4		4wBE	CE	5712	1969	
16 82		4wBE	CE	B3132B	1984	

a   one of WR N7606, WR N7620 or WR N7621

DEFENCE INFRASTRUCTURE ORGANISATION, DEFENCE TRAINING ESTATES - NORTH, REDESDALE RANGES MILITARY TARGET RAILWAY, REDESDALE RANGES, BUSHMAN'S ROAD, SILLS, ROCHESTER, NORTHUMBERLAND
(operated by Landmarc Support Services Ltd t/a Landmarc Solutions)

Gauge 2ft 6in   R.T.C.   NT 826999

–		2w-2PM	Wkm	3245	1943	OOU
6	SILLS 1	2w-2PM	Wkm	11686	1990	OOU
7	SILLS 2	2w-2PM	Wkm	11687	1990	OOU
(8)	SILLS 4	2w-2PM	Wkm	11688	1990	OOU
9	SILLS 3	2w-2PM	Wkm	11689	1990	OOU

Gauge 1000mm – White Spot Target Railway (elevated railway)   NT 862045

2337		4wBE	SAAB	108	1986	OOU a
2338		4wBE	SAAB	(133?	19xx)	OOU a

a   stored at Otterburn Camp

LYNEMOUTH POWER LTD, LYNEMOUTH POWER STATION, LYNEMOUTH, ASHINGTON

NE63 9NW

(part of EP UK Investments Ltd, a subsidiary of EPH Energetický a Průmslový Holdings)

Gauge 4ft 8½in   www.lynemouthpower.com   **NZ 302903**

–	2w-2PMR	Bance	057/98	1998
TNS 106	2-2wDMR	Robel		
		56.27-10-AG37		1982

NEXUS TYNE WEAR METRO, NEWCASTLE-UPON-TYNE

Locomotives are kept at :-

Howdon Satellite Depot, Wallsend	NE28 0JH	NZ 333671
Nexus Learning Centre, South Shields, Co. Durham	NE33 1TA	NZ 362674
Nexus Rail Headquarters, South Gosforth	NE3 1XD	NZ 250684
Nexus Rail Track Engineering, Palmersville	NE12 9TA	NZ 297701
South Gosforth Car Sheds, Newcastle-upon-Tyne	NE3 5DG	NZ 250685

Locomotives may also be occasionally found at other locations on the Tyne & Wear Metro network.

Gauge 4ft 8½in   www.nexus.org.uk

BL1 97901	4wBE/WE		HE	9174	1989	
BL2 97902	4wBE/WE		HE	9175	1989	
BL3 97903	4wBE/WE		HE	9176	1989	
–	4wBE	R/R	Vollert		2023	
YF59 YHK 99709 976055-2	4wDMR	R/R	_(Ford	833760	2009	
			(Aquarius		2010	
YR10 FDP	4wDMR	R/R	_(Ford	833517	2010	
			(Aquarius		2011	
T597 OKK	4wDM	R/R	_(DAF	167811	1999	
			(SRS		1999	
F171 DUA	4wDM	R/R	_(Unimog	140729	1988	
			(Zweiweg	1188	1988	OOU
NA52 JNN	4wDM	R/R	Unimog	200083	2003	
NK12 CWZ 99709 979101-1	4wDM	R/R	_(Unimog	228343	2012	
			(LHGroup		2012	
NK16 GUX 99709 942161-9	4wDM	R/R	_(Unimog	239027	2016	
			(LHGroup		2016	
NK16 GUW 99709 942162-7	4wDM	R/R	_(Unimog	239031	2016	
			(LHGroup		2016	
AE66 CPU 99709 977041-1	4wDM	R/R	_(Unimog	242230	2016	
			(LHGroup		2016	
AE66 CPV 99709 977042-9	4wDM	R/R	_(Unimog	242499	2016	
			(LHGroup		2016	
–	2w-2BER		Bance	094	2000	
HCT 035	4wDH		Perm	035	1993	OOU

NORTHUMBRIA RAIL LTD, BEDLINGTON

Adminstration address only.

The following is a FLEET LIST of locomotives owned by (or in the care of) this contractor

Gauge 4ft 8½in

801	Bo-BoDE	Alco	77120	1950	a	
323.674-2 SIMONSIDE / SPLUTTER	4wDH	Gmd	4991	1957	a	
323-539-7 CHEVIOT	4wDH	Gmd	4861	1955	b	
424839	0-4-0DE	RH	424839	1959	c	
1212 HELGA	4w-4DMR	EK		1958	a	

RB 004	4wDMR	Leyland/DerbyC&W			
		RB004	1984	d
MPP 10188	2w-2BER	Bance	098/00	2000	a
HCT 022	4wDH	Perm		022	1988	e

a currently at Nene Valley Railway, Cambridgeshire
b currently at Arlington Fleet Group Ltd, Eastleigh, Hampshire
c currently at M. Fairnington, Wooler, Northumberland
d currently at Waverley Route Heritage Association, Whitrope, Borders
e currently at Scottish Railway Preservation Society, Bo'ness, West Lothian

IAN STOREY ENGINEERING LTD, HEPSCOTT, near MORPETH   NE61 6LN
Gauge 4ft 8½in   NZ 223844

| (DB 965053) | 2w-2PMR | Wkm | 7576 | 1956 | Dsm |

PRESERVATION SITES

ALLENHEADS TRUST LTD,
ALLENHEADS HERITAGE CENTRE, ALLENHEADS   NE47 9HN
Gauge 2ft 0in   NY 859453

| – | 0-4-0BE | WR | | | |

ALN VALLEY RAILWAY, LIONHEART RAILWAY STATION,
LLOYD'S FIELD, LIONHEART ENTERPRISE PARK, ALNWICK   NE66 2EZ
Gauge 4ft 8½in   www.alnvalleyrailway.co.uk   NU 202122

No.9		0-6-0T	OC	HC	1243	1917	
RENISHAW IRONWORKS No.6		0-6-0ST	OC	HC	1366	1919	
48	9103/48	0-6-0ST	IC	HE	2864	1943	
No.60		0-6-0ST	IC	HE	3686	1948	
	PENICUIK	0-4-0ST	OC	HL	3799	1935	
	MERLIN // MYRDDIN	0-4-0ST	OC	P	1967	1939	
(01564)	12088 "SHIRLEY"	0-6-0DE		Derby		1951	
	20/110/711	0-6-0DH		AB	615	1977	
			reb	AB	6719	1987	
	"DRAX"	0-6-0DM		EES	8199	1963	
55804	144004	2-2wDMR		_(BRE(D)		1986	
				(Alex	2785/07	1986	
	rebuilt as	2-2wDHR		AB		1991	
55827	144004	2w-2DMR		_(BRE(D)		1986	
				(Alex	2785/08	1986	
	rebuilt as	2w-2DHR		AB		1991	
55816	144016	2-2wDMR		_(BRE(D)		1987	
				(Alex	2785/31	1987	
	rebuilt as	2-2wDHR		AB		1991	
55839	144016	2w-2DMR		_(BRE(D)		1987	
				(Alex	2785/32	1987	
	rebuilt as	2w-2DHR		AB		1991	

55852	(144016)		2-2wDMR		_(BRE(D)		1988	
					(Alex	1187/3	1988	
		rebuilt as	2-2wDHR		AB		1991	
(DB 965952)	BUZZ		2w-2PMR		Wkm	10648	1972	
(PWM 2807	B170W)		2w-2PMR		Wkm	4985	1949	Dsm a

a   converted to non-powered passenger trailer

CHAINBRIDGE HONEY FARM and TRANSPORT MUSEUM, HORNCLIFFE, BERWICK-UPON-TWEED   TD15 2XT
Gauge   4ft 8½in   www.chainbridgehoney.com   **NT 936507**

2234	(45010)	4+0-4-0VBTR	Derby		1904	Dsm a

a   converted to hauled coaching stock

THE DALES SCHOOL, BLYTHDALE, COWPEN ROAD, BLYTH   NE24 4RE
Gauge   4ft 8½in   www.thedales.northumberland.sch.uk   **NZ 288816**

55802	144002		2-2wDMR	_(BRE(D)		1986	
				(Alex	2785/03	1986	
		rebuilt as	2-2wDHR	AB		1990	
55825	144002		2w-2DMR	_(BRE(D)		1986	
				(Alex	2785/04	1986	
		rebuilt as	2w-2DHR	AB		1990	

DANIEL FARM SHOP & TEA ROOMS, SLED LANE, WYLAM   NE41 8JH
Gauge   2ft 6in   (Closed)   **NT 933385**

CANNONBALL	4wDM	s/o	RH	175403	1935

HEATHERSLAW LIGHT RAILWAY CO LTD, FORD FORGE, HEATHERSLAW MILL, CORNHILL-ON-TWEED   TD12 4TJ
Gauge   1ft 3in   www.heatherslawlightrailway.co.uk   **NZ 123637**

	BUNTY	2-6-0T	OC	_(SmithN		C2005	
				(AK	85R	2010	a
	THE LADY AUGUSTA	0-4-2	OC	TaylorB		1989	
97655	"BINKY"	0-6-0DH		Heatherslaw		2014	
	rebuild of	6wDH		SmithN		1989	b
	CLIVE	4w-4wDH		SmithN		2000	

a   Locomotive part built by SmithN and completed by AK
b   uses parts supplied by AK

H. MEERS, BEAL STATION GOODS YARD, BEAL   TD15 2PB
Gauge   4ft 8½in   **NU 063426**

1611	0-4-0ST	OC	P	1611	1923

NORTH TYNESIDE COUNCIL,
TYNE & WEAR ARCHIVES & MUSEUMS, STEPHENSON STEAM RAILWAY,
MIDDLE ENGINE LANE, Near CHIRTON   NE29 8DX

The railway is operated by the North Tyneside Steam Railway Association

Gauge 4ft 8½in   www.stephensonsteamrailway.org.uk   **NZ 323693**

A.No.5	0-6-0PT	IC	K	2509	1883	
ASHINGTON No.5						
JACKIE MILBURN 1924-1988	0-6-0ST	OC	P	1970	1939	
"BILLY"	0-4-0	VC	GS(K)		1816	
(No.69)	0-6-0ST	IC	HE	3785	1953	
No.1 TED GARRETT JP, DL, MP	0-6-0T	OC	RSHN	7683	1951	
401	0-6-0ST	OC	WB	2994	1950	
(D2078) 03078	0-6-0DM		Don		1959	
(D4145) 08915	0-6-0DE		Hor		1962	
3267 (DE 900730)	4w-4wRER		York		1904	
10	0-6-0DM		Consett		1958	
E4	Bo-BoWE/BE		Siemens	457	1909	

M. FAIRNINGTON, AGRICULTURAL ENGINEERS,
UNIT A, BERWICK ROAD, WOOLER   NE71 6AH

Gauge 4ft 8½in   **NT 995287**

424839 NORTHUMBRIA	0-4-0DE		RH	424839	1959	a	

a   property of Northumbria Rail Ltd, Bedlington, Northumberland
     locomotives for restoration occasionally present

WOODHORN NARROW GAUGE RAILWAY,
WOODHORN COLLIERY MUSEUM, ASHINGTON   NE63 9YF

Gauge 2ft 0in   www.woodhornrailway.com   **NZ 287884**

No.1 BLACK DIAMOND / HUNSLET 6348 /					
SEAHAM VANE TEMPEST 1993	4wDM	HE	6348	1975	
3 RIO GEN	4wDH	HE	9353	1995	
No.2 EDWARD STANTON	4wDH	Schöma	5240	1991	

NOTTINGHAMSHIRE

INDUSTRIAL SITES

BALFOUR BEATTY RAIL LTD,
OLD STATION YARD, DERBY ROAD, SANDIACRE NG10 5AG
Gauge 4ft 8½in   www.balfourbeatty.com   SK 483360

XTU 2104	H867 ARB		4wDM	R/R	Unimog 166648	1991

BLYTH PLANT & TOOL HIRE,
REDBRIDGE HOUSE, WORKSOP ROAD, BLYTH S81 0TX
Gauge 4ft 8½in   SK 622857

F511 LRR	4wDM	R/R	_(Multicar		1989
			(Perm		1989

BODEN RAIL ENGINEERING LTD,
COLWICK TRACKTION MAINTENANCE DEPOT,
MALLARD ROAD, NETHERFIELD, NOTTINGHAM (NG4 2PE)
Gauge 4ft 8½in   SK 631404

97301	(D6800 37100)		Co-CoDE		_(EE	3229	1962	
					(VF	D754	1962	OOU
2150	MARDY MONSTER		0-6-0ST	OC	P		2150	1954
NCB No.11	BIRCH COPPICE		4wDH		TH		134C	1964
		a rebuild of	4wVBT	VCG	S		9578	1954

mainline and preserved locomotives for maintenance/overhaul usually present

DB CARGO UK LTD, TOTON TRACTION MAINTENANCE DEPOT,
off STATION ROAD, SANDIACRE NG10 1HA
Gauge 4ft 8½in   www.uk.dbcargo.com   SK 484354

	99709 979112-8	4wDH	R/R	Zephir	2925	2021	
000118		4wBE		_(Express	E602	2016	
				(Hegenscheidt	101896	2016	

HARRY NEEDLE RAILROAD COMPANY LTD, RAIL MAINTENANCE DEPOT,
WORKSOP DEPOT, BABBAGE WAY, off SANDY LANE, WORKSOP S80 1UJ
Gauge 4ft 8½in   www.harry-needle.co.uk   SK 578799

(D1945	47502)	47715 HAYMARKET	Co-CoDE	BT	707	1966	
(D3000)	13000		0-6-0DE	Derby		1952	a
(D3536	08421)	09201	0-6-0DE	Derby		1958	
(D3655)	08500		0-6-0DE	Don		1958	
(D3657)	08502		0-6-0DE	Don		1958	
(D3745)	08578		0-6-0DE	Crewe		1959	
(D3790)	08623		0-6-0DE	Derby		1959	
(D3867)	08700		0-6-0DE	Hor		1960	

(D3881)	08714		0-6-0DE	Crewe		1960	
(D3967)	08799	IAN GODDARD 1938 - 2016				
			0-6-0DE	Derby		1960	
(D3986)	08818) 4 MOLLY		0-6-0DE	Derby		1960	
(D4033)	08865 GILLY		0-6-0DE	Dar		1960	
(D4045)	08877) WIGAN 1		0-6-0DE	Dar		1961	
(D4134)	08904		0-6-0DE	Hor		1962	
(D4155)	08925		0-6-0DE	Hor		1962	b
D4164	08934		0-6-0DE	Dar		1962	b
(D5207)	25057		Bo-BoDE	Derby		1963	
D7633	(25283 25904)		Bo-BoDE	BP	8043	1965	
(D7663	25213 25313)		Bo-BoDE	Derby		1966	
(D8042	20042) 20312		Bo-BoDE	_(EE	2764	1959	
				(VF	D489	1959	
(D8069)	20069		Bo-BoDE	_(EE	2975	1961	
				(RSHD	8227	1961	
(D8095	20095) 20305		Bo-BoDE	_(EE	3001	1961	
				(RSHD	8253	1961	
(D8101	20101) 20901		Bo-BoDE	_(EE	3007	1961	
				(RSHD	8259	1961	c
(D8187	20187) 20308		Bo-BoDE	_(EE	3668	1967	
				(EEV	D1063	1967	
(D8325	20225) 20905		Bo-BoDE	_(EE	3706	1967	
				(EEV	D1101	1967	c
(E6010	73104)		Bo-BoDE/RE	_(EE	3572	1965	
73951				(EEV	E342	1965	
			reb	RVEL		2014	
(E6019	73113 73211)		Bo-BoDE/RE	_(EE	3581	1966	
73952				(EEV	E351	1966	
			reb	RVEL		2014	
3102	#BRUMSTAR		Bo-BoWE/RE	GEC-Alsthom		1992	
29			0-6-0DH	HE	7017	1971	
7189	GAFFER		0-6-0DH	HE	7189	1970	
2			0-6-0DH	Jung	12347	1956	
			reb	Newag	270	2015	
4			0-6-0DH	Jung	13286	1962	
			reb	Newag		2016	
1			0-6-0DH	Jung	13289	1962	
			reb	Newag	269	2015	
–			0-6-0DH	S	10186	1964	
			reb	HAB	6459	1989	
01545	(253)		4wDH	TH	271V	1977	
01546	(255)		4wDH	TH	273V	1977	

a	privately owned
b	property of GB Railfreight Ltd
c	on hire to Balfour Beatty

Additionally, vehicles are stored at Worksop Up Yard (SK 571804)

mainline and preserved vehicles for maintenance/overhaul/storage usually present

HARSCO RAIL LTD, UNIT 1, CHEWTON STREET, EASTWOOD,
NOTTINGHAM (part of the Harsco Corporation)   www.harscorail.com   **NG16 3HB**

Road/Rail vehicles under construction, repair, or for hire, are usually present.   **SK 475462**

NOTTINGHAM TRAMS LTD, NOTTINGHAM EXPRESS TRANSIT, WILKINSON STREET DEPOT, NOTTINGHAM

Gauge 4ft 8½in   www.thetram.net

NG7 7NW
SK 553423

–	4wBE		ACI	20699	2013
–	4wBE		Niteq	B183	2003
FG52 WCC	4wDM	R/R	_(Unimog	199635	2002
			(Zagro	3139	2002
GK03 ACX	4wDM	R/R	_(Unimog	201624	2003
			(Zweiweg		2003

SIMS METAL MANAGEMENT LTD, HARRIMANS LANE, DUNKIRK,
NOTTINGHAM (member of the Sims Group Ltd)   www.simsmm.co.uk

Locomotives for scrap are occasionally present at this location.

NG7 2SD
SK 549337

PRESERVATION SITES

BILSTHORPE HERITAGE MUSEUM,
rear of VILLAGE HALL, CROSS STREET, BILSTHORPE

Gauge 3ft 0in   www.bilsthorpemuseum.co.uk

NG22 8QY
SK 646608

(E2)	4wBEF		CE	B1504A	1977
		reb	CE	B3156	1984

K. & J. BOWNES & SONS, 1 CORNMILL FARM BUNGALOW,
off OWDAY LANE, WALLINGWELLS, WORKSOP

Gauge 4ft 8½in   www.kjbownes.co.uk

S81 8DA
SK 574838

–	0-6-0F	OC	WB	3019	1952	Pvd

Visitors by appointment only

CLIPSTONE HEADSTOCKS (operated by Nottingham Mining Heritage Centre CIC)
CLIPSTONE COLLIERY, MANSFIELD ROAD, CLIPSTONE, MANSFIELD

Gauge 2ft 6in   www.clipstoneheadstocks.com

NG21 9EH
SK 595632

No.5	4wBEF		_(EE	2300	1956
			(Bg	3436	1956

Gauge 2ft 0in

6 TURBO TED	4wDM		MR	9543	1950

GREAT CENTRAL RAILWAY (NOTTINGHAM) LTD,
NOTTINGHAM TRANSPORT HERITAGE CENTRE,
MERE WAY, RUDDINGTON, NOTTINGHAM

Gauge 4ft 8½in   www.gcrn.co.uk

NG11 6JS
SK 575322

61264 (1264)	4-6-0	OC	NBQ	26165	1947

–		2-8-0	OC	Alco	(70610	1943?)	Dsm
(411.09)		2-8-0	OC	BLW	69621	1943	Dsm
	"JULIA"	0-6-0ST	IC	HC	1682	1937	a
	(DOLOBRAN)	0-6-0ST	IC	MW	1762	1910	
No.15	(8310/41) (RHYL)	0-6-0ST	IC	MW	2009	1921	a
5	(ABERNANT)	0-6-0ST	IC	MW	2015	1921	a
(D417)	50017 ROYAL OAK	Co-CoDE		_(EE	3787	1968	
				(EEV	D1158	1968	
(D2118)	03118	0-6-0DM		Sdn		1959	
D3290	(08220)	0-6-0DE		Derby		1956	
(D3861)	08694	0-6-0DE		Hor		1959	
(D3952)	08784	0-6-0DE		Derby		1960	
(D5580)	31162	A1A-A1ADE		BT	180	1960	
(D5634)	31210	A1A-A1ADE		BT	234	1960	
(D5662)	31235	A1A-A1A DE		BT	262	1960	
(D6709)	37009 (37340)	Co-CoDE		_(EE	2872	1960	
				(VF	D588	1960	
D8154	(20154)	Bo-BoDE		_(EE	3625	1966	
				(EEV	D1024	1966	
43025	(253012)	Bo-BoDE		Crewe		1976	
43044	(253022) EDWARD PAXMAN	Bo-BoDE		Crewe		1977	
–		4wDM		RH	263001	1949	
D2959		0-4-0DE		RH	449754	1961	
H 014		0-6-0DH		RR	10262	1967	
(50645)	M53(645) (T061)	2-2w-2w-2DMR		DerbyC&W		1958	
50926	(W53926 977814)	2-2w-2w-2DMR		DerbyC&W		1959	
W51138	T003	2-2w-2w-2DMR		DerbyC&W		1958	
W51151	T004	2-2w-2w-2DMR		DerbyC&W		1958	
55803	144003	4wDMR		_(BRE(D)		1986	
				(Alex	2785/05	1986	
	rebuilt as	4wDHR		AB		1991	
55826	144003	4wDMR		_(BRE(D)		1986	
				(Alex	2785/06	1986	
	rebuilt as	4wDHR		AB		1991	
XLR 8104	R649 JFE	4wDM	R/R	_(Landrover		1997	
				(Raynesway		1997	
DX 68807		4wDMR		Perm	008	1986	
HCT 005	XOP 2298	4wDH		Perm	005	1987	

a   currently off site for restoration

SHERWOOD FOREST RAILWAY, SHERWOOD FOREST FARM PARK, LAMB PENS LANE, EDWINSTOWE, MANSFIELD   NG21 9HL

Gauge 1ft 3in   www.sherwoodforestrailway.com   **SK 586655**

No.1	SMOKEY JOE	0-4-0STT	OC	HardyK	01	1991	
2	PET	0-4-0ST	OC	HardyK	02	1998	
3	ANNE	4wBE		HardyK	E3	1993	
1		2-2wPM		Rosewall		1947	
			reb	Vanstone		1980	OOU
4	"GEORGE"	4wDH		SherwoodF		2005	
	"MILLIE"	4wDH		SherwoodF		2015	
	a rebuild of	2w-2DM		Longleat?			

SUNDOWN ADVENTURELAND,
TRESWELL ROAD, RAMPTON, near RETFORD   **DN22 0HX**
Gauge 1ft 6in   www.sundownadventureland.co.uk   **SK 793792**

–	2-4-0RE	s/o	SL	546.4.93	1993

NEIL WHITE, PRIVATE LOCATION
Gauge 4ft 8½in   **Private Site**

1677	2-2w-2w-2RER	Metcam		c1960

OXFORDSHIRE

INDUSTRIAL SITES

MINISTRY OF DEFENCE, DEFENCE STORAGE & DISTRIBUTION CENTRE,
BICESTER MILITARY RAILWAY
Gauge 4ft 8½in

(M61183) 977349 (936003) No.3	4w-4wRER	Afd/Elh		1957	a

a   preserved vehicle currently stored here

See also Section 6 for details.

SCIENTIFIC RESEARCH LABORATORY, PRIVATE LOCATION
Gauge 5ft 6in   **Private Site**

RLJ 7	2w-2CE	R/R	HU	1958
		reb	CE	B4611	2015

PRESERVATION SITES

BLENHEIM PALACE RAILWAY, PLEASURE GARDENS, WOODSTOCK   **OX20 1PP**
Gauge 1ft 3in   www.blenheimpalace.com   **SP 444163**

SIR WINSTON CHURCHILL	0-6-2DH	s/o	AK	39	1992
WINSTON	0-6-0DH	s/o	AK	94	2014

CHINNOR & PRINCES RISBOROUGH RAILWAY ASSOCIATION,
THE ICKNIELD LINE, STATION APPROACH, STATION ROAD, CHINNOR   **OX39 4ER**
Gauge 4ft 8½in   www.chinnorrailway.co.uk   **SP 756002**

4555 WARRIOR	2-6-2T	OC	Sdn	1924
D2069 (03069)	0-6-0DM		Don	1959
D3018 (13018 08011) HAVERSHAM	0-6-0DE		Derby	1953

(D3993 08825) 97808	0-6-0DE		Derby		1960
(D6927) 37227	Co-CoDE		_(EE	3413	1963
			(EEV	D871	1963
D8059 (20059)	Bo-BoDE		_(EE	2965	1961
			(RSHD	8217	1961
D8568	Bo-BoDE		CE	4365U/69	1963
43054 (253027)	Bo-BoDE		Crewe		1977
IRIS	0-6-0DH		RH	459515	1961
(51375) 977992 (960301)	2-2w-2w-2DMR		PSteel		1960
W55023	2-2w-2w-2DMR		PSteel		1960
W55024 (977858 960010)	2-2w-2w-2DMR		PSteel		1960
(61736) 1198	4w-4wRER		Elh		1960
(61737) 1198	4w-4wRER		Elh		1960
–	2w-2PMR		Bance	AL2025	1995
WD 9024	2w-2PMR		Wkm	7090	1955

CHOLSEY & WALLINGFORD RAILWAY, HITHERCROFT ROAD, WALLINGFORD

Gauge 4ft 8½in   www.cholsey-wallingford-railway.com

OX10 9DQ
SU 600891

12 "ISEBROOK"	4wVBT	VCG	S	6515	1926
(D3030 08022) LION	0-6-0DE		Derby		1953
(D3074 08060) 060 UNICORN	0-6-0DE		Dar		1953
(D3190 08123) "GEORGE MASON"	0-6-0DE		Derby		1955
CARPENTER	0-4-0DM		FH	3270	1948
No.1 WILLY SKUNK	2w-2PMR		Wkm	8774	1960

COTSWOLD WILD LIFE PARK, BRADWELL GROVE, BURFORD

Gauge 2ft 0in   www.cotswoldwildlifepark.co.uk

OX18 4JP
SP 237084

No.3	0-4-0DH	s/o	AK	17	1985
No.4 BELLA	0-4-0DH	s/o	AK	68	2003

COULSDON OLD VEHICLE & ENGINEERING SOCIETY, BUILDING D8, MOD GRAVEN HILL SITE, BICESTER

Gauge 4ft 8½in   (Closed)

Private Site

WD 40 FREDDY	4wDM		RH	458961	1962

R. DIXON, PRIVATE SITE, BICESTER

Gauge 2ft 6in

Private Site

(10)	4wDHF		HE	9053	1981	Dsm

Gauge 2ft 0in

No.3 CDC	0-4-2PT	OC	HE	3758	1952
–	4wDM		LB	54181	1964
DORIS	4wDM		RH	462365	1960
EL 9	4wBE		CE	5961C	1972
MBS 236	4wBE		LMM	1066	1950

GREAT WESTERN SOCIETY, DIDCOT RAILWAY CENTRE (OX11 7NJ)

Gauge 7ft 0¼in   www.didcotrailwaycentre.org.uk   SU 524906

	FIRE FLY	2-2-2	IC	FFP		1989	
	IRON DUKE	4-2-2	IC	Resco		1985	a

Gauge 4ft 8½in

1014	COUNTY OF GLAMORGAN	4-6-0	OC	_(Llangollen		2006	
				(GWS		2006	
	a rebuild of 7927 WILLINGTON HALL			Sdn		1950	
1338		0-4-0ST	OC	K	3799	1898	
1340	TROJAN	0-4-0ST	OC	AE	1386	1897	
1363		0-6-0ST	OC	Sdn	2377	1910	
1466	(4866)	0-4-2T	IC	Sdn		1936	
2999	LADY OF LEGEND	4-6-0	OC	GWS/Riley		2006	
	a rebuild of 4942 MAINDY HALL			Sdn		1929	b
3650		0-6-0PT	IC	Sdn		1939	
3738		0-6-0PT	IC	Sdn		1937	
3822		2-8-0	OC	Sdn		1940	
4079	PENDENNIS CASTLE	4-6-0	4C	Sdn		1924	
4144		2-6-2T	OC	Sdn		1946	
5051	EARL BATHURST	4-6-0	4C	Sdn		1936	
5227		2-8-0T	OC	Sdn		1924	
ROD 5322		2-6-0	OC	Sdn		1917	
5572		2-6-2T	OC	Sdn		1929	
5900	HINDERTON HALL	4-6-0	OC	Sdn		1931	
6023	KING EDWARD II	4-6-0	4C	Sdn		1930	
6106		2-6-2T	OC	Sdn		1931	
6697		0-6-2T	IC	AW	985	1928	
6998	BURTON AGNES HALL	4-6-0	OC	Sdn		1949	
7202		2-8-2T	OC	Sdn		1934	
7808	COOKHAM MANOR	4-6-0	OC	Sdn		1938	
No.5		0-4-0WT	OC	GE		1857	
No.2409	KING GEORGE	0-6-0T	IC	GWS		2022	
	a rebuild of	0-6-0ST	IC	HE	2409	1942	
	SIR ROBERT McALPINE & SONS (LONDON) LIMITED No.31						
		0-6-0ST	IC	HC	1026	1913	
No.1	BONNIE PRINCE CHARLIE	0-4-0ST	OC	RSHN	7544	1949	
93		4 + 0-4-0VBTR	OC	Sdn		1908	
			reb	_(TyseleyLW		2007	
				(Llangollen		2009	
18000		A1A-A1A	GTE	_(BBC(S)	4559	1949	
				(SLM	3977	1949	
D1023	WESTERN FUSILIER	C-CDH		Sdn		1963	
(D3771	08604) 604 PHANTOM	0-6-0DE		Derby		1959	
(D5800)	31270 ATHENA	A1A-A1ADE		BT	301	1961	
D9516		0-6-0DH		Sdn		1964	
(W22W)	No.22	Bo-BoDMR		Sdn		1940	
(5)		0-4-0PM		GWS		c1985	
	a rebuild of	0-4-0WE		DerbyC&W		1960	
DL26		0-6-0DM		HE	5238	1962	
(A 21 W)		2w-2PMR		Wkm	6892	1954	DsmT

(B 37 W	PWM 3963)	2w-2PMR		Wkm	6948	1955	DsmT
68007	PWM 4303	2w-2PMR		Wkm	7506	1956	

a   incorporates parts of RSHN 7135 / 1944
b   carries works plate Great Western Railway Co, Didcot Works 3540 / 2018

P. ROGERS, PRIVATE LOCATION
Gauge 4ft 8½in   **Private Site**

(A144	PWM 2176)	2w-2PMR		Wkm	4153	1946	OOU

JIM SHACKELL, WITNEY
Gauge 1ft 3in   **Private Site**

No.103	JOHN	4-4-2	OC	Barnes	103	1921	
5751	PRINCE WILLIAM	4-6-2	OC	G&S		1949	a
–		2-6-2	OC	LemonB		1967	u/c
5127		4-4-0	OC	McGarigle		c1924	
712		0-4-0	OC	Morse		1951	
–		4-4-0PM	OC	Morse		1939	

a   carries plate G&S 9/1946

RUTLAND

INDUSTRIAL SITES

HEIDELBERG MATERIALS CEMENT (HANSON CEMENT), KETTON WORKS, KETCO AVENUE, KETTON (part of the Hanson Heidelberg Cement Group)   PE9 3SX
Gauge 4ft 8½in   www.heidelbergmaterials.com   **SK 987057**

(D3789)	08622	H 028 19	0-6-0DE	Derby	1959	a
(D3977)	08809	24	0-6-0DE	Derby	1960	a

a   property of Rail Management Services Ltd, Chesterfield, Derbyshire

PRESERVATION SITES

RUTLAND RAILWAY MUSEUM LTD,
t/a ROCKS BY RAIL : THE LIVING IRONSTONE MUSEUM, COTTESMORE IRON ORE SIDINGS, ASHWELL ROAD, COTTESMORE, near OAKHAM   LE15 7FF
Gauge 4ft 8½in   www.rocks-by-rail.org   **SK 887137**

(RRM 2)		0-4-0ST	OC	AB	1931	1927
(RRM 6)	SIR THOMAS ROYDEN	0-4-0ST	OC	AB	2088	1940
(RRM 196)	BELVOIR	0-6-0ST	OC	AB	2350	1954
	(STAMFORD)	0-6-0ST	OC	AE	1972	1927
	"RHOS"	0-6-0ST	OC	HC	1308	1918

(RRM 5)	SINGAPORE	0-4-0ST	OC	HL	3865	1936		
RRM 1	UPPINGHAM	0-4-0ST	OC	P	1257	1912		
RRM 4	(ELIZABETH)	0-4-0ST	OC	P	1759	1928		
(RRM 183)	CRANFORD No.2	0-6-0ST	OC	WB	2668	1942		
9	EXTON PARK	0-6-0ST	OC	YE	2521	1952		
RRM 175		0-4-0DH		JF	4220007	1960		
(DS 1169)	IMP	4wDM		RH	207103	1941		
RRM 12		4wDM		RH	306092	1950		
(RRM 176)	ELIZABETH	0-4-0DE		RH	421436	1958		
RRM 150	ERIC TONKS	0-4-0DE		RH	544997	1969		
RRM 21	BETTY	0-4-0DH		RR	10201	1964		
	JEAN	0-4-0DH		RR	10204	1965		
(RRM 186)	GRAHAM	0-4-0DH		RR	10207	1965		
No.8		4wDH		TH	178V	1967		
(RRM 177)	No.4 MR.D	4wDH		TH	186V	1967		
(RRM 16)	DE 5	0-6-0DE		YE	2791	1962		
1382		0-6-0DE		YE	2872	1962		
68/038		2w-2DMR		Wkm	1947	1935		
	rebuilt as	2w-2DMR/BER		DonM		1983	DsmT	
(DB 965072)		2w-2PMR		Wkm	7587	1957		
–		2w-2PMR		Wkm		1956	DsmT	a

a   one of Wkm 7445/1956 or Wkm 7581/1956

SHROPSHIRE

INDUSTRIAL SITES

TELFORD and WREKIN COUNCIL, TELFORD INTERNATIONAL RAILFREIGHT PARK, DONNINGTON, TELFORD   **TF1 7GA**
Gauge 4ft 8½in   **SJ 693132**

–	4wBE	R/R	Niteq	B305	2010
	refurbished		BEAZ	M2011	2017

VLR TECHNOLOGIES LTD, REVOLUTION VERY LIGHT RAILWAY IRONBRIDGE, BUILDWAS ROAD, IRONBRIDGE   **TF8 7BL**
Gauge 4ft 8½in   www.revolutionvlr.com   **SJ 693132**

–	4w-4wBE/DER	TDI	2020

PRESERVATION SITES

CAMBRIAN HERITAGE RAILWAYS
Cambrian Railways Society, Oswestry
Locomotives can be found at :	Blodwell Junction, Llanyblodwel	SY10 8LZ	SJ 252230
	Oswestry Cycle & Railway Museum, Station Yard, Oswestry	SY11 1RE	SJ 294297
	Weston Wharf, Western Road, Morda	SY10 9ES	SJ 298276

Gauge 4ft 8½in   www.cambrianrailways.com

–		0-6-0ST	OC	AB	885	1900
–		0-4-0ST	OC	AB	2261	1949
	NORMA	0-6-0ST	IC	HE	3770	1952
	ADAM	0-4-0ST	OC	P	1430	1916
	OLIVER VELTOM	0-4-0ST	OC	P	2131	1951
E6036	(73129)	Bo-BoDE/RE		_(EE	3598	1965
				(EEV	E368	1965
7	HO37	0-6-0DH		EEV	D1201	1967
–		4wDM		FH	3541	1952
	ALPHA	4wDM		FH	3953	1960
–		0-4-0DM		HC	D843	1954
	SCOTTIE	4wDM		RH	412427	1957
11517	ALUN EVANS	0-4-0DE		RH	458641	1963
55806	144006	2-2wDMR		_(BRE(D)		1986
				(Alex	2785/11	1986
	rebuilt as	2-2wDHR		AB		1990
55829	144006	2w-2DMR		_(BRE(D)		1986
				(Alex	2785/12	1986
	rebuilt as	2w-2DHR		AB		1990
55807	144007	2-2wDMR		_(BRE(D)		1986
				(Alex	2785/13	1986
	rebuilt as	2-2wDHR		AB		1990
55830	144007	2w-2DMR		_(BRE(D)		1986
				(Alex	2785/14	1986
	rebuilt as	2w-2DHR		AB		1990
98205		4wDMR		Plasser	52760A	1985

Cambrian Railways Society, Weston Wharf, Morda   SY10 9ES
Gauge 2ft 0in   SJ 298276

496038	BARNY	4wDM	RH	496038	1963	Pvd
	LLANFFORDA	4wDM	RH	496039	1963	Pvd
7	VULCAN	4wDM	RH	7002/0967/6	1967	Pvd

Cambrian Railway Trust, Llynclys Goods Yard, Llynclys   SY10 8BX
Gauge 4ft 8½in   SJ 285239

D3019	(13019)	0-6-0DE	Derby		1953
–		0-6-0DH	EEV	D1230	1969
	TELEMON	0-4-0DM	_(VF	D295	1955
			(DC	2568	1955
W51187		2-2w-2w-2DMR	MetCam		1958
W51205		2-2w-2w-2DMR	MetCam		1958
W51512		2-2w-2w-2DMR	MetCam		1959
28690		4w-4wRER	DerbyC&W		c1939
–		2w-2PMR	Geismar	98/28	1998	a

a   stored at private site in Weston Rhyn

See also **Oswestry and District Narrow Gauge Group** (below)

G.FAIRHURST, c/o ENGLISH NATURE, MANOR HOUSE WORKS, near WHIXALL

Gauge 2ft 0in   SJ 505366

–		4wPM	MR	1934	1919	a
–		4wDM	RH	191679	1938	a

a   locomotives stored at a private location

HADLEY LEARNING COMMUNITY,
CRESCENT ROAD, HADLEY, TELFORD   TF1 5JU

Gauge 4ft 0in   www.hadleylearningcommunity.org.uk

–		4wG	TU		1987

IRONBRIDGE GORGE MUSEUM TRUST LTD
- IRONBRIDGE BIRTHPLACE OF INDUSTRY

Blists Hill Victorian Town, Legges Way, Madeley, Telford   TF7 5UD

Gauge 4ft 8½in   www.ironbridge.org.uk   SJ 693033

–	0-6-0ST	OC	AB	782	1896

Gauge 3ft 0in

–	4wG	OC	CastleGKN	1990	a

a   working replica of Trevithick loco, operates on plateway-type flanged rails.

Gauge 2ft 0in

SIR PETER GADSDEN	4wBE		AK	84	2008

Coalbrookdale Museum of Iron, Coach Road, Coalbrookdale   TF8 7QD

Locomotives on static display.

Gauge 4ft 8½in   SJ 668047

–	0-4-0VBT	VCG	S	6155	1925
a rebuild of	0-4-0ST	OC	MW	437	1873
–	0-4-0VBT	VCG	S	6185	1925
a rebuild of (6)	0-4-0ST	OC	Coalbrookdale	c1865	

Enginuity, Coach Road, Coalbrookdale   TF8 7DX

Gauge 4ft 8½in   SJ 667048

5	0-4-0ST	OC	Coalbrookdale	c1865	

PRIVATE OWNER, PRIVATE LOCATION

Gauge 4ft 8½in   **Private Site**

68019 (DE 320501)	2w-2PMR	Wkm	7598	1957	

OSWESTRY & DISTRICT NARROW GAUGE GROUP,
LLYNCLYS GOODS YARD, LLYNCLYS   SY10 8BX

Locomotives stored at various locations

Gauge 3ft 0in   SJ 285239

A	4wDMF	RH	433388	1959	Dsm

Gauge 2ft 6in

(ND 3051)		0-4-0DM		HE	2022	1939	Dsm

Gauge 2ft 0in

	IORWERTH	0-4-0VBT	VC	ACAB		2008		
	BUSTA	0-4-0PM		ACAB	1	2004	Dsm	
8		4wPM		KC		c1926		
89		4wDM		MR	4565	1928		
	(WEDHOLME)	4wDM		MR	8885	1944	Dsm	
–	"WARRIOR"	4wDM		RH	191680	1938		
–		4wDM		RH	452294	1960		
	(RATTY)	2w-2PM		BarberA			Dsm	a

a   currently stored off site

SEVERN VALLEY RAILWAY CO LTD

Locomotives are kept at :-

Bewdley, Worcestershire	DY12 1BG	SO 793753
Bridgnorth, Shropshire	WV16 5DT	SO 715926
Highley, Shropshire	(WV16 6NU)	SO 749831
Kidderminster, Worcestershire	DY10 1QX	SO 836757

Gauge 4ft 8½in   www.svr.co.uk

813		0-6-0ST	IC	HC	555	1900
1501		0-6-0PT	OC	Sdn		1949
2857		2-8-0	OC	Sdn		1918
4150		2-6-2T	OC	Sdn		1947
4930	HAGLEY HALL	4-6-0	OC	Sdn		1929
7325		2-6-0	OC	Sdn		1932
7714		0-6-0PT	IC	KS	4449	1930
7812	ERLESTOKE MANOR	4-6-0	OC	Sdn		1939
34027	TAW VALLEY	4-6-2	3C reb	Bton Elh		1946 1957
42968		2-6-0	OC	Crewe	136	1934
43106		2-6-0	OC	Dar	2148	1951
60532	BLUE PETER	4-6-2	3C	Don	2023	1948
75069		4-6-0	OC	Sdn		1955
82045		2-6-2T	OC	SVR		2015
	WARWICKSHIRE	0-6-0ST	IC	MW	2047	1926
(71516)	WELSH GUARDSMAN	0-6-0ST	IC	RSHN	7170	1944
	DUNROBIN	0-4-4T	IC	SS	4085	1895
	CATCH ME WHO CAN	2-2-0	IC	SVR		2008
D182 (46045) (97404)		1Co-Co1DE		Derby		1962
(D306) 40106		1Co-Co1DE		_(EE	2726	1960
	ATLANTIC CONVEYOR			(RSHD	8136	1960
(D407) 50007 HERCULES		Co-CoDE		_(EE	3777	1967
				(EEV	D1148	1967
(D431) 50031 HOOD		Co-CoDE		_(EE	3801	1968
				(EEV	D1172	1968
(D433) 50033 GLORIOUS		Co-CoDE		_(EE	3803	1968
				(EEV	D1174	1968
(D435) 50035 (50135)		Co-CoDE		_(EE	3805	1968
	(ARK ROYAL)			(EEV	D1176	1968

(D444)	50044 EXETER	Co-CoDE	_(EE	3814	1968	
			(EEV	D1185	1968	
(D449)	50049 (50149) DEFIANCE	Co-CoDE	_(EE	3819	1968	
	50014 WARSPITE		(EEV	D1190	1968	
D821	GREYHOUND	B-BDH	Sdn		1960	
D1013	WESTERN RANGER	C-CDH	Sdn		1962	
D1015	(89416) WESTERN CHAMPION	C-CDH	Sdn		1962	
D1048	WESTERN LADY	C-CDH	Crewe		1962	
D1062	WESTERN COURIER	C-CDH	Crewe		1963	
D3022	(08015)	0-6-0DE	Derby		1953	
(D3201	08133) 13201	0-6-0DE	Derby		1955	
D3586	(08471)	0-6-0DE	Crewe		1958	
(D3802	08635) H3802	0-6-0DE	Derby		1959	
(D4013	08845) 09107	0-6-0DE	Hor		1961	
D4100	(09012) DICK HARDY	0-6-0DE	Hor		1961	
(D4126)	08896 STEVEN DENT	0-6-0DE	Hor		1962	
(D6608	37274) 37308	Co-CoDE	_(EE	3568	1965	
			(EEV	D997	1965	
D7029		B-BDH	BPH	7923	1962	
(D8007)	20007	Bo-BoDE	_(EE	2354	1957	
			(VF	D382	1957	a
(D8048)	20048	Bo-BoDE	_(EE	2770	1959	
			(VF	D495	1959	
(D8305)	20205	Bo-BoDE	_(EE	3686	1967	
			(EEV	D1081	1967	a
D9551		0-6-0DH	Sdn		1965	
12099		0-6-0DE	Derby		1952	
D2957		0-4-0DM	RH	319290	1953	
D2960	SILVER SPOON	0-4-0DM	RH	281269	1950	
D2961		0-4-0DE	RH	418596	1957	
M50933		2-2w-2w-2DMR	DerbyC&W		1960	
M51941		2-2w-2w-2DMR	DerbyC&W		1960	
E52064		2-2w-2w-2DMR	DerbyC&W		1961	
(PPM 50	999900) 12	2w-2FER	PPM	12	2000	
(LMS 017)		2w-2PMR	Lesmac	LMS017	2006	
(DX 68811)	99709 018044-6 ETI 41	4wDHR	Perm	001	1987	
(PT 2P)		2w-2PMR	Wkm	1580	1934	
PWM 3189		2w-2PMR	Wkm	5019	1948	DsmT
DB 965054		2w-2PMR	Wkm	7577	1957	DsmT
(PT 1P	TP 49P)	2w-2PMR	Wkm	7690	1957	
8085	(9021)	2w-2PMR	Wkm	8085	1958	

a   property of Class 20189 Ltd, Ripley, Derbyshire

The Engine House, Highley   (WV16 6NU)
Gauge 4ft 8½in   SO 748829

4566		2-6-2T	OC	Sdn		1924
5764		0-6-0PT	IC	Sdn		1929
7819	HINTON MANOR	4-6-0	OC	Sdn		1939
46443		2-6-0	OC	Crewe		1950
47383		0-6-0T	IC	VF	3954	1926
48773	(8233 WD 70307)	2-8-0	OC	NBH	24607	1940
80079		2-6-4T	OC	Bton		1954

| WD 600 | GORDON | | 2-10-0 | OC | NBH | 25437 | 1943 | |
| | THE LADY ARMAGHDALE | | 0-6-0T | IC | HE | 686 | 1898 | |

M. SHILL, PRIVATE SITE, near SHREWSBURY

Gauge 4ft 8½in　　　　　　　　　　　　　　　　　　　　　　**Private Site**

| | BOB DARVILL | | 4wDM | | MR | 9921 | 1959 | |

SHREWSBURY STEAM TRUST, c/o SHROPSHIRE COUNCIL, SHREWSBURY MUSEUMS SERVICE, COLEHAM PUMPING STATION, LONGDEN COLEHAM, SHREWSBURY

SY3 7DN

Gauge 2ft 0in　　　www.shropshiremuseums.org.uk　　　　**SJ 496121**

| | – | | 2w-2BE | | TargettR | | 2017 | |

SHROPSHIRE MINES TRUST, SNAILBEACH LEAD MINES

Locomotives are kept elsewhere at a private location.　　　　**Private Site**

Gauge 1ft 11½in　　　www.shropshiremines.org.uk　　　　**SJ 375022**

	–		0-4-0BE		WR	S7950	1978	Dsm
	RED DWARF		0-4-0BE		WR	5655	1956	
		rebuilt as	4wBE		ShepherdFG		1980	

THE TANAT VALLEY LIGHT RAILWAY CO LTD, NANTMAWR VISITOR CENTRE, NANTMAWR BANK, NANTMAWR

SY10 9HW

Gauge 4ft 8½in　　　www.tanatvalleyrailway.co.uk　　　　**SJ 254244**

EE8416			4wDM		RH	338416	1953	
ARMY 251	FRANCIS BAILY OF THATCHAM		0-4-0DM		RH	390772	1956	
	–		4wDM		RH	416568	1957	
55642	143601		2-2wDMR		_(AB	669	1985	
					(Alex	1784/1	1985	
		rebuilt as	2-2wDHR		RFSD		1988	
55667	143601		2w-2DMR		_(AB	670	1985	
					(Alex	1784/2	1985	
		rebuilt as	2w-2DHR		RFSD		1988	
55657	143616		2-2wDMR		_(AB	709	1985	
					(Alex	1784/31	1985	
		rebuilt as	2-2wDHR		RFSD		1988	
55682	143616		2w-2DMR		_(AB	710	1985	
					(Alex	1784/32	1985	
		rebuilt as	2w-2DHR		RFSD		1988	
55660	143619		2-2wDMR		_(AB	713	1985	
					(Alex	1784/37	1985	
		rebuilt as	2-2wDHR		RFSD		1988	
55685	143619		2w-2DMR		_(AB	714	1985	
					(Alex	1784/38	1985	
		rebuilt as	2w-2DHR		RFSD		1988	
61937	(309616 977963) 960101		4w-4wWER		York		1962	
99709	901004-0 YELLOW PERIL		2w-2PMR		Lesmac	LMS014	2006	

Gauge 2ft 0in

14005	STEAM TRAM		4wVBT	G		1969
	rebuilt from		4wDM	L	14005	1940
39005			4wPM	L	39005	1952
(C37)			2w-2PMR	Locospoor B7281E		Dsm
	RAIL TAXI		4-2-0PMR	MorrisRP		1967
No.21			4wDM	RH	373359	1954
VP1200			4wPM	TeasdaleS		2011
PWM 2788	6887 (A16W)		2w-2PMR	Wkm	6887	1954
	rebuilt as		2w-2DMR	ENG/GEM		1994

The Rail Trolley Trust
Gauge 4ft 8½in   www.railtrolleytrust.co.uk

–		2w-2BER		Bance	054	1998	
MPP 9832		2w-2BER		Bance	00101	1996	
–		2w-2PMR		Geismar ST/99/37		1999	
–		2w-2PMR		Geismar			
	RTU 7417	2w-2BER		Geismar 5E/0001		2003	
–		2w-2PMR		Lesmac LMS005		2004	Dsm
99709	901003-2	2w-2PMR		Lesmac LMS013		2006	
(T003)	CEPS 68097 105099	4wDHR		Perm	T003	1988	
(DX 68805)	BRIAN	4wDMR		Perm	006	1986	
(DX 68806)		4wDMR		Perm	007	1986	
V205	(54187) JERRY M	2w-2DH		Perm	205	1980	
			reb	RVR		c2005	
TNS 105		2w-2DMR		Robel 56.27-10-AG36		1982	
–		2w-2PMR		Syl	14384		
(DE) 950021		2w-2PMR		Wkm	590	1932	
(DX 68088 966034)		2w-2DMR		Wkm	10842	1975	DsmT

Gauge monorail – The Monorail Collection

1795	HULL	1w1PM		RM	1795	1952	
2910	HAZLEMERE	1w1PM		RM	2910	1953	
3250	BOLTON (HEAP CLOUGH)	1w1PM		RM	3250	1953	
3906	WANTAGE	1w1PM		RM	3906	1954	
3981	(BISHOPS SUTTON)	1w1PM		RM	3981	1955	
4114	BOSTON 'A' (FOLDHILL)	1w1PM		RM	4114	1955	
4810	(BINFIELD)	1w1PM		RM	4810	1955	
4904	SACRISTON	1w1PM		RM	4904	1956	
(4989)	1989 STEYNING	1w1PM		RM	4989	1956	
5013	STROUD	1w1PM		RM	5013	1956	a Dsm
5041	BODMIN	1w1PM		RM	5041	1956	
(5074)		1w1PM		RM	5074	1956	b
(6560)		1w1PM		RM	6560	1957	b
6572	FELBRIDGE	1w1PM		RM	6572	1957	
7050	COOKHAM	1w1PM		RM	7050	1957	
7182	ESHER 'A'	1w1PM		RM	7182	1958	
7481	EDENBRIDGE	1w1PM		RM	7481	1958	
7493	HENFIELD	1w1PM		RM	7493	1958	
7498	HASLINGDEN	1w1PM		RM	7498	1958	
–		1w1PM	s/o	RM	7824	c1959	

8071	CHIPPING NORTON	1w1PM	RM	8071	1959		
	UCKFIELD 'A'	1w1PM	RM	8073	1959		Dsm
8118	(GODSTONE)	2wPH	RM	8118	1959		
8415	BINFIELD 'B'	1w1PM	RM	8415	1959		Dsm
–		2wDH	RM	8423	1959	a	
BATTERY ELECTRIC MONORAIL TRAM LOCO		1w1PM	RM	8564	1959		
	rebuilt as	1w1BE	PilbeamA				Dsm
8583	SHENFIELD	2wPH	RM	8583	1959		
–		1w1PM	RM	8611	1959		Dsm
8633	ESHER 'B'	1w1PM	RM	8633	1959		
8638	SLEAFORD	1w1PM	RM	8638	1960		
8653	(CHERTSEY)	2wPH	RM	8653	1959		
8663	(LLANGEFNI)	2wPH	RM	8663	1960		
8862	"UCKFIELD 'B'"	2wPH	RM	8862	1960		
8992	FAVERSHAM	1w1PM	RM	8992	1960		
9391	(WANSTEAD)	2wPH	RM	9391	1960		
9392	(WOODFORD)	2wPH	RM	9392	1960		
9536	CREWKERNE	1w1PM	RM	9536	1961		
9537	(KIDLINGTON)	1w1PM	RM	9537	1961		
9795	(NAYLAND)	2wPH	RM	9795	1960		
9811	(ESHER 'C')	2wPH	RM	9811	1961		
9812	(ESHER 'D')	2wPH	RM	9812	1961		
9852	TADLEY	2wPH	RM	9852	1961		
9853	"INGRAVE 'A'"	2wPH	RM	9853	1961		
9869	UCKFIELD 'C'	2wPH	RM	9869	1961		
9925	LLANGOLLEN	2wPH	RM	9925	1961		
10051	GILLINGHAM	1w1PM	RM	10051	1961		
10067	CAMBERLEY	2wPH	RM	10067	1961		
10069	(LIVERPOOL 'A')	2wPH	RM	10069	1961		
–		2wPH	RM	10070	1961	a	Dsm
	(SPALDING 'A')	2wGasH	RM	10248	1961	c	
10258	HESWALL	2wPH	RM	10258	1961		
10260	"EARLEY"	2wPH	RM	10260	1961		
	OWSTON FERRY	1w1PM	RM	10268	1961		
10422	HALE	2wPH	RM	10422	1961		
10593	SPALDING 'B'	2wGasH	RM	10593	1962	c	
10744		1w1PH	RM	10744	1962		
10782	HARWELL	1w1PH	RM	10782	1962		
10809	STORRINGTON	1w1PH	RM	10809	1962		OOU
–	(HULL 'B')	1w1PH	RM	10894	1962		Dsm
10900	BOURNEMOUTH 'A'	1w1PH	RM	10900	1962		OOU
10952	(HAVERHILL)	1w1PH	RM	10952	1962		OOU
11097	(AYLESFORD)	1w1PH	RM	11097	1962		OOU
11204	OXFORD	1w1PH	RM	11204	1963		OOU
11206	(EYNSHAM)	1w1PH	RM	11206	1963		OOU
–		1w1PH	RM	11338	1963		Dsm
–		1w1PH	RM	11451	1963		Dsm
11457	(WEST CALDER)	1w1PH	RM	11457	1963		OOU
"11459"	(TANKERSLEY)	1w1PH	RM	11459	1963		OOU
"11460"	ASHINGTON	1w1PH	RM	11460	1963		OOU
11809	HECKINGTON	1w1PH	RM	11809	1963	d	
11836	(BOWBURN)	1w1PH	RM	11836	1963		OOU

12124	WICKHAM	1w1PH	RM	12124	1964		OOU
12126	CHELTENHAM	1w1PH	RM	12126	1964		
12432	(HOWDEN)	1w1PH	RM	12432	1964		
"12438"	(DURHAM)	1w1PH	RM	12438	1964		OOU
–		1w1DH	RM	12625	1964		Dsm
–		1w1DH	RM	12626	1964		Dsm
12634	(HULL 'C')	1w1DH	RM	12634	1964		OOU
12649	MOSTYN	1w1GasH	RM	12649	1964		OOU
12713	ABINGDON	1w1PM	RM	12713	1964		
12869	(SWINDON)	1w1PH	RM	12869	1964		OOU
	LONDON 'A'	2wPH	RM	13300	1965	d	
	LONDON 'B'	2wPH	RM	13301	1965	d	
13313	SAUNDERSFOOT	2wPH	RM	13313	1965	d	
13318	"LIVERPOOL 'B'"	2wPH	RM	13318	1965		OOU
"13626"	BOURNEMOUTH 'B'	2wPH	RM	13626	1965		OOU
13629	LANCHESTER	2wPH	RM	13629	1965		
"13911"	(ESHER 'E')	2wPH	RM	13911	1965		OOU
13913	SWANWICK	2wPH	RM	13913	1965		OOU
13929	(INGRAVE 'B')	2wPH	RM	13929	1965		OOU
14753	WITTON GILBERT	2wPH	RM	14753	1966		OOU
14766	(WHARNCLIFFE-SIDE)	2wPH	RM	14766	1966		
	BANBURY	2wDH	RM	14769	1966		Dsm
14789	LEESWOOD	2wPH	RM	14789	1966		
14806	TIPTREE	2wPH	RM	14806	1966		OOU
15152	"WESTBURY"	2wDH	RM	15152	1967		OOU
15773	PENTRE HALKYN	2wPH	RM	15773	1967	d	
15778	(HOLME UPON SPALDING MOOR)	2wPH	RM	15778	1967		
15784	(CONINGSBY)	2wDH	RM	15784	1967	e	
No.2	MEXBOROUGH	2wDH	Metalair	20002	1968		
3	LINCOLN 'A'	2wDH	Metalair	20003	1968		
4	(BRENTWOOD 'A')	2wPH	Metalair	20004	1968		
5	(BRENTWOOD 'B')	2wPH	Metalair	20005	1968		
20007	FARINGDON	1w1DH	Metalair	20007	1968		
17	(MONO TANKER 'A')	2wDH	Metalair	20017	1968		
18	MONO TANKER 'B'	2wDH	Metalair	20018	1968		
19	LINCOLN 'B'	2wDH	Metalair	20019	1968		
25	SKELLINGTHORPE	2wDH	Metalair	20025	1968		
No.39	BROUGHTON ASTLEY	2wPH	Metalair	20039	1968		Dsm
44	CHIRK	2wPH	Metalair	20044	1968		
65		2wDH	Metalair	20065	1968	f	
	SHEPSHED	2wDH	Metalair	20067	1968		
69		2wDH	Metalair	20069	1968	g	
20091	PARTINGTON	1w1DH	Metalair	20091	1968		
No.140	RAILCAR A	2wDH	Metalair	20140	1969	e	
No.141	RAILCAR B	2wDH	Metalair	20141	1969	e	
No.146	MONOLOCO	0-2-0ST	OC	CM	11003	1998	h
161	LLANGEFNI (WITHERNSEA)	2wPH	Metalair	20161	1969	d	
No.212	POOLE	2wDH	Metalair	20212	1970		
222	TREFNANT	2wPH	Metalair	20222	1971		
20228		2wDH	Metalair	20228	1970		
20240	(SALISBURY)	2wDH	Metalair	20240	1971		

	DOWNTON	2wDH		Metalair	20242	1971	
265	BOSTON 'B'	2wDH		Metalair	20265	1973	i

a   converted to unpowered wagon
b   converted to permanent way manrider
c   horticultural flat power wagon
d   hydraulic tip power wagon
e   power wagon
f   converted to passenger carriage
g   converted to barrel carrying wagon
h   built on chassis of trailer wagon Metalair 20146/1969
i   converted to railcar 2002

TELFORD HORSEHAY STEAM TRUST, TELFORD STEAM RAILWAY, THE OLD LOCO SHED, BRIDGE ROAD, HORSEHAY, TELFORD   TF4 2NF

Gauge 4ft 8½in   www.telfordsteamrailway.co.uk   SJ 675073

5619		0-6-2T	IC	Sdn		1925	
"139	BEATTY"	0-4-0ST	OC	HL	3240	1917	OOU
	ROCKET	0-4-0ST	OC	P	1722	1926	
3		0-4-0ST	OC	P	1990	1940	
(D2051)		0-6-0DM		Don		1959	
(D3925)	08757 EAGLE	0-6-0DE		Hor		1961	
(D6963)	37263	Co-CoDE		_(EE	3523	1964	
				(EEV	D952	1964	
27414	TOM	0-4-0DH		NBQ	27414	1954	
183062	FOLLY	4wDM		RH	183062	1937	Dsm
VL 6	HECTOR	4wDM		RH	382824	1955	
525947		0-4-0DH		RH	525947	1968	
55545	142004	4wDMR		_(BRE(D)		1985	
				(Leyland R5.12	1985	
	rebuilt as	4wDHR		AB		1990	
55595	142004	4wDMR		_(BRE(D)		1985	
				(Leyland R5.11	1985	
	rebuilt as	4wDHR		AB		1990	
55708	142058	4wDMR		_(BRE(D)		1986	
				(Leyland R5.152	1986	
	rebuilt as	4wDHR		AB		1990	
55754	142058	4wDMR		_(BRE(D)		1986	
				(Leyland R5.155	1986	
	rebuilt as	4wDHR		AB		1990	
55813	144013	4wDMR		_(BRE(D)		1986	
				(Alex 2785/25	1986	
	rebuilt as	4wDHR		BRE(D)		1988	
55836	144013	4wDMR		_(BRE(D)		1986	
				(Alex 2785/26	1986	
	rebuilt as	4wDHR		BRE(D)		1988	
T111725	99709 975006-6	2-4wDMR	R/R	_(John Deere 029255			
				(Harsco DP270/005			

Gauge 2ft 0in - Telford Town Tramway   SJ 674072

–		4wVBTram	VCG	_(Kierstead		1979	
				(AK		1979	
–		4wDM		RH	222101	1943	Dsm
–		4wDM		RH	229633	1944	

SOMERSET

INDUSTRIAL SITES

AGGREGATE INDUSTRIES UK LTD,
MEREHEAD STONE TERMINAL, TORR WORKS, SHEPTON MALLET   BA4 4RA
(part of Holcim Group; operated by Mendip Rail Ltd)

Gauge 4ft 8½in   www.aggregate.com   **ST 693426**

(D3810)	08643		0-6-0DE	Hor		1959	
(D3955	08787)	08296	0-6-0DE	Derby		1960	+
(D4163)	08933		0-6-0DE	Dar		1962	+
277			6wDE	GECT	5475	1978	a
44	WESTERN YEOMAN II		Bo-BoDE	GM	798033-1	1980	

    +   currently at Hunslet Ltd, Barton-under-Needwood, Staffordshire
    a   property of Ed Murray & Sons Ltd, Hartlepool, Teesside

HEIDELBERG MATERIALS AGGREGATES,
(HANSON AGGREGATES), WHATLEY QUARRY, near FROME   BA11 3LF
(part of the Heidelberg Cement Group; operated by Mendip Rail Ltd)

Gauge 4ft 8½in   www.heidelbergmaterials.com   **ST 733479**

(D3817)	08650	0-6-0DE	Hor		1959
(D4177)	08947	0-6-0DE	Dar		1962
120	KENNETH JOHN WITCOMBE	4w-4wDE	GM	37903	1972

PRESERVATION SITES

BLATCHFORD LIGHT RAILWAY,
EMBOROUGH QUARRY, EMBOROUGH, near WELLS
This is a private railway, not open to the public.

Gauge 2ft 6in   **Private Site**

2	YARD No. B6	0-4-0DM		HE	2266	1940
ND 3060		0-4-0DM		HE	2398	1941
	YARD No.1075	4wDH		HE	7447	1976
	YARD No.1074	4wDH		HE	7451	1976
10		4wDHF	RACK	HE	9057	1981
–		4wDM		RH	398101	1956

Gauge 750mm

–		4wDH?	GIA	55134	1966

Gauge 2ft 4in

	CORRIS	4wDM	RH	398102	1956

Mr. BOND, YEOVIL AREA

Gauge 3ft 0in   **Private Site**

| – | | 4wDM | | JF | 3930048 | 1951 |

Stored at unknown location

PAUL CHANT, PILTOWN FARM, WEST PENNARD, GLASTONBURY   BA6 8NQ

Gauge 4ft 8½in   ST 556390

| 2 | | 4wDH | | TH | 136C | 1964 |
| | a rebuild of | 4wVBT | VCG | S | | |

CLEVEDON MINIATURE RAILWAY, SALTHOUSE FIELDS, CLEVEDON   BS21 7RH

Gauge 1ft 3in   Closed   ST 397710

| 5305 | | 4-6-0BE | s/o | MossAJ | | 1999 |
| | | | reb | LSLL | | 2021 |

Railway is currently closed with the locomotive stored off site at an unknown location

EAST SOMERSET RAILWAY CO LTD,
WEST CRANMORE RAILWAY STATION, SHEPTON MALLET   BA4 4QP

Locomotives are kept at :-   West Cranmore Station   ST 664429
                                                     Merryfield Lane Station   ST 654425

Gauge 4ft 8½in   www.eastsomersetrailway.com

4110		2-6-2T	OC	Sdn		1936	
4247		2-8-0T	OC	Sdn	2637	1916	
46447		2-6-0	OC	Crewe		1950	
1719	LADY NAN	0-4-0ST	OC	AB	1719	1920	
No.31		0-6-0T	IC	RSHN	7609	1950	Dsm
(51909	L231)	2-2w-2w-2DMR		DerbyC&W		1960	
DH16		4wDH		S	10175	1964	
199		4wDH		RR	10199	1964	
39	"CABOT"	0-6-0DH		RR	10218	1965	
–		0-6-0DH		RR	10221	1965	
2	JOAN	0-4-0DH		S	10165	1964	
(W51947)		2-2w-2w-2DMR		DerbyC&W		1961	Dsm
–		2-2wBER		Cranmore		c2015	

GARTELL LIGHT RAILWAY,
COMMON LANE, YENSTON, near TEMPLECOMBE   BA8 0NB

Gauge 2ft 0in   www.newglr.weebly.com   ST 718218

| No.6 | MR.G | 0-4-2T | OC | NDLW | 698 | 1998 |
| 9 | JEAN | 0-4-0TT | OC | NDLW | | 2008 |

8	FAITH	0-4-2T	OC	UphillJ		2016	
No.5	ALISON	4wDH		AK	No.10	1983	
No.2	ANDREW	4wDH		BD	3699	1973	
No.1	AMANDA	Bo-BoDH		Gartell		2003	a
–		4wDM		RH	193984	1939	b

a incorporates parts from SL 7323 1973
b converted into a brake van

O. JANES, PRIVATE LOCATION
Gauge 1ft 3in   **Private Site**

	MOUNTAINEER	0-4-0TT	OC	WVanHeiden		1972
–		0-6-0DH	s/o	MossAJ		2022

SANDFORD STATION RAILWAY HERITAGE CENTRE LTD,
ST. MONICA TRUST RETIREMENT VILLAGE,
STATION ROAD, SANDFORD, WINSCOMBE   **BS25 5AA**
Gauge 4ft 8½in   www.sandfordstation.co.uk   **ST 416595**

	MARJORIE	4wVBT	VCG	S	9387	1948

SOMERSET & DORSET RAILWAY HERITAGE TRUST, MIDSOMER NORTON
SOUTH STATION, SILVER STREET, MIDSOMER NORTON   **BA3 2EY**
Gauge 4ft 8½in   www.sdjr.co.uk   **ST 664536**

	JOYCE	4wVBT	VCG	S	7109	1927
D4095	(08881) 881	0-6-0DE		Hor		1961
D1120	DAVID J. COOK	0-6-0DH		EEV	D1120	1966
(Sc52006)		2-2w-2w-2DMR		DerbyC&W		1961
(Sc52025)		2-2w-2w-2DMR		DerbyC&W		1961
B40W	(PWM 4301 TR18)	2w-2PMR		Wkm	7504	1956

SOUTH WEST MAIN LINE STEAM COMPANY plc,
YEOVIL RAILWAY CENTRE, YEOVIL JUNCTION STATION, YEOVIL   **BA22 9UU**
Gauge 4ft 8½in   www.yeovilrailway.freeservers.com   **ST 570140**

	LORD FISHER	0-4-0ST	OC	AB	1398	1915
	PECTIN	0-4-0ST	OC	P	1579	1921
–		0-4-0DM		JF	22898	1940	Dsm
	SAM	0-4-0DM		JF	22900	1941
44	COCKNEY REBEL	0-4-0DM		JF	4000007	1947
DS1174	RIVER YEO	4wDM		RH	458959	1961

THIRTY INCH RAILWAYS LTD, CHARD, SOMERSET
Gauge 1ft 0in   **Private site**

D1	GREEN DRAGON	4w-4wDH		WhalleyB	2009

WEST SOMERSET RAILWAY plc

Locomotives are kept at :-

Bishops Lydeard	TA4 3RU	ST 164290
Dunster	TA24 6PJ	SS 996447
Minehead	TA24 5BG	SS 975463
Williton Goods Yard	TA4 4RQ	ST 085416

Gauge 4ft 8½in   www.west-somerset-railway.co.uk

4561		2-6-2T	OC	Sdn		1924	
7828	ODNEY MANOR	4-6-0	OC	Sdn		1950	
9351		2-6-0	OC	WSR		2004	
	rebuilt from 5193	2-6-2T	OC	Sdn		1934	
9466		0-6-0PT	IC	RSHN	7617	1952	
80064		2-6-4T	OC	Bton		1953	
	FORESTER	0-4-0ST	OC	AB	1260	1911	
–		0-4-0F	OC	AB	1984	1930	
	LLANTANAM ABBEY	0-6-0ST	OC	AB	2074	1939	
	VICTORY	0-4-0ST	OC	AB	2201	1945	
D1010	WESTERN CAMPAIGNER	C-CDH		Sdn		1962	
D2133		0-6-0DM		Sdn		1960	
D4107	(09019)	0-6-0DE		Hor		1961	
(D)6566	(33048)	Bo-BoDE		BRCW	DEL170	1961	
D6575	(33057)	Bo-BoDE		BRCW	DEL179	1961	
D7017		B-BDH		BPH	7911	1961	
D7018		B-BDH		BPH	7912	1961	
(D9518)	9312/95 No.7	0-6-0DH		Sdn		1964	
D9526		0-6-0DH		Sdn		1964	
–		0-4-0DH		AB	578	1972	
–		0-4-0DH		AB	579	1972	
200793	GOWER PRINCESS	4wDM		RH	200793	1940	
W51354		2-2w-2w-2DMR		PSteel		1960	
51663		2-2w-2w-2DMR		DerbyC&W		1960	a Dsm
51859	859	2-2w-2w-2DMR		DerbyC&W		1960	
51880	880	2-2w-2w-2DMR		DerbyC&W		1960	
(51887)		2-2w-2w-2DMR		DerbyC&W		1960	
4162		2w-2PMR		Geismar ST/02/27		2002	
4163		2w-2PMR		Geismar ST/02/28		2002	

a   frame used by Permanent Way Department as a flat wagon

WESTONZOYLAND ENGINE TRUST, WESTONZOYLAND PUMPING STATION, MUSEUM OF STEAM POWER & LAND DRAINAGE, HOOPERS LANE, WESTONZOYLAND, near BRIDGWATER   TA7 OLS

Gauge 2ft 0in   www.wzlet.org   ST 340328

–		4wVBT	VCG	DonnellyN		2018
	BOOTHBY	4wPM		FH	1830	1933
–		4wPM		L	6299	1935
–		4wDM		L	34758	1949
–		4wDM		L	42319	1956
87030		4wDM		MR	40S310	1968
–		4wDM		RH	235711	1945
SP 88		4wBE		CE	B0402A	1974

STAFFORDSHIRE

INDUSTRIAL SITES

ALSTOM TRANSPORT (part of Alstom Holdings S.A.),
CENTRAL RIVERS DEPOT, BARTON-UNDER-NEEDWOOD   **DE13 8ES**
Gauge 4ft 8½in   www.alstom.com   **SJ 203177**

(D4173) 08943		0-6-0DE	Dar	1962	a

     a   on hire from Harry Needle Railroad Co Ltd, Derbyshire

CLAYTON EQUIPMENT LTD,
SECOND AVENUE, CENTRUM 100, BURTON-UPON-TRENT   **DE14 2WF**
New CE locomotives under construction / overhaul / repair usually present.
Gauge 2ft 0in   www.claytonequipment.co.uk   **SK 225221**

	DIANE	4wBE	CE	RD001	2016	a

     a   demonstration locomotive

DB CARGO MAINTENANCE LTD, WHEILDON ROAD, STOKE-ON-TRENT   **ST4 4HP**
Gauge 4ft 8½in   www.uk.dbcargo.com   **SJ 880439**

P403D	DENISE	4wDH	S		10029	1960	a
99709 979117-7		4wDH	R/R	Zephir	2930	2021

     a   property of Railway Support Services Ltd, Wishaw, Warwickshire

ELECTRO-MOTIVE DIESEL LTD, LONGPORT GOODS YARD, BROOKSIDE
INDUSTRIAL ESTATE, off STATION STREET, LONGPORT   **ST6 4NF**
(Subsidiary of Progress Rail; A Caterpillar Company)
Gauge 4ft 8½in   www.progressrail.com   **SJ 855495**

(D3685)	08523	H 061	0-6-0DE	Don		1958	a
(66048)			Co-CoDE	GMC	968702-48	1998	Dsm

     a   property of Rail Management Services Ltd, Chesterfield, Derbyshire

F.M.B. / T.G.S., PRIVATE SITE
Locomotives under repair or restoration are usually present.
Gauge 2ft 6in   **Private Site**

(2	LENA)	0-4-2ST	OC	KS	1098	1910
(6	LUCY)	0-4-2ST	OC	KS	1313	1916
–		0-4-2ST	OC	KS	3025	1917
–		4wDM		Plymouth		1939
	a rebuild of	4wPM		Plymouth	513	1918

HUNSLET LTD t/a HUNSLET ENGINE COMPANY (part of the Ed Murray Group),
GRAYCAR BUSINESS PARK, BARTON-UNDER-NEEDWOOD   **DE13 8EN**
Locomotives under construction / repair / overhaul / for resale and hire usually present.
The following is a FLEET LIST of the locomotives owned by this contractor.
Gauge 4ft 8½in   www.hunsletengine.co.uk   **SK 208187**

(D3516)	08401		0-6-0DE		Derby		1958	
(D3560)	08445		0-6-0DE		Derby		1958	a
(D3587)	08472		0-6-0DE		Crewe		1958	b
(D3782)	08615	UNCLE DAI	0-6-0DE		Derby		1959	c
(D3991)	08823	KEVLA	0-6-0DE		Derby		1960	c
10	AVON		0-6-0DH		AB	600	1976	d
11	FORTH		0-6-0DH		AB	649	1980	
				reb	HE	750215	2006	d
	(SAM)		0-6-0DH		AB	659	1982	
				reb	HAB	6768	1990	
				reb	HE		2004	e f
	(EMILY)		0-6-0DH		AB	660	1982	
				reb	HAB	6769	1990	
				reb	HE		2004	e g
	RACHAEL		0-6-0DH		HE	7003	1971	
	LAURA		0-6-0DH		HE	8805	1978	h
	EMMA		0-6-0DH		HE	8902	1978	j
	BLUEBIRD		0-6-0DH		HE	8998	1981	e
	REDWING		0-6-0DH		HE	8999	1981	j
	PATRICK D. DUGGAN		0-6-0DH		HE	9386	2015	
		rebuild of			EEV	D1226	1967	k
	GRAHAM LEE JNR		0-6-0DH		HE	9387	2015	
		rebuild of			EEV	_(D1249	1968	
					(	3947	1968	k
DH50-2	(TAZ)		0-6-0DHF		TH	246V	1973	
				reb	HE	9377	2011	l
	EDWARD		4wDH		TH	267V	1976	m
DH50-1	(SAMMY)		0-6-0DH		TH	278V	1978	
				reb	HE	9376	2011	l
	–		0-6-0DH		TH	290V	1980	
				reb	HE	9382	2012	
293V	(01575)		0-6-0DH		TH	293V	1980	n
	STEELMAN DH75		6wDH		TH	V325	1987	

a   currently at Prologis RFI, DIRFT (Daventry), Crick, Northamptonshire
b   currently at Aggregate Industries UK Ltd, Bardon Hill, Leicestershire
c   currently at Tata Steel Europe, Shotton Works, Flintshire
d   currently at INEOS, Grangemouth, Scotland
e   currently at ICL UK (Cleveland) Ltd, Tees Dock, Teesside
f   carried ABG 660 plates from 2004 to 2007
g   carried ABG 659 plates from 2004 to 2007
h   currently at Nemesis Rail, Burton-upon-Trent, Staffordshire
j   currently at Reid Freight Services Ltd, Cinderhill IE, Stoke-on-Trent, Staffordshire
k   currently at Tarmac plc, Tunstead Quarry, Buxton, Derbyshire
l   currently at Tata Steel Europe, Trostre Works, Llanelli, South Wales
m   currently at Puma Energy (UK) Ltd, Milford Haven, South Wales
n   currently at Reid Freight Services Ltd, Cockshute Sidings, Stoke-on-Trent,

Locomotives present for overhaul :
Gauge 4ft 8½in

(D3692)	08530		0-6-0DE		Dar		1959	$	**M**

| (D3955 | 08787) | 08296 | 0-6-0DE | Derby | 1960 | + |
| (D4163) | 08933 | | 0-6-0DE | Dar | 1962 | + |

$ property of Freightliner Group Ltd
+ property of Mendip Rail Ltd

LAND RECOVERY LTD, COPP LANE, LONGTON, STOKE-ON-TRENT

ST6 4NU

Gauge 4ft 8½in   www.landrecovery.co.uk

SJ 851507

| (01581) LRR 041 | BUTTERCUP | 4wDM | R/R | Trackmobile | | |
| | TM 4150 MAGNUM | | | LGN971310798 | 1998 | |

LEESE BROS. (ECCLESHALL) LTD,
SCRAP METAL MERCHANTS, PLATT BRIDGE, ECCLESHALL

ST21 6EN

Gauge 2ft 0in

SJ 810292

1	4wDH	Schöma	6154	2007
2	4wDH	Schöma	6155	2007
3	4wDH	Schöma	6156	2007
4	4wDH	Schöma	6157	2007
5	4wDH	Schöma	6158	2007

Gauge 600mm

| – | 4wDH | Schöma | 6299 | 2008 |
| – | 4wDH | Schöma | 6300 | 2008 |

NEMESIS RAIL LTD, BURTON RAIL DEPOT,
DERBY ROAD, BURTON-UPON-TRENT

DE14 1RS

Locomotives for maintenance, overhaul and storage usually present

Gauge 4ft 8½in   www.nemesisrail.com

SK 250244, SK 251244

(D61)	45112	ROYAL ARMY ORDNANCE CORPS				
		1Co-Co1DE	Crewe		1962	
(D1713)	47488		Co-CoDE	BT	475	1964	
(D1921)	47244	47640 UNIVERSITY OF STRATHCLYDE				
		Co-CoDE	BT	683	1965	
(D1927)	47250	47600) 47744	Co-CoDE	BT	689	1966	
(D1932)	47493)	47701	Co-CoDE	BT	694	1966	
(D2324)	2324		0-6-0DM	_(RSHD	8183	1961	
				(DC	2705	1961	a
D3236	(08168 | 13236)	0-6-0DE	Dar		1956	
(D3557 | 08442)	0042	0-6-0DE	Derby		1958	Dsm
(D3588)	08473		0-6-0DE	Crewe		1958	Dsm
(D3610)	08495		0-6-0DE	Hor		1958	
(D3670)	09006		0-6-0DE	Dar		1959	a
(D3693)	08531		0-6-0DE	Dar		1959	b
(D3742)	08575		0-6-0DE	Crewe		1959	b OOU
(D3878)	08711		0-6-0DE	Crewe		1960	a
(D3973)	08805	HUNSLET	0-6-0DE	Derby		1960	c
(D4102)	09014		0-6-0DE	Hor		1961	a
(D4148)	08918		0-6-0DE	Hor		1962	a
D5217	(25067)		Bo-BoDE	Derby		1963	

(D5304)	26004	Bo-BoDE	BRCW	DEL49	1958	
(D5311)	26011	Bo-BoDE	BRCW	DEL56	1959	
(D5546)	31128 CHARYBDIS	A1A-A1ADE	BT	145	1959	
(D5547	31129) 31461	A1A-A1ADE	BT	146	1959	
(D5581	31163) 97205	A1A-A1ADE	BT	181	1960	
(D5654	31228) 31454 (31554)	A1A-A1ADE	BT	254	1960	
(D6514)	33103 SWORDFISH	Bo-BoDE	BRCW	DEL106	1960	d
(D6574)	33019 GRIFFON	Bo-BoDE	BRCW	DEL126	1960	
(D6790	37090) 37508 (37606)	Co-CoDE	_(EE	3217	1962	
			(RSHD	8336	1962	
(D6955)	37255	Co-CoDE	_(EE	3512	1964	
			(EEV	D943	1964	
(D7615)	25265	Bo-BoDE	Derby		1966	
(D8041	20041) 20904	Bo-BoDE	_(EE	2763	1959	
			(VF	D488	1959	a
(D8083	20083) 20903	Bo-BoDE	_(EE	2989	1961	
			(RSHD	8241	1961	a
(D8127	20127) 20303	Bo-BoDE	_(EE	3033	1962	
			(RSHD	8284	1962	a
(E6020)	73114	Bo-BoDE/RE	_(EE	3582	1965	
			(EEV	E352	1965	
X7202	KEMIRA 2 52	0-6-0DH	EEV	D1233	1968	a
33		0-6-0DH	EEV-AEI	3998	1970	a
–		0-6-0DH	EEV-AEI	5352	1971	a
	–	4wDE	_(Express	ES402	2009	
			(Hegenscheidt	101470	2009	
X7215	KEMIRA 1 51	0-6-0DH	GECT	5380	1972	a
–		0-6-0DH	HE	7041	1971	a
28	3D 63/000/316	0-6-0DH	HE	7181	1970	a
	LAURA	0-6-0DH	HE	8805	1978	e
H 013		0-4-0DH	S	10137	1962	a
9		0-6-0DH	TH	237V	1971	a
(51993	977834 S003 960933)	2-2w-2w-2DMR	DerbyC&W		1961	
SC52005	(977832)	2-2w-2w-2DMR	DerbyC&W		1961	
(52012	977835 S003 960933)	2-2w-2w-2DMR	DerbyC&W		1960	
SC52031		2-2w-2w-2DMR	DerbyC&W		1961	
(W79976)		4wDMR	ACCars		1958	
62043	1753 CHRIS GREEN	4w-4wRER	York		1965	
(65302)	5703 (930 204 977874)	4w-4RER	Afd/Elh		1954	
(65304)	5705 (930 204 977875)	4w-4RER	Afd/Elh		1954	
(S65373)	5759	4w-4RER	Elh		1956	
	99709 909137-0	2w-2PMR	Geismar	ST/04/06	2004	

a   property of Harry Needle Railroad Co Ltd, Derbyshire
b   property of Freightliner Group Ltd
c   property of West Midlands Trains, Soho, Birmingham
d   currently at Wyvernrail plc, Ecclesbourne Valley Railway, Derbyshire
e   property of Hunslet Ltd, Barton-under-Needwood, Staffordshire

REID FREIGHT SERVICES LTD,
Newpark Works, Cinderhill Industrial Estate,
Weston Coyney Road, Longton, Stoke-On-Trent   **ST3 5LB**

Locomotives in transit or for storage may be present in yard.

Gauge 4ft 8½in   www.reidfreight.co.uk   **SJ 925435**

(D3763)	08596	0-6-0DE		Derby		1959	a
	ALEX	0-6-0DH		AB	614	1977	a OOU
	EMMA	0-6-0DH		HE	8902	1978	b
	REDWING	0-6-0DH		HE	8999	1981	c

Cockshute Sidings, Shelton New Road, Stoke-on-Trent   **ST4 7AB**

Locomotives in transit or for storage may be present in yard.

Gauge 4ft 8½in   **SJ 864473**

(411.388)		2-8-0	OC	Alco	70284	1942	Dsm
(01560)		4wDH		RR	10229	1965	c OOU
(01575)	293V	0-6-0DH		TH	293V	1980	b OOU
(11229)	229 483009	2-2w-2w-2RER		MetCam		1938	OOU
(65379)	930 206 977925	4w-4RER		Afd/Elh		1954	OOU
(65382)	930 206 977924	4w-4RER		Afd/Elh		1954	OOU
MPP 13731	99709 901028-9	2w-2BER		Geismar 5E/0006		2006	

a   property of Harry Needle Railroad Co Ltd, Derbyshire
b   property of Hunslet Ltd, Barton-under-Needwood, Staffordshire
c   property of Ed Murray & Sons Ltd, Hartlepool, Teesside

STATFOLD ENGINEERING LTD,
STATFOLD BARN FARM, ASHBY ROAD, TAMWORTH   **B79 0BU**

Locomotives under restoration usually present

Gauge 4ft 8½in   www.statfoldengineering.co.uk   **SK 240064**

S.134	WHELDALE	0-6-0ST	IC	HE	3168	1944	

Gauge 1067mm

5	2-6-2T	OC	Thunes	4	1902	

J. WATSON & SONS LTD, SCRAP METAL MERCHANTS,
COMMON ROAD, STAFFORD (Part of the Watson Group)   **ST16 3DG**

Locomotives for scrap or resale occasionally present.   www.jwatsonandsons.co.uk   **SJ 923249**

A.P. WEBB PLANT HIRE LTD, COMMON ROAD, STAFFORD   **ST16 3DQ**

Gauge 4ft 8½in   www.apwebbplanthire.co.uk   **SJ 922249**

024	99709 979030-2	4wDM	R/R	_(Unimog 164499	1990
				(Zweiweg 1402	1990
041	99709 979029-4	4wDM	R/R	_(Unimog 170264	1991
				(Zweiweg	1991

PRESERVATION SITES

CHASEWATER LIGHT RAILWAY AND MUSEUM COMPANY, CHASEWATER RAILWAY, CHASEWATER COUNTRY PARK, POOL ROAD, near BROWNHILLS   WS8 7NL

Gauge 4ft 8½in   www.chasewaterrailway.co.uk   SK 034070

431		0-6-0ST	OC	HC	431	1895	
	HOLLY BANK No.3	0-6-0ST	IC	HE	3783	1953	
4	ASBESTOS	0-4-0ST	OC	HL	2780	1909	
917		0-4-0ST	OC	P	917	1902	
11	CYNTHIA	4wVBT	VCG	S	9366	1945	
5		4wVBT	VCG	S	9632	1957	
No.6	DUNLOP	0-4-0ST	OC	WB	2648	1941	
"34"		0-4-0DE		BBT	3097	1956	
MARSTON THOMPSON EVERSHED		0-4-0DM		Bg	3410	1955	
–		0-4-0DM		Bg	3590	1962	
(251)		6wDE		GECT	5414	1976	
–		0-6-0DM		HC	D615	1938	
6678	4/33	0-4-0DH		HE	6678	1969	
		reb		HE		1982	
6		0-6-0DH		HE	9000	1983	
		reb		HE	9288	1987	
		reb		YEC	L117	1992	
–		0-4-0DM		JF	4100013	1948	
21		4wDM		KC	1612	1929	
No.2	UBIQUE	4wPM		MR	1930	1919	
15099	MORRIS	4wDM		MR	2026	1920	a
RRM 106		0-4-0DH		NBQ	27656	1957	
D2911		0-4-0DH		NBQ	27876	1959	
–		4wDM		RH	305306	1952	
R.O.F. PEMBREY No.10 // MYFANWY		0-4-0DH		_(RSHD	8366	1962	
				(WB	3211	1962	
TH 103C	"MEGAN"	4wDH		TH	103C	1960	
	a rebuild of	4wVBT	VCG	S	9390	1949	
(01568)	HELEN	4wDH		TH	264V	1976	
	(HEM HEATH 3D)	0-6-0DM		WB	3119	1956	
–		4wDH		WB	3208	1961	
(AD 9118)		4wDHR		BD	3707	1975	
55568	142027	2-2wDMR		_(BRE(D)		1985	
				(Leyland R5.74		1985	
	rebuilt as	2-2wDHR		RFSD		1988	OOU
55618	142027	2w-2DMR		_(BRE(D)		1985	
				(Leyland R5.73		1985	
	rebuilt as	2w-2DHR		RFSD		1988	
55570	142029	2-2wDMR		_(BRE(D)		1985	
				(Leyland R5.20		1985	
	rebuilt as	2-2wDHR		HAB		1990	
55620	142029	2w-2DMR		_(BRE(D)		1985	
				(Leyland R5.19		1985	
	rebuilt as	2w-2DHR		HAB		1990	
55571	142030	2-2wDMR		_(BRE(D)		1985	
				(Leyland R5.21		1985	
	rebuilt as	2-2wDHR		HAB		1990	

55621	142030		2w-2DMR	_(BRE(D)		1985
				(Leyland R5.22		1985
		rebuilt as	2w-2DHR	HAB		1990
900331			2w-2PMR	Wkm	496	1932
TR 1	PWM 2773		2w-2PMR	Wkm	6872	1954

a   plate reads MR 2028

Gauge 2ft 0in - Chasewater Narrow Gauge Railway   **SK 034070**

8		4wDHF	HE	7385	1976
	BOLTON FELL	4wDM	LB	52726	1961
SO 30		4wDM	MR	5609	1931
–		4wDM	RH	174535	1936
	YD No.988	4wDM	RH	235729	1944
	YKSMM 2004/3004	4wDMF	RH	441424	1960
–		4wDMF	RH	480678	1961

The Rail Trolley Trust, Brownhills West Station   SK 034070
Gauge 4ft 8½in   www.railtrolleytrust.co.uk

900312	YORK 21	2w-2PMR	Wkm		1931	a	
(900332)	LNER 59	2w-2PMR	Wkm	497	1932		
–		2w-2PMR	Wkm	4164	1948	a	Dsm
A159	PWM 2191	2w-2PMR	Wkm	4168	1948		
(9033)		2w-2PMR	Wkm	6857	1954		
(1)		2w-2PMR	Wkm	6952	1955	a	
9036		2w-2PMR	Wkm	8196	1958	b	
–		2w-2BER	Bance		c1990	a	

a   stored off site
b   currently at Keighley Bus museum, Keighley, West Yorkshire

CHATTERLEY WHITFIELD MINING MUSEUM, TUNSTALL
Site opens on selected dates, advertised on the website.
Gauge 2ft 6in   www.chatterleywhitfieldfriends.org.uk   **SJ 883531**

3	TOM	4wBEF	Bg	3555	1961
2	CW 1986 111 JERRY	4wBEF	Bg	3578	1961
–		4wBEF	_(EE	3223	1962
			(RSHD	8344	1962

CHURNET VALLEY RAILWAY (1992) plc
Locomotives are kept at :-   Cheddleton   ST10 2HA SJ 983519
    Oakamoor   SK 046450
Gauge 4ft 8½in   www.churnet-valley-railway.co.uk

44422		0-6-0	IC	Derby		1927
(48173	8173)	2-8-0	OC	Crewe		1943
3278	FRANKLIN D. ROOSEVELT	2-8-0	OC	Alco	71533	1944
6046	(411.144)	2-8-0	OC	BLW	72080	1945
Tkh 2871		0-6-0T	OC	Chrz	2871	1951
Tkh 2944	HOTSPUR	0-6-0T	OC	Chrz	2944	1952
5197		2-8-0	OC	Lima	8856	1945
	KATIE	0-4-0ST	OC	AB	2226	1946

(D1994)	47292		Co-CoDE		Crewe		1966	
D3800	(08633)		0-6-0DE		Derby		1959	
(D6513)	33102	SOPHIE	Bo-BoDE		BRCW	DEL105	1960	
(D6539)	33021	EASTLEIGH	Bo-BoDE		BRCW	DEL131	1961	
(D7672)	25322	(25912) TAMWORTH CASTLE						
			Bo-BoDE		Derby		1967	
(D8057)	20057		Bo-BoDE		_(EE	2963	1961	
					(RSHD	8215	1961	
	BRIGHTSIDE		0-4-0DH		YE	2672	1959	
No.6	ROGER H. BENNETT		0-6-0DE		YE	2748	1959	a
RT1			4w-4wDHR		Balfour Beatty		1993	b
(68706	98706)		4wDHR		Perm	010	1986	b
(DX) 98710			4wDHR		Perm	011	1986	b
DX 98707			4wDHR		Perm	012	1986	b
(DX) 68801			4wDHR		Perm	002	1985	b
(DX) 68803	(RTU 8803)		4wDHR		Perm	004	1985	b
	C965 YOR		4wDMR	R/R	Bruff	523	1986	

a carries worksplate dated 1956
b property of B & R Track Services

DRAYTON MANOR PARK, FAZELEY, TAMWORTH

B78 3TW

Gauge 2ft 0in   www.draytonmanor.co.uk

SK 194016

1	THOMAS	4-4wDH	s/o	MetallbauE 081309	2008
6	PERCY	4-4wDH	s/o	MetallbauE	2008
	ROSIE	4-4wDH	s/o	MetallbauE	2009
	POLPERRO EXPRESS	4-4-0+4w-4wDH	s/o	SL	2151	2003

JAN FORD, BREWOOD HALL, SPARROWS END LANE, BREWOOD

ST19 9DB

Gauge 2ft 0in   Private site - visitors by appointment only

Private Site

	PHOENIX	0-4-0ST	OC	Ferndale	21	2001	a

a carries worksplate H.K. Porter 18635 1896

FOXFIELD LIGHT RAILWAY SOCIETY, FOXFIELD STEAM RAILWAY, BLYTHE BRIDGE, near STOKE-ON-TRENT

Locomotives are kept at :-   Blythe Bridge (Caverswall Road) Station   ST11 9BG   SJ 957421
                                 Site of Foxfield Colliery, near Dilhorne   SJ 976446

Gauge 4ft 8½in   www.foxfieldrailway.co.uk

No.2		0-6-2T	IC	Stoke		1923	
–		0-4-0ST	OC	AE	1563	1908	
No.1		0-4-0ST	OC	BP	1827	1879	
4101		0-4-0CT	OC	D	4101	1901	
	BELLEROPHON	0-6-0WT	OC	Haydock	C	1874	a
	WHISTON	0-6-0ST	IC	HE	3694	1950	
6		0-4-0ST	OC	Heath		1885	
			reb	CW		1934	
–		0-4-0ST	OC	HL	3581	1924	
	MOSS BAY	0-4-0ST	OC	KS	4167	1920	
–		0-4-0ST	OC	KS	4388	1926	

	Name	Type		Builder	Works No.	Date	Note
	"THE WELSHMAN"	0-6-0ST	IC	MW	1207	1890	
	HENRY CORT	0-4-0ST	OC	P	933	1903	
		reb		EV		1920	
		reb		EV		1933	
	"ACKTON HALL NO.3"	0-6-0ST	IC	P	1567	1920	
	"IRONBRIDGE"	0-4-0ST	OC	P	1803	1933	
No.11		0-4-0ST	OC	P	2081	1947	
–		4wVBT	VCG	S	9535	1952	
	LEWISHAM	0-6-0ST	OC	WB	2221	1927	
	HAWARDEN	0-4-0ST	OC	WB	2623	1940	
	FLORENCE No.2	0-6-0ST	OC	WB	3059	1953	
	CLIVE	0-6-0DH		AB	486	1964	
	WD 820	0-4-0DM		_(DC	2157	1940	
				_(EEDK	1188	1940	
				(VF		1940	
–		4wBE/WE		EEDK	1130	1939	
–		6wDM		KS	4421	1929	
	HELEN	4wDM		MR	2262	1923	
	"HERCULES"	4wDM		RH	242915	1946	b
(No.1	MERRY TOM)	4wDM		RH	275886	1949	
D2971		0-4-0DM		RH	313394	1952	
–		4wDM		RH	408496	1957	
–		0-4-0DE		RH	424841	1960	
	WOLSTANTON No.3	0-6-0DM		WB	3150	1959	
	BAGNALL	4wDH		WB	3207	1961	
–		4wDH		TH	111C	1961	
	a rebuild of	4wVBT	VCG	S			
	LUDSTONE	0-6-0DE		YE	2868	1962	
55705	142055	2-2wDMR		_(BRE(D)		1986	
				(Leyland R5.156		1986	
	rebuilt as	2-2wDHR		RFSD		1990	
55751	142055	2w-2DMR		_(BRE(D)		1986	
				(Leyland R5.153		1986	
	rebuilt as	2w-2DHR		RFSD		1990	
PWM 3764	(68065 B3 TR38)	2w-2PMR		Wkm	6643	1953	

a   carries works plate D
b   rebuilt from 2ft 0in gauge

Gauge 1ft 6in

	Name	Type		Builder	Works No.	Date
	JMLM7	0-4-0BE		WR	M7548	1972

M. HAMBLY, PRIVATE LOCATION, TAMWORTH
Gauge 4ft 8½in   **Private Site**

	Name	Type	Builder	Works No.	Date	Note
	–	2w-2PMR	Fairmont			a
	–	2w-2PMR	Wkm	4146	1947	
(DX 68010	DB965987)	2w-2PMR	Wkm	7073	1955	b
DB 965071		2w-2PMR	Wkm	7586	1957	

a   engine no.114866
b   currently stored off site

LAWRENCE HODGKINSON, 17 BILLINGTON AVENUE, LITTLE HAYWOOD

These locomotives are kept at a private location.

Gauge 2ft 6in Private Site

–		4wDM	RH	441948	1959	
–		4wDM	RH	476112	1962	

Gauge 2ft 0in

–		4wPM	L	962	1928	

A. HODGSON

Gauge 2ft 6in Private Site

5	4wBE	BV	690	1974	
–	4wBE	WR	892	1935	

Gauge 2ft 0in

No.1 BESSIE	4wDM	RH	170374	1934	

LIME KILN WHARF INDUSTRIAL RAILWAY, LIME KILN BASIN, WHITEBRIDGE INDUSTRIAL ESTATE, WHITEBRIDGE LANE, STONE ST15 8LQ

Gauge 2ft 0in Private Site SJ 894345

LOD 758227	4wDM	MR	8813	1943	
LOD 758019	4wDM	MR	8820	1943	
–	4wDM	RH	171901	1934	a

a   carries plate RH 191679

MILL MEECE PUMPING STATION PRESERVATION TRUST LTD, MILL MEECE PUMPING STATION, COTES HEATH, near ECCLESHALL ST21 6QU

Gauge 2ft 0in www.millmeecepumpingstation.co.uk SJ 831339

–	4wDM	L	39419	1953	

MOSELEY RAILWAY TRUST, APEDALE VALLEY LIGHT RAILWAY, APEDALE COMMUNITY COUNTRY PARK, LOOMER ROAD, CHESTERTON, NEWCASTLE-UNDER-LYME ST5 7LB

Gauge 3ft 0in www.avlr.org.uk SJ 823484

06/22/6/2	4wDM	RH	224337	1944	

Gauge 2ft 6in

L8	OLD NICK	4wDH	AB	556	1970	
12	ELECTRA	4wBE	BV	565	1970	
"54"	(YARD No.54 TO 235)	4wDMR	FH	2196	1940	
–		4wDM	SMH	60SD755	1980	
"70"	"CRYSTAL"	4wBE	WR	K7070	1970	

Gauge 2ft 0in

RENISHAW No.2	0-4-0T	OC	AE	1986	1926	
OGWEN	0-4-0T	OC	AE	2066	1933	

104	9	0-6-0WT	OC	HC	1238	1916	
303	1215	4-6-0T	OC	HE	1215	1916	
	STANHOPE	0-4-2ST	OC	KS	2395	1917	
–		0-6-0T	OC	KS	3014	1916	
5662		0-4-0WTT	OC	OK	5662	1912	
"1"	"BILLET"	4wBE		WR	C6717	1963	
"2"	CABLE MILL	4wBE		WR	C6716	1963	
3	81A 136	4wDM		MR	8878	1944	
6	"GENESIX"	4wPM		MR	7066	1938	
7	(Z. & W. WADE)	4wDM		MR	8663	1941	
"9"		4wDM		MR	7522	1948	
"10"		4wPM		MR	9104	1941	
"11"	ALD HAGUE	4wPM		FH	3465	1954	
13	"THE PILK"	4wDM		MR	11142	1960	
14	"KNOTHOLE WORKER"	4wDM		MR	22045	1959	
"15"		4wDM		FH	2514	1941	
"17"		4wDM		MR	8698	1941	Dsm
18	L.C.W.W. 8103	4wDM		HE	6299	1964	
20		4wDM		MR	8748	1942	
21		4wDM		MR	8669	1941	
24		4wDM		HE	1974	1939	
"25"	ND 10448	4wDM		HE	6007	1963	
26	"TWUSK"	4wDM		HE	6018	1961	
27	ANNIE	4wDM		RH	198297	1939	
28	24 "DOROTHY"	4wDM		RH	198228	1940	
29	VANGUARD	4wDM		RH	195846	1939	
"30"	"FRIDEN"	4wDM		RH	237914	1946	a
"31"	"PLUTO"	4wDM		RH	189972	1938	
"32"	LISTER	4wDM		LB	52885	1962	
"33"		4wPM		MR	7033	1936	
34		4wDM		RH	164350	1933	
"35"	DX 68061 PWM 2214 (TR26)	2w-2DMR		Wkm	4131	1947	
37	7	4wDM		RH	260719	1948	
"No.38"	KENNETH	4wDM		RH	223749	1944	
"No.39"	LR 2832	4wPM		MR	1111	1918	
"40"	(739) "SLUDGE"	4wDM		SMH	40SD516	1979	
41		4wDM		MR	5821	1934	
No.42		4wDM		MR	7710	1939	
"43"	THUNDERBIRD 4	4wDM		SMH	104G063	1976	
44	CHAUMONT	4wDH		HU	LX1002	1968	
"45"	87008	4wDM		RH	179870	1936	
"47"	LR 3090	4wPM		MR	1369	1918	
"48"	R12 ND 6458	4wDM		RH	235725	1944	
50	DELTA	0-4-0DM		Dtz	10050	1931	
"52"	LAWR	0-4-0PM	s/o	Bg	1695	1928	
"53"		4wDM		FH	2306	1940	
"56"	13 CAT C	4wDH		HE	9082	1984	
"57"	3 TOE RAG	4wDH		HE	8827	1979	
"58"		4wDM		HC	D558	1930	
"59"	4588	4wPM		OK	4588	1932	
"60"	"LORD AUSTIN"	4wPM		MR	6035	1937	

"61" (LR 3041)	4wDM		MR	1320	1918	
62 P396 81A03 "LBT"	4wDM		RH	497542	1963	
"64"	4wDMF		RH	256314	1949	
"65" 5 42005	4wDM		RH	223667	1943	
"66" 24.8 "PIKROSE"	4wBE		PWR	B0366	1993	
"67" 6 "AMENE"	4wBE		WR	D6912	1964	
"69"	4wBE		CE	B0475	1975	
71	4wBE		CE	5843	1971	
"72 LADY ANN"	4wBE		CE	B0922A	1975	
"74"	4wDM		OK	(3444 1930?)		
"75"	4wDM		MR	22235	1965	
"80"	4wDM		LB	52610	1961	
"81" LR 2573	4wDM		MR	2197	1923	
"82" No.33 "DRACULA"	4wDM		FH	3582	1954	Dsm
"83" "RHIWBACH"	2w-2PM		Rhiwbach		1935	
"84"	4wPM		H	984	1931	
"86" LR 2638	4wDM		FH	2586	1941	
"89" L2 GHOST	4wDM		R&R	84	1938	
"90" DH 887 (ND 10393)	4wDH		BD	3756	1981	
"92"	4wDM		RH	193974	1938	
"93" "PROMETHEUS"	4wBE		CE	B4299	1998	
"94"	4wDH		AK	46	1993	
"95" MERLIN	4wDH	s/o	HU	LX1001	1968	
"96" (74-Lb-00003)	0-4-0DM		BLW			
a rebuild of	0-4-0PM		BLW	48378	1917	
"98" L5 "JOHN STEVENSON"	4wDM		MR	21520	1955	
"98" (LM 21)	4wDM		RH	243392	1946	
"106" Y.W.A. L2	4wDM		RH			
"107"	4wDM		RH	175418	1936	Dsm
–	4wDM		RH	191658	1938	
–	4wDM		AK	No.4	1979	
2	4wDM		LB	50888	1959	
rebuilt as	4wDH					
–	4wDM		LB	52528	1961	
MBS 387	4wBE		LMM	1053	1950	
"35" 9303/507	0-4-0DM		HE	6619	1966	
20 60S317 "VILYA"	4wDM		MR	60S317	1966	
23	4wDM		L	52031	1960	
P 1215	4wDM		MR	5213	1930	
754	4wDM		SMH	60SD754	1980	
–	4wDM		MR	8979	1946	Dsm
4 VULCAN A.M.W. No.197	4wDM		RH	198287	1940	
"101" WoW 3164 RTT/767172	2w-2PM		Wkm	3164	c1943	
"THE CHAIR"	2w-2PMR		SharmanJ		2018	Dsm

a   currently off site for restoration

Gauge 1ft 11½in

"91" (RTT/767156)	2w-2PM		Wkm	3158	c1943

Gauge 1ft 8in

"100" (DL 354013)	4wDM		RH	354013	1953
–	4wDM		RH	375360	1955

Gauge 1ft 6in

"68" S147 263 016 "PORSCHE" 263054   4wBE   CE   5926/2   1972

Apedale Heritage Centre, Loomer Road, Chesterton ST5 7LB
Gauge 2ft 0in   www.apedale.co.uk   SJ 822483

"16" 6 "MARGARET"   4wDHF   HE   9056   1982
"78" CITY OF GLOUCESTER   4wPM   MR   5038   1930
"36" 6 "COMMERCIAL" ORDER No. N/NS/N/146/149
                               4wDM   RH   280865 1949

THE NATIONAL BREWERY CENTRE,
HORNINGLOW STREET, BURTON-ON-TRENT DE14 1NG
Gauge 4ft 8½in   (Closed)   SK 248234

No.9   0-4-0ST OC NR   5907   1901
No.20   4wDM   KC   1926

R. PHILLIPS
Gauge 2ft 0in   Private Site

(4)   4wDM   RH   260716 1949
–   4wDM   MR   9778   1953
–   4wPM   SkinnerD   c1975

PRIVATE OWNER, c/o WILDON (UK) LTD, OLD STATION YARD,
DODSLEIGH LANE, LOWER LEIGH, near UTTOXETER ST10 4SJ
Gauge 4ft 8½in   Private Site   SK 014357

2957   4wDM   RH   512572 1965

PRIVATE SITE, BROUGHTON, near ECCLESHALL
Gauge 1ft 0in   Private site

–   4w-4wDH s/o WhalleyB   c2016

THE STAFFORDSHIRE NARROW GAUGE RAILWAY LTD,
AMERTON RAILWAY, AMERTON FARM & CRAFT CENTRE,
STOWE-BY-CHARTLEY, near STAFFORD ST18 0LA
Gauge 3ft 0in   www.amertonrailway.co.uk   SJ 993278

No.1 (ED 10)   0-4-0ST OC WB   1889   1911

Gauge 2ft 0in

–   0-8-0T   OC Hen   14019   1916
JENNIE   0-4-0ST OC HE   3905   2007 a
No.56 (LORNA DOONE)   0-4-0ST OC KS   4250   1922
ISABEL   0-4-0ST OC WB   1491   1897
A10 RNAD TRECWN   4wDH   BD   3782   1984

GOLSPIE / THE TRENTHAM EXPRESS	0-4-0DM	s/o	Bg	2085	1934		
	DREADNOUGHT	0-4-0DM	s/o	Bg	3024	1939	
–		0-4-0DM	s/o	Bg	3235	1947	
3		0-4-0DM		Dtz	19531	1937	
–		4wDM		FH	2025	1937	
	GORDON	4wDHF		HE	8561	1978	
	rebuilt as	4wDH				2004	
–		4wDM		Jung	5869	1934	
–		4wDM		MR	7471	1940	
87033		4wDM		MR	40SD501	1975	
ND 6507		4wDM		RH	221623	1943	
–		4wDM		RH	506491	1964	
SP 90		4wBE		CE	B0402C	1974	
(SP 972) 72292/1		4wBE		CE	B3686A	1990	

a   carries plate dated 2005

STATFOLD NARROW GAUGE MUSEUM TRUST LTD, STATFOLD COUNTRY PARK, STATFOLD BARN RAILWAY, ASHBY ROAD, TAMWORTH   B79 0BU

Private site hosting regular ticket-only public open days; strictly no public access at any other time.

Gauge 4ft 8½in   www.statfold.com   **SK 240064, 242061**

No.1	GLENFIELD 15	0-4-0CT	OC	AB	880	1902	
	HODBARROW	0-4-0ST	OC	HE	299	1882	
–		0-4-0PM		Bg	680	1916	
–		0-4-0PM		Bg	800	1920	
No.14	CADBURY	0-4-0DM		HC	D1012	1956	Pvd
	(WALTER)	4wBE		CE	B4427A	2006	
"78"	LIBBIE	2w-2PMR		Bg/DC	1097	1920	
14/3	2705-A9	2w-2PMR		Fairmont	252319		
	CN 168-31	2w-2PMR		Fairmont			a

Gauge 3ft 6in

–	4wPMR		Wkm	5864	1951	Dsm

Gauge 3ft 0in

	"HANDYMAN"	0-4-0ST	OC	HC	573	1900	a
LM 110		4wDM		RH	379066	1954	

Gauge 2ft 6in

50		0-6-0DH	HB	D1419	1971	
YARD No.26 (ND 6506) SAM "69"		0-4-0DM	HE	2019	1939	
(51) TOM		0-6-0DH	_(HE	8847	1981	
			(HC	DM1447	1981	
19 "83"	4wDHF	RACK	HE	9294	1990	

Gauge 750mm

1	0-4-0WT	OC	OK	614	1900	
5	0-4-4-0T	4C	OK	1473	1905	

Gauge 2ft 0in

	CEGIN	0-4-0WT	OC	AB	1991	1931
"184"	MARCHLYN	0-4-0T	OC	AE	2067	1933
–		0-4-0	OC	Dav	1650	1918

–		0-6-0T	OC		Dec	1735	1919	
11	FIJI	0-6-0	OC		HC	972	1912	
19		0-4-0ST	OC		HC	1056	1914	
	ALPHA	0-6-0T	OC		HC	1172	1924	
GP 39		0-6-0WT	OC		HC	1643	1930	
	CLOISTER	0-4-0ST	OC		HE	542	1891	
	JONATHAN	0-4-0ST	OC		HE	678	1898	
	SYBIL MARY	0-4-0ST	OC		HE	921	1906	
	"SEAFORTH"	0-4-2T	OC		HE	1026	1910	
No.2	HOWARD	0-4-2ST	OC		HE	1842	1936	
	rebuilt as	0-4-2T	OC					
	rebuilt as	0-4-2ST	OC		Statfold		2014	
No.4	TRANGKIL No.4	0-4-2ST	OC		HE	3902	1971	b
	STATFOLD	0-4-0ST	OC		HE	3903	2005	
	JACK LANE	0-4-0ST	OC		HE	3904	2006	
	SID	0-4-0CA	IC		HE	9902	2009	
	SACCHARINE	0-4-2T	OC		JF	13355	1912	
9	"72" SF. DJATIBARANG	0-4-4-0T	4C		Jung	4878	1930	
No.1	SRAGI No.1 // S.S. TRAM BIMA	0-4-2T	OC		KraussS	4045	1899	
	DIANA	0-4-0T	OC		KS	1158	1917	
	ROGER	0-4-0ST	OC		KS	3128	1918	
	SRAGI 14 MAX	0-6-0WT	OC		OK	10750	1923	
No.1	HARROGATE	0-6-0ST	OC		P	2050	1944	
19.B		0-4-0ST+WT	OC		SS	3518	1888	
			reb		TyseleyLW		2004	
	WENDY	0-4-0ST	OC		WB	2091	1919	
	A. BOULLE	4-4-0T	OC		WB	2627	1940	
	ISIBUTU	4-4-0T	OC		WB	2820	1945	
	ISAAC	0-4-2T	OC		WB	3023	1953	
No.6	HOWARD	0-4-0VB	VC		Wbton	2	2007	
	CHARLES	4wPM			Brookville	3526	1949	
–	(521166)	4wPM			Brookville	3746	1951	
	rebuilt as	4wDM			Statfold		2014	
–		4wPM			FH	(1776	1931 ?)	
1	D4	4wDM			Funkey	1001		
4	D5	4wDM			Funkey	1033		
	rebuilt as	4wDH						
–		0-6-0DMF			HC	DM803	1954	Dsm
2	ATLAS	4wDM			HE	2463	1944	
			reb		ALR	No.2	1983	
S.1985.0055		4wDM			HE	2959	1944	
–		4wDH			HE	8819	1979	
9332	STATFOLD WORKS	4wDH			HE	9332	1994	
–		4wDHF			HE	9351	1994	
W114H	WESTERN REEFS GOLD MINE	4wDH			HT	6720	1965	
D7	N13	4wDH			HT	7588	1968	
–		4wDM			HU	36863	1929	
(8)		4wPM			HU	39924	1924	
	AGWI PET 2	4wPM			MR	4724	1939	
	BRAMBRIDGE HALL	4wPM			MR	5226	1930	
39581		4wDM			MR	8640	1941	c DsmT
	CHARLIE	4wDM			MR	9976	1954	
	20777	0-4-0DM			OK	20777	1936	

2		4wPM		Plymouth		1891	1924	
	rebuilt as	4wDM						
No.7	TINY	4wDM		Plymouth	5800	1954		
No.8	TIM	4wDH		Plymouth	6137	1958		
4	ALISTAIR	4wDM		RH	201970	1940		
	THE GOOSE	4-4wDMR		Statfold		2015		
No.3		4wPM		VIW	4049	1929		
	CONTEX 1	4wBE		CE	5940A	1972		Dsm
JMLM23	280537	4wBE		GB	420253	1970		
			reb	WR		1983		
			reb	CE	B4623	2016		
809	(215 216)	2w-2-2-2wRE		EEDK	809	1931		d
(No.4	THE ECLIPSE)	0-4-0WE		Greaves		1927		
	a rebuild of	0-4-0ST	OC	WB	1445	1895		
	THE COALITION	0-4-0WE		Greaves		1930		
	a rebuild of	0-4-0ST	OC	WB	1278	1890		
"82"		4wBE		WR	6092	1958		
–		4-4wPMR		NemethJ		2009		
	rebuilt as	4-2wPMR		HE	9903	2009		e

Gauge 600mm

(No.11	ESCUCHA)	0-4-0T	OC	BH	748	1883	
2	MINAS DE ALLER 2	0-6-0PT	OC	Corpet	439	1884	f
No.1	CDC	0-4-2PT	OC	HE	3756	1952	
RTT/767178		2w-2PM		Wkm	3170	c1943	

Gauge 1ft 11½in

	No.1K / No.2K	0-4-0+0-4-0T 4C		BP	5292	1909	
	GERTRUDE	0-4-0ST	OC	HE	995	1909	g
	LIASSIC	0-6-0ST	OC	P	1632	1923	
–		2w-2PMR		Wkm	10943	1976	

Gauge 1ft 10¾in

	KING OF THE SCARLETS	0-4-0ST	OC	HE	492	1888
	MICHAEL	0-4-0ST	OC	HE	1709	1932

Gauge 1ft 6in

No. 1	WOOLWICH	0-4-0T	OC	AE	1748	1916
	JACK	0-4-0WT	OC	HE	684	1898
BWR 3	CARNEGIE	0-4-4-0DM		HE	4524	1954

Gauge 1ft 0¼in - Mease Valley Light Railway

	VICTORIA	2-4-2T	OC	ESR	332	2007
–		0-6-0DH		AK	73	2005

Gauge monorail

A4		2wPH	RM	8253	1959

a	currently off site for overhaul
b	convertible to 2ft 6in gauge
c	converted to brake tender
d	built 1931 but originally carried plates dated 1930
e	converted Landrover
f	worksplate shows build date of 1884, but locomotive was completed in 1885
g	locomotive is sectioned

JOHN STRIKE, PRIVATE LOCATION, near TAMWORTH
Gauge 2ft 0in   Private Site
    –   4wDM   RH   175127   1935

SUFFOLK

INDUSTRIAL SITES

FELIXSTOWE DOCK & RAILWAY CO LTD, t/a PORT OF FELIXSTOWE
FELIXSTOWE (member of HPH Group)   IP11 3SY
Freightliner Group Ltd, Central Freightliner Terminal
Gauge 4ft 8½in   www.freightliner.co.uk   TM 274344

    (D3693)   08531   0-6-0DE   Dar   1959 M

GB Railfreight Ltd, South Freightliner Terminal
Gauge 4ft 8½in   www.gbrailfreight.com   TM 285326

    (D3906)   08738   0-6-0DE   Crewe   1960 a

      a   property of Railway Support Services Ltd, Wishaw, Warwickshire

FREIGHTLINER MAINTENANCE LTD, IPSWICH DEPOT,
off RANLEIGH ROAD, IPSWICH   IP2 0AB
Gauge 4ft 8½in   www.freightliner.co.uk   TM 152439

    (D3858)   08691   TERRI   0-6-0DE   Hor   1959 M
    (D4121)   08891   0-6-0DE   Hor   1962 M
      –   4wBE   _(Express   E608   2020
                                   (Hegenscheidt   2020

PRESERVATION SITES

APPLE MOUNT RETREAT,
APPLE MOUNT FARMHOUSE, THORP MORIEUX, SUDBURY   IP30 0NQ
Gauge 4ft 8½in   Private Site   TL 923521

    10289   (10229 129 483009)   2-2w-2w-2RER   MetCam   1938

EAST ANGLIA TRANSPORT MUSEUM SOCIETY, EAST SUFFOLK LIGHT RAILWAY,
CHAPEL ROAD, CARLTON COLVILLE, LOWESTOFT   NR33 8RL
Gauge 2ft 0in   www.eatransportmuseum.co.uk   TM 505903

      –   4wDM   MR   5902   1932 a Dsm
    2   ALDBURGH   4wDM   MR   5912   1934
   No.6 THORPNESS   4wDM   MR   22209 1964

No.5	ORFORDNESS	4wDM		MR	22211	1964	
4	LEISTON	4wDM		RH	177604	1936	

a   converted to a brake van

HALESWORTH TO SOUTHWOLD NARROW GAUGE RAILWAY SOCIETY, HALESWORTH

Gauge 3ft 0in   www.halesworthtosouthwoldrailway.co.uk   **Private Site**

LM 373	4wDH		DunEW		1984	
		reb	BoothWKelly		2008	+

+   rebuilt as non-self propelled generator and compressor unit

Gauge 900mm (operates on 3ft 0in gauge)

RS106	"HOLTON"	4wDH	RFSD	L106	1989

Blythburgh Station, Station Road, Halesworth   **IP19 9LQ**

Gauge 3ft 0in   (operates on advertised open days)   **TM 452754**

–	4wBE	GB		1936

IPSWICH TRANSPORT MUSEUM,
THE OLD TROLLEYBUS DEPOT, COBHAM ROAD, IPSWICH   IP3 9JD

Gauge monorail   www.ipswichtransportmuseum.co.uk   **TM 193428**

–	1w1PH	RM	11345	1963	a

a   not on public display

LEISTON WORKS RAILWAY TRUST, LEISTON   (IP16 4ER)

Gauge 4ft 8½in   www.leistonworksrailway.onesuffolk.net   **TM 443626**

	R.O.F. PURITON No.2	0-4-0DM	AB	349	1941	
(960236)		2w-2PMR	Wkm	1519	1934	a

a   stored off site

LONG SHOP MUSEUM, MAIN STREET, LEISTON   IP16 4ES

Gauge 4ft 8½in   www.longshopmuseum.co.uk   **TM 443626**

"SIRAPITE"	4wT	G	AP	6158	1906

THE MID-SUFFOLK LIGHT RAILWAY COMPANY,
BROCKFORD STATION, WETHERINGSETT, near STOWMARKET   IP14 5PW

Gauge 4ft 8½in   www.mslr.org.uk   **TM 128659**

(68088)	985	0-4-0T	IC	Dar	(1205)	1923	
	CALEDONIA WORKS	0-4-0ST	OC	AB	1219	1910	
–		0-4-0VBT	OC	Cockerill	2525	1907	
(No.4)	"ALSTON"	0-6-0ST	OC	HC	1604	1928	
–		0-4-0DM		JF	20337	1934	
	SIR WILLIAM McALPINE	4wDM		RH	294266	1951	

(5)	ALSTON	0-4-0DM		RH	304470	1951	
	(DE) 960220	2w-2PMR		Wkm	1949	1935	

PLEASUREWOOD HILLS THEME PARK, LEISURE WAY, LOWESTOFT

NR32 4TZ

Gauge 4ft 8½in www.pleasurewoodhills.com

TM 545965

D.2069		4wDM		RH	305315	1952

Gauge 2ft 0in

167 7324 SUFFOLK PUNCH	4w-2-4wPH	s/o	Chance	79.50167.24	1979

S.A. PYE, 109 PAPER MILL LANE, BRAMFORD, near IPSWICH

IP8 4BU

Gauge 4ft 8½in

TM 127472

DUDLEY	BARKING POWER	4wDM		FH	(3294	1948)

S. SMITH, HERRINGFLEET HILLS, HERRINGFLEET

Gauge 2ft 0in

Private Site

–	4wDM	s/o	MR	9774	1953

SOUTHWOLD RAILWAY TRUST, STEAMWORKS, BLYTH ROAD, SOUTHWOLD

IP18 6AZ

Gauge 3ft 0in www.southwoldrailway.co.uk

TM 500765

No.3	BLYTH	2-4-0T	OC	NBRES	(007)	2021	a
	SCALDWELL	0-6-0ST	OC	P	1316	1913	
No.5	MELLS	4wDH		MR	105H006	1969	

a   carries worksplate SS 2850/1879

STEAM TRACTION LTD, c/o WEBB TRUCK EQUIPMENT, MELFORD ROAD, ACTON, near SUDBURY

CO10 0BB

Gauge 5ft 0in www.skiploaders.co.uk

TL 883455

1077	2-8-2	OC	Jung	11787	1953

STONHAM BARNS PARK, PETTAUGH ROAD, STONHAM ASPAL, near STOWMARKET

IP14 6AT

Gauge 4ft 8½in www.stonhambarns.co.uk

TM 127472

–	0-4-0F	OC	RSHN	7803	1954

SURREY

INDUSTRIAL SITES

R. BANCE & CO LTD, COCKROW HILL HOUSE, ST. MARY'S ROAD, SURBITON
Administration address only. Vehicles under construction, for repair or resale, at another, private, location.
Gauge 4ft 8½in   www.bance.com   Private Site

–		2w-2BER	Bance	064/98	1998
–		2w-2BER	Bance	095/00	2000
–		2w-2BER	Bance	255/11	2011
99709 901231-9		2w-2PMR	Bance	278	2016
	reb		Bance	278	2021

PRESERVATION SITES

M. HAYTER,
1 HEATHERVIEW COTTAGES, SHORTFIELD COMMON, FRENSHAM, FARNHAM
Gauge 2ft 0in   Private Site

–	2w-2PM	Wkm	2981	1941	Dsm

MIZENS RAILWAY, BARRS HILL, KNAPHILL, WOKING
Gauge 3ft 6in   www.wokingminiaturerailwaysociety.com

GU21 2JW
SU 967593

–	4-8-2T	OC	D	3819	1899

OLD KILN LIGHT RAILWAY, THE RURAL LIFE CENTRE, OLD KILN MUSEUM,
THE REEDS, REEDS ROAD, TILFORD, near FARNHAM
Gauge 2ft 0in   www.oldkilnlightrailway.co.uk

GU10 2DL
SU 858434

	PAMELA	0-4-0ST	OC	HE	920	1906	
	EMMET	0-4-0T	OC	Moors Valley 20	1995	
	a rebuild of	0-4-0DM		OK	21160	1938	a
M.N.No.1	ELOUISE	0-6-0WT	OC	OK	9998	1922	
	ALTONIA	0-4-0DM	s/o	Bg	1769	1929	
	rebuilt as	0-4-0DH	s/o	BES		1992	
	CORBIERE	4wDM		FH	2528	1941	
HE 1944	STINKER	4wDM		HE	1944	1939	
(AD) 36	CHAMPION	4wDH		HE	7011	1971	
(NG) 37	WEY VALLEY	4wDH		HE	7012	1971	
			reb	HAB	6014	1988	
(NG) 38	WEYFARER	4wDH		HE	7013	1971	
	"WALTER"	4wDM		Moës		c1955	
	(FIDO)	4wPM		MR	5297	1931	Dsm
5713	(EAGLE)	4wDM		MR	5713	1936	
(LOD/758039)	PHOEBE	4wDM		MR	8887	1944	Dsm
–		4wDM		MR	8981	1946	

No.1 PAUL COOPER	4wDM	MR	9655	1951	
SANDROCK	4wDM	RH	177639	1936	
RED DWARF	4wDM	RH	181820	1936	
(AD 22) "SUSAN"	4wDM	RH	211609	1941	
497760	4wDMF	RH	497760	1963	
No.4 L12 (RTT 767093) LIZ	2w-2PM	Wkm	3031	1941	
rebuilt as	4wDM	Hayter		1973	
SUE	2w-2PM	Wkm	3287	1943	
rebuilt as	2w-2PMR	‡		1977	

‡   rebuilt by E.J.Stephens, Wey Valley Light Railway, Farnham, Surrey
a   on loan from Moors Valley Railway, Dorset

The Rustics Timber Group, Woodland Tramway
Gauge 2ft 0in

–	2w-2PM	Wkm	3032	1941	Dsm

P. RAMPTON, PRIVATE SITE, HAMBLEDON
The following locomotives are stored at private locations.
The locomotives are not available for viewing or photography.

Gauge 2ft 0in   Private Site

82	2-6-2+2-6-2T	4C	Hano	10634	1928

Gauge 600mm

RENISHAW 4	0-4-4-0T	VCG	AE	2057	1931
7 SOTILLOS	0-6-2T	OC	Borsig	6022	1906
2 SAMELICES	0-6-0T	OC	Couillet	1209	1898
101	0-4-2T	OC	Hen	16073	1918
102	0-4-0T	OC	Hen	16043	1918
RENISHAW 5	0-4-4-0T	4C	WB	2545	1936
R.P.M.. No.2	0-4-2T	OC	WB	2895	1948

Gauge 550mm

4 SAN JUSTO	0-4-2ST	OC	HC	639	1902
5 SANTA ANA	0-4-2ST	OC	HC	640	1902

FRANK SAXBY, GUILDFORD
Gauge 2ft 0in   Private Site

SUE	4wG	IC	Saxby	1943	1999

EAST SUSSEX

INDUSTRIAL SITES

ST. LEONARDS RAILWAY ENGINEERING LTD,
ST. LEONARDS WEST MARINA DEPOT,
BRIDGE WAY, ST. LEONARDS-ON-SEA
Gauge 4ft 8½in   www.hastingsdiesels.co.uk

TN38 8AP
TQ 775084

D2995	(07011)		0-6-0DE	Resco	L105	1978	
		a rebuild of		RH	480696	1962	
S60000	1001	HASTINGS	4-4wDER	Afd/Elh		1957	
S60001	1001		4-4wDER	Afd/Elh		1957	OOU
(S60016)	S60116	(S60226) 1001 MOUNTFIELD					
			4-4wDER	Afd/Elh		1957	
(S60018)	S60118	1001 TUNBRIDGE WELLS					
			4-4wDER	Afd/Elh		1957	
(S60019)	S60119	1013	4-4wDER	Afd/Elh		1957	
(S60145	977939	930301)	4-4wDER	Afd/Elh		1962	
(S60149	977940	930301)	4-4wDER	Afd/Elh		1962	

SUSSEX POLICE TRAINING CENTRE,
KINGSTANDING, MARESFIELD, UCKFIELD,
Gauge 4ft 8½in   www.sussex.police.uk

TN22 3JJ
TQ 473290

55584	142043		2-2wDMR	_(BRE(D)		1986
				(Leyland R5.92		1986
		rebuilt as	2-2wDHR	RFSD		1988
55634	142043		2w-2DMR	_(BRE(D)		1986
				(Leyland R5.91		1986
		rebuilt as	2w-2DHR	RFSD		1988

PRESERVATION SITES

BLUEBELL RAILWAY CO LTD
Locomotives are kept at :-

Horsted Keynes, West Sussex	RH17 7BB	TQ 372293
Kingscote	RH19 4LD	TQ 367355
Sheffield Park	TN22 3QL	TQ 403238

Gauge 4ft 8½in   www.bluebell-railway.com

(30096)	96	NORMANDY	0-4-0T	OC	9E		1893	
30583	(488)		4-4-2T	OC	N	3209	1885	
(30847)	847		4-6-0	OC	Elh		1936	
(30928)	No.928	STOWE	4-4-0	3C	Elh		1934	
(31027)	No.27	PRIMROSE	0-6-0T	IC	Afd		1910	Dsm
(31065)	No.65		0-6-0	IC	Afd		1896	
(31263)	No.263		0-4-4T	IC	Afd		1905	
(31323)	323	BLUEBELL	0-6-0T	IC	Afd		1910	
(31592)	No.592		0-6-0	IC	Longhedge		1901	
31618	(1618)		2-6-0	OC	Bton		1928	

(31638)	1638		2-6-0	OC	Afd		1931	
32424	BEACHY HEAD		4-4-2	OC	Bluebell		2007	
(32473)	B473 BIRCH GROVE		0-6-2T	IC	Bton		1898	
(32636)	No.672 FENCHURCH		0-6-0T	IC	Bton		1872	
(32655)	55 STEPNEY		0-6-0T	IC	Bton		1875	
(34023)	21 C 123 BLACKMOOR VALE		4-6-2	3C	Bton		1946	
34059	SIR ARCHIBALD SINCLAIR		4-6-2	3C	Bton		1947	
				reb	Elh		1960	
(58850)	27505		0-6-0T	OC	Bow	181	1880	
73082	CAMELOT		4-6-0	OC	Derby		1955	
75027			4-6-0	OC	Sdn		1954	
84030			2-6-2T	OC	Bluebell		2011	
	a rebuild of 78059		2-6-0	OC	Dar		1956	u/c
80100			2-6-4T	OC	Bton		1955	
80151			2-6-4T	OC	Bton		1957	
92240			2-10-0	OC	Crewe		1958	
No.3	CAPTAIN BAXTER		0-4-0T	OC	FJ	158	1877	
No.4	SHARPTHORN		0-6-0ST	IC	MW	641	1877	
D4106	(09018)		0-6-0DE		Hor		1961	
D6570	(33052) ASHFORD		Bo-BoDE		BRCW	DEL174	1961	
E6040	(73133)		Bo-BoDE/RE		_(EE	3712	1966	
	BLUEBELL RAILWAY				(EEV	E372	1966	
	BRITANNIA		4wPM		H	957	1926	a
–			4wDH		RR	10241	1966	
				reb	TH	247C	1973	
S60130	1305 BRIGHTON ROYAL PAVILION							
			4-4wDER		Afd/Elh		1962	
387	(2761 LT512)		4w-4wRER		Ashbury		1898	
				reb			1907	b Dsm
	99709 901120-4		2w-2PMR	R/R	Geismar M44/072		2010	
DXN 68001	(68078)		2w-2DMR		Wkm	10708	1974	
				reb	Ashford		1993	

- a   runs on propane gas
- b   converted to hauled coaching stock

DRUSILLA'S ZOO PARK, SAFARI EXPRESS,
ALFRISTON ROAD, BERWICK, ALFRISTON, near EASTBOURNE   BN26 5QS
Gauge 2ft 0in   www.drusillas.co.uk   TQ 524050

240597	0-4-2+4-4wDH	s/o	MetallbauE	2007	

THE CLAUDE JESSETT TRUST COMPANY, THE GREAT BUSH RAILWAY,
TINKERS PARK, MAIN ROAD, HADLOW DOWN, near UCKFIELD   TN22 4HS
Gauge 2ft 0in   www.tinkerspark.com   TQ 538241

(No.28)	SAO DOMINGOS	0-6-0WT	OC	OK	11784	1928
(31)	(GRÜNWALD)	4wDM		Diema	1600	1953
No.15	OLDE	4wDM		HE	2176	1940
(1)	AMINAL	4wPM		MR	(5361	1933)
	rebuilt as	4wDM	reb	‡		
(No.25)	WOLF	4wDM		MR	7469	1940
(4)	MILD	4wDM		MR	8687	1941

(No.36)	DRUSILLA	4wDM		MR	22236	1965	
10	CAPE	4wDM		OK	5926	1935	
5	(ALPHA)	4wDM		RH	183744	1937	
(14)	(ALBANY)	4wDM		RH	213840	1941	
(29)	24 R. J. BROWN	4wDM		RH	382820	1955	
(No.22)	(LAMA)	4wBE		WR	5033	1953	
(23)		0-4-0BE		WR	M7534	1972	Dsm
(No.24)	TITCH	0-4-0BE		WR	M7535	1972	

‡   rebuilt by Ludlay Brick Co, Berwick, near Eastbourne, Sussex

LAVENDER LINE PRESERVATION SOCIETY,
ISFIELD STATION, STATION ROAD, ISFIELD, near UCKFIELD   TN22 5XB
Gauge 4ft 8½in   www.lavender-line.co.uk   TQ 452171

	LADY LISA	0-4-0VBT	OC	Cockerill	2945	1920	
D4113	09025	0-6-0DE		Hor		1961	
221	QUEENBOROUGH	0-4-0DM		AB	354	1941	
			reb	BIS		1957	
15		4wDM		FH	3658	1953	
			rep	Resco	L112		
830	WEM	0-4-0DM		_(VF	5257	1945	
				(DC	2176	1945	
S60117	205018 1118	4-4wDER		Afd/Elh		1957	
S60122	(205023) (1123)	4-4wDER		Afd/Elh		1959	Pvd
60151	205033 (1133)	4-4wDER		Afd/Elh		1962	
(61928)	309624 977966) 960102	4w-4wWER		York		1962	
999507		4wDMR		Wkm	8025	1958	
	99709 909130-5	2w-2DMR		Geismar	ST/03/56	2003	
	ARMY 9035	2w-2PMR		Wkm	8195	1958	

A. PRAGNELL, PRIVATE LOCATION, near HORSTED KEYNES
Gauge 4ft 8½in   **Private Site**

(900855)		2w-2PMR	Wkm	6967	1954	a

a   carries plate Wkm 7445/1956

RAILWAY RETREATS NORTHIAM,
NORTHIAM STATION, STATION ROAD, NORTHIAM   TN31 6QT
Gauge 4ft 8½in   www.railwayretreats.co.uk   TQ 834265

	"RAILWAY RETREATS"	0-4-0ST	OC	AB	2015	1935	a

a   carries worksplate AB 1609/1918

ROTHER VALLEY RAILWAY LTD, ROBERTSBRIDGE JUNCTION STATION,
STATION ROAD, ROBERTSBRIDGE   TN32 5DG
Gauge 4ft 8½in   www.rvr.org.uk   TQ 734235

D2112	(03112)	0-6-0DM		Don		1960	
(D99)	(11790)	0-4-0DM		_(VF	D77	1947	
				(DC	2251	1947	

1	TITAN	0-4-0DM	_(VF	D140	1951
			(DC	2274	1951

TUNBRIDGE WELLS & ERIDGE RAILWAY PRESERVATION SOCIETY LTD, SPA VALLEY RAILWAY, GROOMBRIDGE STATION, NEWTON WILLOWS, GROOMBRIDGE, near TUNBRIDGE WELLS   TN3 9RD

Gauge 4ft 8½in   www.spavalleyrailway.co.uk   **TQ 532371**

For details of locomotives see under Kent entry.

VOLKS ELECTRIC RAILWAY, MADEIRA DRIVE, BRIGHTON   BN12 1EN

Gauge 2ft 8½in   www.volkselectricrailway.co.uk   **TQ 326035**

9		4wRER		BE		1898	
	reb			BE		c1911	
	reb			BrightonTC		1950	
3		4wRER		VER		1892	Dsm
4		4wRER		VER		1892	
	reb			AK	99R	2017	
6		4wRER		VER		1901	
	reb			AK	101R	2018	
7		4wRER		VER		1901	
8		4wRER		VER		1901	
9		4wRER		VER		1910	
10		4wRER		VER		1926	
	reb			AK	102R	2018	
–		4wDM		AK	40SD530	1987	

WEST SUSSEX

INDUSTRIAL SITES

SIEMENS AG, SIEMENS RAIL SYSTEMS, THREE BRIDGES TRAINCARE FACILITY, THREE BRIDGES, CRAWLEY   RH10 1LY

Gauge 4ft 8½in   www.mobility.siemens.com   **TQ 288365**

–		4wBE		?		c2015
–		4wBE	R/R	Zagro		2019

PRESERVATION SITES

BLUEBELL RAILWAY CO LTD,

East Grinstead Station, Firbank Way, East Grinstead RH19 1DD TQ 385379
Horsted Keynes Station, Station Approach, Horstead Keynes RH17 7BB TQ 372293

Gauge 4ft 8½in   www.bluebell-railway.co.uk

For details of locomotives see under East Sussex entry.

HOLLYCOMBE STEAM & WOODLAND GARDEN SOCIETY,
HOLLYCOMBE STEAM COLLECTION, IRON HILL, LIPHOOK   GU30 7LP

Gauge 4ft 8½in   www.hollycombe.co.uk   SU 852295

50	COMMANDER B	0-4-0ST	OC	HL	2450	1899	

Gauge 1000mm

ND 3646	4wDM	RH	210961	1941	

Gauge 3ft 0in

"EXCELSIOR"	2-2-0WT	G	AP	1607	1880	

Gauge 2ft 6in

BEV YARD No.4B/9B ND 3305 B40	4wBE	WR	3805	1948	

Gauge 2ft 0in – Quarry Railway

70	CALEDONIA	0-4-0WT	OC	AB	1995	1931	
38	JERRY M	0-4-0ST	OC	HE	638	1895	
	COCO	0-4-0WT	OC	OK	3136	1908	
NG 53		4wDH		AB	764	1988	
			reb	HAB		1996	
	FLOATER	4wDM		MR	60S318	1964	a
C.F.V.O.	No.25	4wDM		Plymouth	4848	1945	
16	JACK	4wDM		RH	203016	1940	
	LIZZIE	4wDH		Ruhrthaler	3909	1969	

a   brake column is stamped 11164 in error

HOTHAM PARK MINIATURE RAILWAY, HOTHAM PARK, BOGNOR REGIS PO21 1HN

Gauge 1ft 0¼in   SZ 939995

	BORIS	0-6-0DH	s/o	AK	80	2007
–	"FRED"	4w-4wBE		_(Parkside (D&S Services		

ANDREW JOHNSTON, MIDHURST

Gauge 2ft 0in   Private Site

	TOBY	4wPM		Carter		1986
–		4wDM		FH	2544	1942
–		4wDM	s/o	MR	8727	1941

LITTLEHAMPTON MINIATURE RAILWAY, MEWSBROOK PARK, LITTLEHAMPTON

BN16 2LX

Gauge 1ft 0¼in   www.littlehamptonminiaturerailway.com   **TQ 042016**

	CHRISTOPHER	2-6-2T	OC	SaundersT		2020
	a rebuild of	0-6-0T	OC	ESR	310	2000
	a rebuild of	0-6-2T	OC	ESR	191	1991
1	QUEEN ELIZABETH	4-4-0PM	s/o	Barnard		1937
	ALBERT	4wDE		GluyasC		2009
	DAISY	0-4-0PH	s/o	RVM		2013
	rebuilt as	0-4-0DH	s/o			2021
	PHILLIPPA	0-4-0DH	s/o	SaundersT		2017

SOUTHERN INDUSTRIAL HISTORY CENTRE TRUST, AMBERLEY MUSEUM, HOUGHTON BRIDGE, AMBERLEY, ARUNDEL

BN18 9LT

Gauge 4ft 8½in   www.amberleymuseum.co.uk   **TQ 031122**

No.5	BURT	4wDM		MR	9019	1950

Gauge 3ft 2¼in

1	(TOWNSEND HOOK)	0-4-0T	OC	FJ	172L	1880
	MONTY	4wDM		OK	7269	1936

Gauge 2ft 11in

–		4wDM		MR	10161	1950

Gauge 2ft 6in

–		4wPM		HU	45913	1932

Gauge 2ft 0in

	BARBOUILLEUR	0-4-0WT	OC	Dec	1126	1947
	POLAR BEAR	2-4-0T	OC	WB	1781	1905
	PETER	0-4-0ST	OC	WB	2067	1918
T0001	ND10261 00 NZ 26	4wDH		BD	3751	1980
808		2w-2-2-2wRE		EEDK	808	1931
	CCSW	4wDM		FH	1980	1936
–		4wPM		(FH	3627	1953 ?)
–		0-4-0DMF		HC	DM686	1948
	(THAKEHAM TILES No.3)	4wDM		HE	2208	1941
–		4wDM		HE	3097	1944
	(THAKEHAM TILES No.4)	4wDM		HE	3653	1948
12		4wDH		HE	8969	1980
–		4wDM		HE		
	"PELDON"	4wDM		JF	21295	1936
–		4wPM		L	33937	1949
–		4wPM		L	34521	1949
–		4wPM		MR	872	1918
	reb			MR	3720	1925
3101		4wPM		MR	1381	1918
27		4wDM		MR	5863	1934
–		4wDM		MR	11001	1956
	SONIA	4wDM		OK	4013	1930
6193	REDLAND	4wDM		OK	6193	1935

7741	THE MAJOR	4wDM		OK	7741	1937	
No.2		4wDM		RH	166024	1933	
–		4wDM		RH	187081	1937	
–		4wDM		R&R	80	1937	
–		4wPM		Thakeham		c1946	
–		4wPM		Thakeham		c1950	
3		4wBE		BE	16303	c1917	a
–		4wBE		WR	4998	1953	
–		4wBE		WR	5031	1953	
–		4wBE		WR	(5034	1953?)	
–		0-4-0BE		WR	T8033	1979	
	RTT/767159	2w-2PM		Wkm	3161	c1943	
WD 904		2w-2PMR		Wkm	3403	1943	

a   rebuilt using parts from BE 16306 c1917

Gauge 1ft 10in

23		0-4-0T	IC	Spence	23L	1921	a

Gauge monorail

–		2wPH		RM	9514	1960	

a   worksplate dated 1920

SOUTHERN TRANSIT BUS COMPANY LTD, THE OLD CEMENT WORKS, A283 SHOREHAM ROAD, UPPER BEEDINGARUNDEL

BN44 3TX

Gauge 4ft 8½in   www.southerntransit.co.uk   **TQ 197085**

1018		4w-4wRER	MetCam	1959
1304		2w-2-2-2wRER	MetCam	1961
9305		4w-4wRER	MetCam	1961

TEESSIDE

Our definition of "Teesside" incorporates the area covered by the former County of Cleveland, created in 1974, and disbanded in 1996 into four separate County Boroughs of Hartlepool, Middlesbrough, Redcar & Cleveland and Stockton-on-Tees.

INDUSTRIAL SITES

BRITISH STEEL LTD, (part of the Jingye Group)
SKINNINGROVE WORKS, CARLIN HOW, SALTBURN-BY-THE-SEA   **TS13 4EE**

Gauge 4ft 8½in   www.britishsteel.co.uk   **NZ 708194**

01572	KATHRYN		0-6-0DH		RR	10256	1966	a
			0-6-0DH		TH	150C	1965	
–		a rebuild of	0-6-0VBT	VCG	S	(9650	1957)	a

a   property of Ed Murray & Sons Ltd, Hartlepool, Teesside

BRITISH STEEL LTD, (part of the Jingye Group)
TEESSIDE BEAM MILL, LACKENBY WORKS, MIDDLESBROUGH
Gauge 4ft 8½in   www.britishsteel.co.uk   **NZ 558222**

253		6wDE	GECT	5416	1976	a	OOU
255		6wDE	GECT	5418	1976	a	OOU
257		6wDE	GECT	5426	1977	a	OOU
260		6wDE	GECT	5429	1977	a	OOU
265		6wDE	GECT	5462	1977	a	OOU
266	SHERRIFFS	6wDE	GECT	5463	1977	a	OOU
269		6wDE	GECT	5466	1977	a	OOU
276	SPAWOOD	6wDE	GECT	5474	1978	a	OOU
(MURR 2)		4wDE	Moyse	1464	1979	a	
	LADY POTTER	0-6-0DH	RR	10214	1964	a	OOU
–		0-6-0DH	RR	10255	1966	a	
–		0-6-0DH	S	10166	1963	a	
01569	EMMA	4wDH	TH	281V	1978	a	OOU
–		4wDM	Lackenby		2010		
	a rebuild of		Robel	54.12-107 AD184	1980	a	

a   property of Ed Murray & Sons Ltd, Hartlepool, Teesside

COBRA MIDDLESBROUGH LTD, (part of the A.V. Dawson Group),
MIDDLESBROUGH GOODS YARD, NORTH ROAD, MIDDLESBROUGH   **TS2 1DQ**
Gauge 4ft 8½in   www.portofmiddlesbrough.com   **NZ 488209**

01567	ELIZABETH	4wDH	TH	276V	1977	a

a   property of Ed Murray & Sons Ltd, Hartlepool, Teesside

EDF ENERGY plc, HARTLEPOOL POWER STATION, SEATON CAREW   **TS25 2BZ**
Gauge 4ft 8½in   www.edfenergy.com   **NZ 532270**

–		0-6-0DH	EEV-AEI	4003	1971	a
	CD40 B4618/6	4wDH	CE	B4618.6	2016	b

a   property of Ed Murray & Sons Ltd, Hartlepool, Teesside
b   property of Railway Support Services Ltd, Wishaw, Warwickshire

ICL UK (CLEVELAND) LTD (part of the ICL Fertilizers Group)
Boulby Mine, Loftus   **TS13 4UZ**
Gauge 4ft 8½in   www.icl-uk.uk   **NZ 763183**

CD40 B4618/1 POLLY	4wDH	CE	B4618.1	2016	a
CD40 B4618/2	4wDH	CE	B4618.2	2016	a

a   property of Railway Support Services Ltd, Wishaw, Warwickshire

Gauge 1ft 6in — Underground   **NZ 765189**

–	0-4-0BE	WR	7654	1974

Tees Dock Terminal, Grangetown

Gauge 4ft 8½in   www.icl-uk.uk

(TS6 6UD)

NZ 549235

–		0-6-0DH	AB	659	1982	
		reb	HAB	6768	1990	
		reb	HE		2004	a
EMILY		0-6-0DH	AB	660	1982	
		reb	HAB	6769	1990	
		reb	HE		2004	a
BLUEBIRD		0-6-0DH	HE	8998	1981	a
(01571) 40-008		0-6-0DH	TH	261V	1976	
		reb	YEC	L124	1996	b

a   property of Hunslet Ltd, Barton-under-Needwood, Staffordshire
b   property of Railway Support Services Ltd, Wishaw, Warwickshire

LIBERTY STEELS, LIBERTY PIPES DIVISION, LIBERTY STEEL HARTLEPOOL, SAW PIPE MILLS (42in and 84in Mills), BRENDA ROAD, HARTLEPOOL   TS25 2EF

(part of the Liberty House Group; part of GFG Alliance)

Gauge 4ft 8½in   www.libertysteelgroup.com

NZ 508288

(MURR 1)	4wDE	Moyse	1364	1976	a

a   property of Ed Murray & Sons Ltd, Hartlepool, Teesside

ED MURRAY & SONS LTD, BRENDA ROAD, HARTLEPOOL   TS25 2BW

The following is a FLEET LIST of the locomotives owned by this contractor.

Gauge 4ft 8½in   www.edmurrays.com

–	"ZIPPY"	0-6-0DH	EEV-AEI	4003	1971	a	
306		0-6-0DH	GECT	5383	1973	b	
253		6wDE	GECT	5416	1976	c	
(425 60 460)		6wDE	GECT	5425	1977	d	
257		6wDE	GECT	5426	1977	c	
260		6wDE	GECT	5429	1977	c	
261		6wDE	GECT	5430	1977	d	
EM 1	(262)	6wDE	GECT	5431	1977	e	
264	PORT MULGRAVE	6wDE	GECT	5461	1977	f	
265		6wDE	GECT	5462	1977	c	
266		6wDE	GECT	5463	1977	c	
268		6wDE	GECT	5465	1977	g	
269		6wDE	GECT	5466	1977	c	
276		6wDE	GECT	5474	1978	c	
277		6wDE	GECT	5475	1978	h	
MURR 1		4wDE	Moyse	1364	1976	j	
(MURR 2)		4wDE	Moyse	1464	1979	c	
3	(2)	4wDE	Moyse	1365	1976	b	Dsm
–		0-6-0DH	S	10107	1963	b	
–		0-6-0DH	S	10166	1963	c	
LADY POTTER		0-6-0DH	RR	10214	1964	c	
(01560)		4wDH	RR	10229	1965	k	
–		0-6-0DH	RR	10255	1966	c	
01572	MS 5482 "KATHRYN"	0-6-0DH	RR	10256	1966	l	

–		0-6-0DH		TH	150C	1965	
	a rebuild of	0-6-0VBT	VCG	S	(9650	1957)	l
(01567)	ELIZABETH	4wDH		TH	276V	1977	m
(01569)	EMMA	4wDH		TH	281V	1978	c
	LOCO No.9	4wDH		TH	287V	1980	b
(01555)	JAMES	4wDH		TH	288V	1980	b
–		0-6-0DH					e +
(1)		4wDM		Lackenby		2010	
	a rebuild of			Robel 54.12-107 AD184	1980	c	
(2)		4wDM		Robel 54.12-107 AD183	1980	b	

a   currently at EDF energy plc, Hartlepool Power Station, Teesside
b   currently at Ed Murray & Sons Ltd, Casebourne Road, Hartlepool (NZ 511313)
c   currently at British Steel Ltd, Lackenby, Teesside
d   currently at DIRFT, Daventry, Northamptonshire
e   currently at Tata Strip Pruducts UK, Llanwern Works, South Wales
f   currently at Tata Steel Europe, Hartlepool, Teesside
g   currently at Hunslet Ltd, Barton-under-Needwood, Staffordshire
h   currently at Aggregate Industries UK Ltd, Merehead, Somerset
j   currently at Liberty Steels, Hartlepool, Teesside
k   currently at Reid Freight Services Ltd, Cockshute Sidings, Stoke, Staffs
l   currently at British Steel Ltd, Skinningrove Works, Teesside
m   currently at Cobra (Middlesbrough) Ltd, Middlesbrough, Teesside
+   one of GECT 5379/1972 or EEV D1248-3946/1968

NORTHERN STEAM ENGINEERING LTD,
LONSDALE HOUSE, ROSS ROAD, PORTRACK, STOCKTON-ON-TEES   TS18 2NH
Gauge 4ft 8½in   www.northernsteamengineering.co.uk   NZ 457189

3814		2-8-0	OC	Sdn		1940
(69523	4744) No.1744	0-6-2T	IC	NBH	22600	1921

vehicles for overhaul / repair occasionally present

PD PORTS LTD, TEESPORT, TEES DOCK, GRANGETOWN   TS6 6UD
(owned by Brookfield Asset Management Inc)
Gauge 4ft 8½in   www.pdports.co.uk   NZ 546232

(D3780)	08613	H 064	LOCO 3	0-6-0DE	Derby	1959	a
(D3956)	08788	H 068		0-6-0DE	Derby	1960	a
(D4015)	08847		LOCO 1	0-6-0DE	Hor	1961	a

a   property of Rail Management Services Ltd, Chesterfield, Derbyshire

PORT OF MIDDLESBROUGH (an A.V. Dawson facility)
Locomotives kept at :   Automotive Coil Store, Depot Road, Middlesbrough   TS2 1LE   NZ 491213
                         Ayrton Store and Railhead, Forty Foot Road, Middlesbrough   TS2 1HG   NZ 493215
                         NSSB Wharf, Riverside Park Road, Middlesbrough   TS2 1UT   NZ 491216
                         Tees Riverside Intermodal Park, Forty Foot Road, Middlesbrough   TS2 1PU   NZ 489212
Gauge 4ft 8½in   www.portofmiddlesbrough.com

(D3525)	08410	0-6-0DE	Derby	1958
(D3765)	08598	0-6-0DE	Derby	1959
(D3767)	08600	0-6-0DE	Derby	1959

(D3942)	08774	ARTHUR VERNON DAWSON				
		0-6-0DE	Derby		1960	
(D4142)	08912	0-6-0DE	Hor		1962	Dsm
	ELEANOR DAWSON	0-4-0DM	Bg/DC	2725	1963	Pvd
7900/58		0-4-0DM	RSHN	7900	1958	Pvd

SEMBCORP UTILITIES TEESSIDE LTD, (part of Sembcorp Industries Group)
WILTON INTERNATIONAL MANUFACTURING SITE, MIDDLESBROUGH
Gauge 4ft 8½in R.T.C. www.sembcorp.com   **NZ 564218**

(D3911)	08743	BRYAN TURNER	0-6-0DE	Crewe		1960	
(D4133)	08903	JOHN W. ANTILL	0-6-0DE	Hor		1962	
RO.005	"WILTONIA"		2w-2DMR	Wkm	7591	1957	
		rebuilt as	2w-2DHR	YEC	L112	1992	
–			2w-2PMR	Wkm	7603	1957	DsmT

SOUTH TEES DEVELOPMENT CORPORATION t/a TEESWORKS,
TEESSIDE WORKS, LACKENBY WORKS, MIDDLESBROUGH
(parts of site are shared by British Steel Ltd)
Gauge 4ft 8½in   R.T.C. www.teesworks.co.uk   **NZ 564223**

8.711	Bo-BoDE	MaK	1600.011	1996	Dsm

STRABAG SE, ANGLO AMERICAN plc,
ANGLO AMERICAN WOODSMITH PROJECT,
WILTON INTERNATIONAL MANUFACTURING SITE, MIDDLESBROUGH
Wilton to Woodsmith Mine MTS Tunnel : Drive 1 – Wilton to Lockwood Beck
Drive 2 – Lockwood Beck to Woodsmith Mine
Drive 3 – Woodsmith Mine to Lockwood Beck

Gauge 900mm   www.angloamerican.com   Wilton Site   **NZ 580227**

-	4wBE	Schöma	6827	2015
-	4wBE	Schöma	7084	2019
-	4wBE	Schöma	7085	2019
-	4wBE	Schöma	7086	2019
-	4wBE	Schöma	7087	2019
-	4wBE	Schöma	7103	2019
-	4wBE	Schöma	7104	2019

TATA STEEL EUROPE, TATA STEEL TUBES EUROPE, NORTH EAST PIPE MILLS,
HARTLEPOOL 20in MILL, BRENDA ROAD, HARTLEPOOL   **TS25 2EF**
(Part of the Tata Group)
Gauge 4ft 8½in   www.tatasteeleurope.com   **NZ 505276, 505278**

264	PORT MULGRAVE	6wDE	GECT	5461	1977	a	OOU

a   property of Ed Murray & Sons Ltd, Hartlepool, Teesside

PRESERVATION SITES

LAND OF IRON, (CLEVELAND IRONSTONE MINING MUSEUM) OFF MILL LANE, DEEPDALE, SKINNINGROVE   TS13 4AP
Gauge 2ft 0in   www.landofiron.org.uk   NZ 712192

No.9	4wDM	RH	182137	1936	Dsm

SALTBURN MINIATURE RAILWAY LTD, VALLEY GARDENS, SALTBURN-BY-THE-SEA   TS13 4SD
Gauge 1ft 3in   www.saltburn-miniature-railway.org.uk   NZ 668214

SALTBURN 150	4-6-2gas	s/o	_(Artisair	1972
			(RADev	1972
rebuilt as	4-6-2DH	s/o	SMR	2011
PRINCE CHARLES	4-6-2DE	s/o	Barlow	1953
BLACKLOCK R.	4-4-2	OC	SMR	2015
	a rebuild of		MossAJ	2001
GEORGE OUTHWAITE	0-4-0DH	s/o	_(SMR	1994
			(WiltonICI	1994

STOCKTON-ON-TEES BOROUGH COUNCIL, PRESTON PARK MUSEUM and GROUNDS, EAGLESCLIFFE, STOCKTON-ON-TEES   TS18 3RH
Gauge 4ft 8½in   www.prestonparkmuseum.co.uk   NZ 429158

–	0-4-0VBT	VCG	HW	21	1870

WARWICKSHIRE

INDUSTRIAL SITES

ALLELYS HEAVY HAULAGE LTD, THE SLOUGH, STUDLEY   B80 7EN
Locomotives in transit are also stored in the yard for short periods of time.
Gauge 4ft 8½in   www.allelys.co.uk   SP 056636

TEUCER	0-4-0DM	_(VF	D294	1955
		(DC	2567	1955

ASSOCIATED BRITISH PORTS HOLDINGS LTD, ABP HAMS HALL RAIL FREIGHT TERMINAL, EDISON ROAD, COLESHILL, BIRMINGHAM B46 1DA

(operated by Maritime Transport Ltd)

Gauge 4ft 8½in www.abports.co.uk **SP 198911**

(D4129)	08899	MIDLAND COUNTIES RAILWAY 175	1839-2014			
		0-6-0DE	Hor		1962	a

a   property of Railway Support Services Ltd, Wishaw, Warwickshire

COSTAIN GROUP plc, PLANT YARD, WATLING STREET, SHAWELL LE17 6AR

Gauge 900mm Closed www.costain.com **SP 538793**

CP20T.1		4wBE	CE	B4536.1	2011
CP20T.2	2 7	4wBE	CE	B4536.2	2011
CP20T.3	1	4wBE	CE	B4536.3	2012
4	1	4wBE	CE	B4536.4	2012
-		4wBE	CE	B4536.5	2012

Gauge 750mm - (currently at Tideway East - Thames Tideway Tunnel contract)

2		4wBE	CE	B4539.1	2012
-		4wBE	CE	B4539.2	2012
	reb	CE		B4573	2012
3		4wBE	CE	B4539.4	2012
-		4wBE	CE	B4539.6	2012
- DUKE OF YORK		4wBE	CE	B4540	2012

Gauge 610mm

1 C		4wBE	CE	B3686C	1990
-		4wBE	CE	B3686D	1990
-		4wBE	CE	B4010A	1994
3		4wBE	CE	B4010B	1994
SP 1122		4wBE	CE	B4057B	1994
SP 1176	2	4wBE	CE		a
SP 1177		4wBE	CE		a
SP 1178		4wBE	CE		a
SP 1179	LOCO 7	4wBE	CE		a
SP 1180	4	4wBE	CE		a
SP 1181	LOCO 8	4wBE	CE		a

a   individual identities unknown - from CE 5590/1, CE 5590/2, CE 5590/8, CE 5949B, CE 5949G and CE B0129

plant yard with locomotives present between contracts - currently stored at unknown location

EUROPEAN METAL RECYCLING LTD, METAL MERCHANTS AND PROCESSORS, TRINITY ROAD, KINGSBURY B78 2LB

Locomotives for scrap are occasionally present at this location.

Gauge 4ft 8½in www.uk.emrgroup.com **SP 219969**

(D3898)	08730	0-6-0DE	Crewe		1960	a
7		4wDH	S	10040	1960	OOU

a   property of Railway Support Services Ltd, Wishaw, Warwickshire

MINISTRY OF DEFENCE, DEFENCE MUNITIONS, KINETON

See Section 6 for details

PORTERBROOK LEASING CO LTD, RAIL INNOVATION CENTRE, STATION ROAD, LONG MARSTON, near STRATFORD-UPON-AVON   CV37 8PL

Locomotives and rolling stock for storage and testing usually present.

Gauge 4ft 8½in   www.porterbrook.co.uk   **SP 152472, 155476, 158474**

WD 70047	MULBERRY	0-4-0DM		AB	362	1942	
01529	(268)	4wDH		TH	310V	1984	a
01543	(263)	4wDH		TH	303V	1982	a
01544	(252)	4wDH		TH	270V	1977	a
–		4wDH	R/R	Minilok	134	1986	a
–		8wDH	R/R	Minilok	160	1991	a
TNS 104		2w-2DMR		Robel 56.27-10-AG35		1983	a

    a   property of Harry Needle Railroad Co Ltd, Barrow Hill, Derbyshire

PRISON SERVICE COLLEGE, NEWBOLD REVEL, STRETTON-UNDER-FOSSE, near RUGBY   CV23 0TH

Gauge 2ft 0in   **SP 455808**

–		4wDM	L	33651	1949	Pvd

R.S.S. - RAILWAY SUPPORT SERVICES LTD, c/o JOHN WATTS FARMS, RYE FARM, RYEFIELD LANE, WISHAW   B76 9QA

Administration address : Montpellier House, Montpellier Drive, Cheltenham

The following is a FLEET LIST of the locomotives owned by this contractor.

Railway Support Services Ltd - Hire Fleet

Gauge 4ft 8½in   www.railwaysupportservices.co.uk   **SP 180945**

(D3520)	08405		0-6-0DE	Derby	1958	a	
(D3526)	08411		0-6-0DE	Derby	1958		OOU
(D3556)	08441		0-6-0DE	Derby	1958	b	
(D3575)	08460	SPIRIT OF THE OAK	0-6-0DE	Crewe	1958	c	
(D3595)	08480		0-6-0DE	Hor	1958	d	
(D3599)	08484	CAPTAIN NATHANIEL DARELL					
			0-6-0DE	Hor	1958	b	
(D3662)	08507		0-6-0DE	Don	1958		
(D3673)	08511		0-6-0DE	Dar	1958	e	
(D3700)	08536		0-6-0DE	Dar	1959		OOU
(D3720)	09009		0-6-0DE	Dar	1959	f	+
(D3735)	08568	ST ROLLOX	0-6-0DE	Crewe	1959		OOU
(D3747)	08580		0-6-0DE	Crewe	1959	g	
(D3760)	08593		0-6-0DE	Crewe	1959		OOU
(D3772)	08605		0-6-0DE	Derby	1959	h	
(D3796)	08629		0-6-0DE	Derby	1959	j	
(D3799)	08632		0-6-0DE	Derby	1959	k	

(D3819)	08652		0-6-0DE	Hor		1959	
(D3830)	08663	ST. SILAS	0-6-0DE	Hor		1959	l
(D3837)	08670		0-6-0DE	Crewe		1960	c
(D3850)	08683		0-6-0DE	Hor		1959	e
(D3870)	08703	STEVE BLICK (CONCRETE BOB) SHUNTERSPOT					
			0-6-0DE	Hor		1960	g
(D3873)	08706		0-6-0DE	Crewe		1960	
(D3876)	08709		0-6-0DE	Crewe		1960	
(D3898)	08730		0-6-0DE	Crewe		1960	m
(D3906)	08738		0-6-0DE	Crewe		1960	n
(D3920)	08752		0-6-0DE	Crewe		1960	o
(D4014)	08846		0-6-0DE	Hor		1961	p
(D4129)	08899	MIDLAND COUNTIES RAILWAY 175 1839-2014					
			0-6-0DE	Hor		1962	q
(D4151)	08921	PONGO	0-6-0DE	Hor		1962	OOU
(D4157)	08927		0-6-0DE	Hor		1962	r
(D4169)	08939		0-6-0DE	Dar		1962	s
CD40	B4618/1	POLLY	4wDH	CE	B4618.1	2016	t
CD40	B4618/2	HAILEY	4wDH	CE	B4618.2	2016	t
CD40	B4618/3	GARY	4wDH	CE	B4618.3	2016	u
CD40	B4618/4	PUSHY-PULLY	4wDH	CE	B4618.4	2016	u
CD40	B4618/5		4wDH	CE	B4618.5	2016	v
CD40	B4618/6		4wDH	CE	B4618.6	2016	w
CD40	B4618/7		4wDH	CE	B4618.7	2016	v
–			4wDH	RR	10252	1966	
P403D	DENISE		4wDH	S	10029	1960	x
10150	01574 P406D	ISAAC	0-6-0DH	S	10150	1963	
1			4wDH	TH	133C	1963	
		rebuild of	4wVBT VCG	S			
–			4wDH	TH	184V	1967	y
(01571)	40-008		0-6-0DH	TH	261V	1976	
				reb	YEC	L124	1996	z

a	currently at Northern Trains Ltd, Neville Hill Depot, Leeds, West Yorkshire
b	currently at Abellio Greater Anglia Ltd, Norwich Crown Point, Norwich
c	currently at GB Railfreight Ltd, Bescot Yards, Walsall
d	currently at London North Eastern Railway Ltd, Craigentinny Depot, Edinburgh, Scotland
e	currently at GB Railfreight Ltd, Eastleigh Yards, Hampshire
f	currently at Gemini Rail Services UK Ltd, Wolverton, Buckinghamshire
g	currently at GB Railfreight Ltd, Whitemoor Yard, Cambridgeshire
h	currently at BBVS JV / HS2 Ltd, Willesden, Greater London
j	currently at Prologis R.F.I., D.I.F.R.T., Daventry, Northamptonshire
k	currently at GB Railfreight Ltd, Peterborough Depot, Cambridgeshire
l	currently at Hitachi Rail Europe, Newton Aycliffe, Co. Durham
m	currently at European Metal Recycling Ltd, Kingsbury, Warwickshire
n	currently at GB Railfreight Ltd, Felixstowe, Essex
o	currently at Imerys Minerals Ltd, Rock Works, near Bugle, Cornwall
p	currently at Tyseley T&RSMD, Birmingham, West Midlands
q	currently at ABP Holdings Ltd, Hams Hall, Warwickshire
r	currently at Avon Valley Railway, Bitton, Gloucestershire
s	currently at Direct Rail Services Ltd, Garston, Merseyside
t	currently at ICL UK (Cleveland) Ltd, Loftus, Teesside
u	currently at Network Rail RIDC, Asfordby, Leicestershire
v	currently at Total UK Ltd, Lindsey Oil Refinery, Lincolnshire
w	currently at EDF Energy, Hartlepool Power Station, Teesside
x	currently at DB Cargo Maintenance Ltd, Stoke-on-Trent, Staffordshire

Vehicles in storage or transit

Locomotives in transit are also stored in the yard for short periods of time.

Gauge 4ft 8½in

45	COLWYN		0-6-0ST	IC	K	5470	1933	Dsm
(D)1662	47484		Co-CoDE		Crewe		1965	
(D3951)	08783		0-6-0DE		Derby		1960	
(D6823	37123)	37679	Co-CoDE		_(EE	3268	1963	
					(EES	8383	1963	
–			0-6-0DH		EEV	3870	1969	
	(TOBY)		4wDM	s/o	RH	235513	1945	
				reb	Shackerstone		2000	
	HAMBLE-LE-RICE		0-6-0DH		TH	294V	1981	a
55658	143603		2-2wDMR		_(AB	711	1985	
					(Alex	1784/33	1985	
		rebuilt as	2-2wDHR		RFSD		1989	
55669	143603		2w-2DMR		_(AB	673	1986	
					(Alex	1784/6	1986	
		rebuilt as	2w-2DHR		RFSD		1989	
(10221	122	483002)	2-2w-2w-2RER		MetCam		1938	
(11142	225	483002)	2-2w-2w-2RER		MetCam		1938	
PWM 2189 (TR 13 DX 68053)			2w-2PMR		Wkm	4166	1948	

a   property of Harry Needle Railroad Co Ltd, Barrow Hill, Derbyshire

SEVERN LAMB UK LTD,
ARDEN INDUSTRIAL ESTATE, TYTHING ROAD, ALCESTER   **B49 6ET**

New miniature locomotives under construction, and locomotives in for repair, usually present.
www.severn-lamb.com   **SP 097587**

UNIPART LOGISTICS LTD, BIRCH COPPICE BUSINESS PARK,
UNIT 2, DANNY MORSON WAY, DORDON, TAMWORTH   **B78 1SE**

Gauge 4ft 8½in   www.unipart.com   **SP 251994**

9210		4wDM	R/R	_(Unimog 199146	2001
				(Zweiweg 1934	2001

PRESERVATION SITES

J.W. LEWIS, BEACON VIEW, BENTLEYS LANE, MAXSTOKE   **B46 2QR**
Gauge 4ft 8½in   **SP 244875**

AD 9124		4wDHR	BD	3713	1975
RS/140		4wDM	FH	3892	1958

JOHN PAYNE, PRIVATE LOCATION
Gauge 4ft 8½in   Private Site

| "11720" | ND10490 | 0-4-0DH | NBQ | 27648 | 1959 |
| – | | 0-4-0DE | YE | 2856 | 1961 |

WEST MIDLANDS

INDUSTRIAL SITES

ALSTOM TRANSPORT, WOLVERHAMPTON (MIDLAND) TRAINCARE CENTRE,
JONES ROAD, OXLEY, WOLVERHAMPTON (part of Alstom Holdings S.A.)   **WV10 6JQ**
Gauge 4ft 8½in   www.alstom.com   **SJ 906010**

| (D3784) | 08617 STEVE PURSER | 0-6-0DE | Derby | 1959 |

BLACK COUNTRY INNOVATIVE MANUFACTURING ORGANISATION
and DUDLEY METROPOLITAN BOROUGH COUNCIL,
VERY LIGHT RAIL INNOVATION CENTRE,
ZOOLOGICAL WAY, off TIPTON ROAD, DUDLEY   **DY1 4SQ**
Gauge 4ft 8½in   www.bcimo.co.uk   **SO 949909**

(D4122)	08892		0-6-0DE	Hor		1962	a
PPM 35			2w-2F/BER	PPM	10	1997	
–			4wBE	R/R	Zephir	2950	2020

a   property of Harry Needle Railroad Co Ltd, Barrow Hill, Derbyshire

GB RAILFREIGHT LTD (part of the Hector Rail Group),
BESCOT LDC UP YARD, SANDY LANE, WEDNESBURY, WALSALL   **(WS10 xxx)**
Gauge 4ft 8½in   www.gbrailfreight.com   **SP 011958**

| (D3575) | 08460 SPIRIT OF THE OAK 0-6-0DE | | Crewe | 1958 | a |
| (D3837) | 08670 | 0-6-0DE | Crewe | 1960 | a |

a   property of Railway Support Services Ltd, Wishaw, Warwickshire

MIDLAND METRO LTD, t/a WEST MIDLANDS METRO,
WEDNESBURY DEPOT, POTTERS LANE, WEDNESBURY
(a wholly owned subsidiary of West Midlands Combined Authority)   **WS10 0AR**
Gauge 4ft 8½in   www.westmidlandsmetro.com   **SO 984945**

Q179 VOH	4wDM	R/R	Unimog	166200	1991
BX64 VMA	4wDM	R/R	Hako		c2015
ERNIE	4wBE		SET	WN1021/1	2007
(A513484)	2w-2PMR		Geismar	ST/08/03	2008

SALFORD METALS LTD, 475 LICHFIELD ROAD, ASTON, BIRMINGHAM — B6 7SP

Gauge 4ft 8½in   www.salfordmetals.co.uk   SP 093899

THE SIR HENRY TARONI EXPRESS	0-4-0DH		HE	7259	1971	
		reb	HE	9286	1987	

TATA STEEL EUROPE, TATA STEEL ROUND OAK RAIL, STEELPARK, THE GATEWAY, BRIERLEY HILL (part of the Tata Group) — DY5 1LJ

Gauge 4ft 8½in   www.tatasteeleurope.com   SO 925879

–	4wBE	R/R	Zephir	2562	2015	a

a   property of DB Cargo (UK) Ltd

T.M.A. ENGINEERING LTD, TYBURN ROAD, ERDINGTON, BIRMINGHAM — B24 8NQ

New locomotives under construction, and for repair, occasionally present.   www.tmaeng.co.uk

WEST MIDLANDS TRAINS, t/a WEST MIDLANDS RAILWAY and LONDON NORTHWESTERN RAILWAY, TYSELEY DEPOT, WARWICK ROAD, TYSELEY, BIRMINGHAM — B11 2HL

(an Abellio, JR East and Mitsui Joint Venture)

Gauge 4ft 8½in   www.londonnorthwesternrailway.co.uk   SP 105840

(D)3783	(08616)	TYSELEY 100 BAM BAM				
		0-6-0DE	Derby		1959	**M**
(D3532)	08417	0-6-0DE	Derby		1958	a
(D4014)	08846	0-6-0DE	Hor		1961	a
	–	4wBE	Windhoff	101007142	2009	

a   property of Harry Needle Railroad Co Ltd, Barrow Hill, Derbyshire
b   property of Railway Support Services Ltd, Wishaw, Warwickshire

WILLENHALL COMMERCIALS LTD, ASHMORE LAKE WAY, WILLENHALL WV12 4LF

Gauge 610mm   www.willenhall-commercials.co.uk   SP 965995

2	VENTILLATA	0-6-0DH		OK	25996	1960

Gauge 600mm

4	0-4-0WT	OC	OK	9922	1922
5	0-4-0WT	OC	OK	10701	1924
3	0-4-0DM		Ruhrthaler	3526	1958

Private site. No casual visitors

PRESERVATION SITES

THE BLACK COUNTRY LIVING MUSEUM TRUST,
THE BLACK COUNTRY LIVING MUSEUM, TIPTON ROAD, DUDLEY **DY1 4SQ**
Gauge 4ft 8½in www.bclm.com **SO 950913**

| 2025 | WINSTON CHURCHILL | 0-6-0ST | IC | MW | | 2025 | 1923 | |

DENVER LIGHT RAILWAY LTD, UNIT 5, FRYERS CLOSE, BLOXWICH **WS3 2XQ**
Gauge 1ft 3in www.denverlightrailway.co.uk **Private Site**

	COUNT LOUIS	4-4-2	OC	BL	32	1923	a
	"PRINCESS CORONATION"	4-6-2	OC	MaxEng		2008	b
a	property of Count Louis Trust						
b	incomplete loco						

L.C.P. PROPERTIES LTD, PENSNETT TRADING ESTATE, SHUT END **DY6 7NA**
Gauge 4ft 8½in www.lcpproperties.co.uk **SO 900898**

| – | | 4wDM | | RH | 215755 | 1942 | |

J.E. SELWAY, near WEDNESBURY
Gauge 1ft 6in **Private Site**

| – | | 4-4-2 | OC | CurwenD | | 1951 | a |
| a | currently stored off site | | | | | | |

THINKTANK – BIRMINGHAM SCIENCE MUSEUM,
MILLENNIUM POINT, CURZON STREET, BIRMINGHAM **B4 7AP**
Gauge 4ft 8½in www.birminghammuseums.org.uk/thinktank **SP 084875**

| 46235 | CITY OF BIRMINGHAM | 4-6-2 | 4C | Crewe | | 1939 | |

TYSELEY LOCOMOTIVE WORKS LTD and VINTAGE TRAINS LTD,
THE STEAM DEPOT, WARWICK ROAD, TYSELEY, BIRMINGHAM **B11 2HL**
Gauge 4ft 8½in www.vintagetrains.co.uk **SP 105841**

2885		2-8-0	OC	Sdn	1938	
3840	COUNTY OF MONTGOMERY	4-4-0	OC	TyseleyLW	2021	u/c
4121		2-6-2T	OC	Sdn	1937	
4588		2-6-2T	OC	Sdn	1927	
4709		2-8-0	OC	Llangollen	2012	u/c
4936	KINLET HALL	4-6-0	OC	Sdn	1929	
4965	ROOD ASHTON HALL	4-6-0	OC	Sdn	1929	
5043	EARL OF MOUNT EDGCUMBE	4-6-0	4C	Sdn	1936	
5080	DEFIANT	4-6-0	4C	Sdn	1939	

5164		2-6-2T	OC	Sdn		1930	
5952	COGAN HALL	4-6-0	OC	Sdn		1935	
6880	BETTON GRANGE	4-6-0	OC	Llangollen		2011	u/c
7027	THORNBURY CASTLE	4-6-0	4C	Sdn		1949	Dsm
7029	CLUN CASTLE	4-6-0	4C	Sdn		1950	
(7752)	L94	0-6-0PT	IC	NBQ	24040	1930	
7760		0-6-0PT	IC	NBQ	24048	1930	
7802	BRADLEY MANOR	4-6-0	OC	Sdn		1938	
7822	FOXCOTE MANOR	4-6-0	OC	Sdn		1950	
9600		0-6-0PT	IC	Sdn		1945	
34053	SIR KEITH PARK	4-6-2	3C	Bton		1947	
			reb	Elh		1958	
45551	5551 THE UNKNOWN WARRIOR						
		4-6-0	3C	Llangollen		2013	u/c
(45593)	5593 KOLHAPUR	4-6-0	3C	NBQ	24151	1935	
46521		2-6-0	OC	Sdn		1953	
71000	DUKE OF GLOUCESTER	4-6-2	3C	Crewe		1954	
80104		2-6-4T	OC	Bton		1955	
–	(CADBURY)	0-4-0T	OC	AE	1977	1925	
No.65		0-6-0T	OC			1994	
	a rebuild of	0-6-0ST	OC	HC	1631	1929	
No.1		0-4-0ST	OC	P	2004	1941	
–		0-6-0ST	IC	RSHN	7289	1945	
No.670		2-2-2	IC	TyseleyLW		1989	u/c
789		2-4-2T	IC	TyseleyLW		2021	u/c
D1755	(47161 47541) 47773	Co-CoDE		BT	483	1964	
(D3029	08021) 13029	0-6-0DE		Derby		1953	
(D6940)	37240	Co-CoDE		_(EE	3497	1964	
				(EEV	D928	1964	
No.6	PRINCESS MARGARET	0-4-0DM		AB	376	1948	
–		4wDM		RH	299099	1950	
–		0-6-0DM		RH	347747	1957	
55819	144019	2-2wDMR		_(BRE(D)		1987	
				(Alex	2785/37	1987	
	rebuilt as	2-2wDHR		(AB		1990	
55842	144019	2-2wDMR		_(BRE(D)		1987	
				(Alex	2785/38	1987	
	rebuilt as	2-2wDHR		(AB		1990	
55855	(144019)	2-2wDMR		_(BRE(D)		1988	
				(Alex	1187/6	1988	
	rebuilt as	2-2wDHR		(AB		1990	
55823	144023	2w-2DMR		_(BRE(D)		1987	
				(Alex	2785/45	1987	
	rebuilt as	2w-2DHR		(AB		1991	
55846	144023	2-2wDMR		_(BRE(D)		1987	
				(Alex	2785/46	1987	
	rebuilt as	2-2wDHR		(AB		1991	
55859	(144023)	2w-2DMR		_(BRE(D)		1988	
				(Alex	1187/10	1988	
	rebuilt as	2w-2DHR		(AB		1991	

Gauge 1ft 3in - The Sutton Miniature Railway

(No.1 SUTTON BELLE)	4-4-2	OC	Cannon		1933
		reb	HuntTG		1953

No.2	SUTTON FLYER	4-4-2	OC	_(Cannon		
				(HuntTG	1950	
No.4		4-4wPMR		G&S	1946	

WILTSHIRE

INDUSTRIAL SITES

DB CARGO UK LTD,
WESTBURY YARD, STATION APPROACH, WESTBURY   BA13 4HP
Gauge 4ft 8½in   www.uk.dbcargo.com   ST 860516

99709 979115-1		4wDH	R/R	Zephir	2928	2021

MINISTRY OF DEFENCE, DEFENCE COMMUNICATIONS SERVICES AGENCY,
BASIL HILL SITE, PARK LANE, CORSHAM
Gauge 600mm - Underground   ST 851694

WD No.1	M.W. 11025	4wDM	RH	179009	1936	OOU
28	M.W. 11026	4wDM	RH	(?)	1938	OOU

MINISTRY OF DEFENCE, LUDGERSHALL RAILHEAD
See Section 6 for details

PRESERVATION SITES

J. BISHOP, PRIVATE SITE, BRADFORD-ON-AVON
Gauge 2ft 6in   Private Site

–	4wDM	RH	359169	1953	

LONGLEAT SAFARI PARK & ADVENTURE PARK,
LONGLEAT RAILWAY, LONGLEAT, WARMINSTER   BA12 7NW
Gauge 1ft 3in   www.longleat.co.uk   ST 808432

7	FLYNN	0-6-0DH	AK	79	2007
8	JOHN THYNN	4w-4wDH	AK	104	2018
	HENRY THYNN	4w-4wDH	AK	112	2021

McARTHUR GLEN, DESIGNER OUTLET SWINDON, KEMBLE DRIVE, SWINDON SN2 2DY

(part of the McArthurglen Group)

Gauge 4ft 8½in   www.mcarthurglen.com   SU 138847

7821	DITCHEAT MANOR	4-6-0	OC	Sdn		1950

NEW COLLEGE SWINDON, DEPARTMENT OF ENGINEERING, NORTH STAR CAMPUS, NORTH STAR AVENUE, SWINDON SN2 1DY

Gauge 1ft 2in   www.newcollege.ac.uk   SU 148855

NORTH STAR	2-2-2	IC	SdnCol		1978

JOHN PAYNE, PRIVATE LOCATION, near SALISBURY

Gauge 4ft 8½in   Private Site

YD No.26653	TANGO	4wDH	BD	3730	1977	
00 NZ 66		4wDH	BD	3732	1977	
–		0-4-0DH	NBQ	27941	1961	
–		4wDM	RH	210481	1941	Dsm

SCIENCE MUSEUM GROUP AT WROUGHTON, NATIONAL COLLECTIONS CENTRE, SCIENCE and INNOVATION PARK, RED BARN GATE, RED BARN, WROUGHTON, near SWINDON SN4 9LT

Not on public display; visitors by appointment only.

Gauge 1ft 11½in   www.sciencemuseumgroup.org.uk   SU 131790

–	0-4-0DM	HE	4369	1951

PAUL STANFORD, CHIPPENHAM

Gauge 2ft 0in   Private Site

3414	2w-2PM	Wkm	3414	1943	
	rebuilt as 2w-2DM			1992	

STEAM - MUSEUM OF THE GREAT WESTERN RAILWAY, SWINDON HERITAGE CENTRE, FIRE FLY AVENUE, SWINDON SN2 2FY

Gauge 7ft 0¼in   www.steam-museum.org.uk   SU 143849

NORTH STAR	2-2-2	IC	Sdn		1925	a

a   replica, incorporating parts of the original, RS 150/1837

Gauge 4ft 8½in

2516		0-6-0	IC	Sdn	1557	1897
2818		2-8-0	OC	Sdn	2122	1905
(3440)	3717 CITY OF TRURO	4-4-0	IC	Sdn	2000	1903
4073	CAERPHILLY CASTLE	4-6-0	4C	Sdn		1923
4248		2-8-0T	OC	Sdn		1916
6000	KING GEORGE V	4-6-0	4C	Sdn		1927

| 9400 | | 0-6-0PT | IC | Sdn | | 1947 |
| A38W | | 2w-2PMR | | Wkm | 8505 | 1960 |

SWINDON & CRICKLADE RAILWAY SOCIETY, near SWINDON

Locomotives are kept at:– Blunsdon Station SN25 2DA SU 110897
Hayes Knoll SU 106907

Gauge 4ft 8½in   www.swindon-cricklade-railway.org

5637		0-6-2T	IC	Sdn		1925	
6695		0-6-2T	IC	AW	983	1928	
35011	GENERAL STEAM NAVIGATION	4-6-2	3C	Elh		1944	
			reb	Elh		1959	
No. 8	FAMBRIDGE	0-4-0ST	OC	AB	2157	1943	
	RICHARD TREVITHICK	0-4-0ST	OC	AB	2354	1954	
3135	SPARTAN	0-6-0T	OC	Chrz	3135	1953	
70		0-6-0T	IC	HC	1464	1921	
	"WILLY"	0-4-0WT	OC	KS	3063	1918	
(D2022)	03022	0-6-0DM		Sdn		1958	
D2152	(03152)	0-6-0DM		Sdn		1960	
D3261	(13261)	0-6-0DE		Derby		1956	
E6003	(73003) SIR HERBERT WALKER	Bo-BoRE/DE		Elh		1962	
PWM 651	(97651)	0-6-0DE		RH	431758	1959	
	WOODBINE	0-4-0DM		JF	21442	1936	
–		0-4-0DM		JF	4210137	1958	a
	BLUNSDON	0-4-0DH		JF	4220031	1964	
–		0-6-0DH		JF	4240015	1962	
51074	595	2-2w-2w-2DMR		GRC&W		1959	
51104	119 021	2-2w-2w-2DMR		GRC&W		1959	
60127	1302 (207203)	4-4wDER		Afd/Elh		1962	
E79978		4wDMR		ACCars		1958	
DR 98504		4wDHR		Plasser	52792	1985	
	99709 901140-2	2w-2PMR		Geismar	M44/090	2011	
68009	(PWM 4305)	2w-2PMR		Wkm	7508	1956	
–		2w-2PMR		Wkm	8089	1958	

a   incorporates parts of JF 4210082 / 1953

P.S. WEAVER, NEW FARM, LACOCK, near CORSHAM

Private Site
ST 899691

Gauge 1ft 9in

| – | | 0-4-0VBT | VCG | WeaverP | | 1978 |

WORCESTERSHIRE

INDUSTRIAL SITES

GRINSTY RAIL LTD,
ARROWVALE BUSINESS PARK, SHAWBANK ROAD, REDDITCH   **B98 8YN**
www.grinstyrail.co.uk   vehicles for overhaul or restoration may occasionally be present

UK LOCO LTD,
UNITS 2 & 3, HEATH PARK, MAIN ROAD, CROPTHORNE, PERSHORE   **WR10 3NE**
www.uk-loco.com   **SO 987446**

New UK Loco vehicles under construction or repair occasionally present

PRESERVATION SITES

THE CHURCHILL (8F) LOCOMOTIVE COMPANY LTD, MALVERN
Gauge 4ft 8½in   Private Site

(45160 8476) 8274	2-8-0	OC	NBH	24648	1941	

DINMORE MANOR LOCOMOTIVE LTD, HONEYBOURNE
Gauge 4ft 8½in   www.dinmoremanor.co.uk   Private Site

3845		2-8-0	OC	Sdn	1942
(55025) 977859 960011 PANDORA	2-2w-2w-2DMR	PSteel			1960

VALLEY RAILWAY ADVENTURE, (EVESHAM VALE LIGHT RAILWAY),
THE VALLEY, (EVESHAM COUNTRY PARK), TWYFORD, EVESHAM   **WR11 4TP**
Gauge 1ft 3in   www.eveshamvalelightrailway.co.uk   **SP 044465**

	MONTY	0-4-2T	OC	ESR	300	1996
	ST EGWIN	0-4-0TT	OC	ESR	312	2003
3	DOUGAL	0-6-2T	OC	SL		1970
	"THE BISCUIT"	4wPMR		Eddy/Knowell		2004
No.4	SLUDGE	4wDM		L	41545	1955
	CROMWELL	4wDM		RH	452280	1960
	rebuilt as	4wDH		AK	13R	1984

LONGLANDS RAILWAY, PRIVATE LOCATION
Gauge 2ft 0in   Private Site

–	4wDM	RH	200512	1940

ROSS-ON-WYE STEAM ENGINE SOCIETY – GREAT WELLAND RAILWAY,
WELLAND STEAM RALLY, WOODSIDE FARM, WELLAND, MALVERN   **WR13 6LN**
Gauge 4ft 8½in   www.wellandsteamrally.co.uk   **SO 803410**

Visiting locomotives may be present on rally days

SEVERN VALLEY RAILWAY CO LTD,
KIDDERMINSTER and BEWDLEY STATIONS
Gauge 4ft 8½in   **SO 836757, 793753**

For details of locomotives see under Shropshire entry.

D. TURNER, "FAIRHAVEN", WYCHBOLD   **Private Site**
Gauge 2ft 0in   **SO 922660**

–	4wDM	MR	8600	1940	

TYTHE HOUSE LIGHT RAILWAY
Gauge 2ft 0in   **Private Site**

–	4wBE	WR			
"No.2"	4wDM	MR	8860	1944	
"No.3"	4wBE	WR	D6905	1964	
–	2w-2PM	Wkm			Dsm

WEST MIDLAND SAFARI PARK, SPRING GROVE, BEWDLEY   **DY12 1LF**
Gauge 400mm - Simba Kiddies Train   www.wmsp.co.uk   **SO 801755**

–	4-6wRE	s/o	_(Dotto	1991	
			(SL	365.3.91	1991

EAST YORKSHIRE

INDUSTRIAL SITES

SOUTH CAVE TRACTORS LTD, t/a SCT-RAIL,
COMMON LANE, NEWPORT, BROUGH, near KINGSTON UPON HULL   **HU15 2RD**
Gauge 4ft 8½in   www.sct-rail.co.uk   **SE 872311**

–	4wBE	R/R	Zagro	80969	2022	a

a   currently at Locomotion, Shildon, Co. Durham
Unimog/ Zagro vehicles for overhaul/resale/refurbishment occasionally present.

SIEMENS AG,
SIEMENS MOBILITY TRAIN MANUFACTURING SITE, GOOLE
Gauge 4ft 8½in   www.mobility.siemens.com

DN14 6XF
SE 732231

78400	332001		4w-4wWER	_(CAF	1997	
				(Siemens	1997	Pvd

WANSFORD TROUT FARM, LTD,
WHIN HILL, WANSFORD, near DRIFFIELD
Gauge 2ft 0in

(YO25 5NW)
TA 051568

–	4wPH	(? Denmark)
–	4wPH	(? Denmark)
–	4wDH	(? Driffield)

PRESERVATION SITES

EAST WRESSLE AND BRIND RAILWAY,
WRESSLE BRICKYARD FARM, WRESSLE, near HOWDEN
Gauge 4ft 8½in

Private Site
SE 724313

JANE	4wDM	RH	371971	1954
PWM 4313 (TR 23 B52)	2w-2PMR	Wkm	7516	1956

HULL CITY MUSEUM & ART GALLERIES, 'STREETLIFE' – MUSEUM OF TRANSPORT,
40 HIGH STREET, KINGSTON UPON HULL
Gauge 4ft 8½in   www.hullmuseums.co.uk

HU1 1PS
TA 103287

	FRANK GALBRAITH	4wVBT	VCG	S		9629	1957

Gauge 3ft 0in

	1	0-4-0Tram	OC	K	2551 T56	1882

MALTON DODGER LTD, t/a YORKSHIRE WOLDS RAILWAY,
FIMBER HALT, BEVERLEY ROAD, FIMBER, near WETWANG
Gauge 4ft 8½in   www.yorkshirewoldsrailway.org.uk

YO25 3HG
SE 911607

5576	SIR TATTON SYKES	0-4-0DH	GECT	5576	1979
	"PATRICIA"	0-6-0DH	JF	4240017	1966

NORTH YORKSHIRE

Locomotives which are located within the County Boroughs of Middlesbrough and Redcar & Cleveland are listed under a separate heading of Teesside.

INDUSTRIAL SITES

AQUARIUS RAILROAD TECHNOLOGIES LTD, OLD SLENINGFORD FARM, MICKLEY, RIPON

HG4 3JB
SE 263769

www.aquariusrail.com   **Private site**   Permission is required prior to visiting.

Road/Rail vehicles under conversion/repair or for hire/resale usually present.

ATKINSON VOS LTD, WENNING AVENUE, HIGH BENTHAM

LA2 7LW
SD 664688

www.unimogs.co.uk

Unimog vehicles for resale, repair or conversion occasionally present.

G.C.S. JOHNSON LTD, BARTON PARK, BARTON, RICHMOND

DL10 6NF
NZ 217079

www.gcsjohnson.co.uk

Locomotives in transit may be stored in this yard for short periods of time

MCL RAIL LTD, GASCOIGNE WOOD SIDINGS, off HAGG LANE, SELBY

LS25 6ES
SE 525315

Gauge 4ft 8½in   www.mclrail.co.uk

PLANT No.11079	SPRITE	0-4-0DH		HC	D1341	1966	
1	TWEETY	4wBE	R/R	Niteq	B259	2007	
		refurbished		BEAZ	M2013	2017	
	TWEETIE	4wBE	R/R	BEAZ	M001	2017	a

a   currently at DB Cargo UK Ltd, Belmont Yard, Doncaster, South Yorkshire
     Other vehicles for storage usually present

NETWORK RAIL, HOLGATE DEPOT, YORK OLD WORKS, HOLGATE ROAD, YORK

YO24 4EH
SE 585518

Gauge 4ft 8½in   www.networkrail.co.uk

99709 979119-3	4wBE	R/R	Colmar		2022	
–	4wBE		Windhoff	101005675/40	2008	a
–	4wBE	R/R	Zephir	2412	2012	

a   located off site at an unknown location

PLASMOR LTD, CONCRETE BLOCK MANUFACTURERS, BLOCK WORKS RAIL TERMINAL, HECK WORKS, GREEN LANE, GREAT HECK

DN14 0BZ
SE 597213

Gauge 4ft 8½in   www.plasmor.co.uk

5481	0-4-0DH	JF	4220038	1966	
5482	4wDH	RR	10280	1968	

A.C. PRICE (ENGINEERING) LTD,
INGLETON INDUSTRIAL ESTATE, NEW ROAD, INGLETON

www.acprice.co.uk

LA6 3NU
SD 691724

Unimog vehicles for resale, repair or conversion occasionally present.

TARMAC plc – A CRH COMPANY,
SWINDEN QUARRY, LINTON, near SKIPTON

Gauge 4ft 8½in   www.tarmac.com

BD23 6BE
SD 983614

CRACOE	6w-6wDH	RFSD	
		067/GA/57000/001	1994

PRESERVATION SITES

TIM ACKERLEY, PRIVATE LOCATION

Gauge 4ft 8½in

Private Site

D3255	0-6-0DE	Derby	1956

DERWENT VALLEY LIGHT RAILWAY SOCIETY, YORKSHIRE MUSEUM OF FARMING, MURTON PARK, MURTON LANE, YORK

Gauge 4ft 8½in   www.dvlr.org.uk

YO19 5UF
SE 650524

–		0-4-0ST	OC	MW	1795	1912
(D2079)	03079	0-6-0DM		Don		1960
D2245	(11215)	0-6-0DM		_(RSHN	7864	1956
				(DC	2577	1956
D3690	(08528)	0-6-0DE		Dar		1959
	DEPARTMENTAL	0-4-0DM		_(DC	2164	1941
	LOCOMOTIVE No.16			_(EEDK	1195	1941
				(VF		1941
	"CHURCHILL"	0-4-0DM		JF	4100005	1947
No.165	BRITISH SUGAR YORK	0-4-0DM		RH	327964	1953
(No.1)		4wDM		RH	417892	1959
(No.2)		4wDM		RH	421419	1958
No.3	KEN COOKE	4wDM		RH	441934	1960
088		4wDM		RH	466630	1962
–		4wWE		KS	1269	1912
–	(DB 965050)	2w-2PMR		Wkm	7565	1957

EMBSAY & BOLTON ABBEY STEAM RAILWAY,
EMBSAY STATION, EMBSAY, near SKIPTON

BD23 6QE

Locomotives are also kept at :–   Bolton Abbey BD23 6AF   SE 060533

Gauge 4ft 8½in   www.embsayboltonabbeyrailway.org.uk

SE 007533

52322	(12322 1300)	0-6-0	IC	Hor	420	1896
	ILLINGWORTH	0-6-0ST	OC	HC	1208	1916

SLOUGH ESTATES LTD No.5		0-6-0ST	OC	HC		1709	1939	
No.140		0-6-0T	OC	HC		1821	1948	
(AIREDALE No.3)		0-6-0ST	IC	HE		1440	1923	
No.7 BEATRICE		0-6-0ST	IC	HE		2705	1945	
S.134 WHELDALE		0-6-0ST	IC	HE		3168	1944	a
(MONCKTON No.1)		0-6-0ST	IC	HE		3788	1953	
CUMBRIA		0-6-0ST	IC	HE		3794	1953	
–		0-4-0ST	OC	RSHN		7661	1950	
D1524	(47004)	Co-CoDE		BT		419	1963	
(D3067)	08054	0-6-0DE		Dar			1953	
(D3941)	08773	0-6-0DE		Derby			1960	
(D5537)	31119	A1A-A1ADE		BT		136	1959	
D5600	(31435)	A1A-A1ADE		BT		200	1960	
(D6994)	37294	Co-CoDE		_(EE		3554	1965	
				(EEV		D983	1965	
(D8169	20169)	Bo-BoDE		_(EE		3640	1966	
				(EEV		D1039	1966	
D8110	(20110)	Bo-BoDE		_(EE		3016	1962	
				(RSHD		8268	1962	
55744	142094	2-2wDMR		_(BRE(D)			1987	
				(Leyland R5.114			1987	
	rebuilt as	2-2wDHR		RFSD			1989	
55790	142094	2w-2DMR		_(BRE(D)			1987	
				(Leyland R5.133			1987	
	rebuilt as	2w-2DHR		RFSD			1989	
3170		4-4wDMR		EBAR/GCR			2016	
	a rebuild of	4-4wPER		NER			1903	
(MEAFORD POWER STATION LOCO No.1)		0-4-0DH		AB		440	1958	
–		0-6-0DM		HC		D1037	1958	
– (H.W.ROBINSON)		0-4-0DM		JF		4100003	1946	
(GR 5093)		4wDM		Robel	21.12 RN5		1973	

a   currently at Statfold Engineering Ltd, Tamworth, Staffordshire

Embsay Narrow Gauge Project, Embsay   BD23 6QE
Gauge 2ft 0in   SE 007533

–	4wPM	L		9993	1938	
–	4wPM	L		10225	1938	OOU

FLAMINGO LAND LTD, FLAMINGO LAND RESORT,
KIRBY MISPERTON, near PICKERING   YO17 6UX
Gauge 2ft 0in – Daktari Express   www.flamingoland.co.uk   SE 778800

97 C.P.HUNTINGTON	4w-2-4wPH s/o	Chance			
		73 5097-24		1973	
1863	4w-2-4wPH s/o	Chance			
		76 50141 24		1976	

Gauge monorail – Zoo monorail   SE 780796

–	2w-2PH	BCM		1984

LIGHTWATER VALLEY LEISURE LTD, LIGHTWATER VALLEY FARM, RIPON

Gauge 1ft 3in   www.lightwatervalley.co.uk

HG4 3HT
SE 285756

278	7		2-8-0DH	s/o	SL	1/84	1984

J.LLOYD, CASTLETON LIGHT RAILWAY, CASTLETON

Gauge 2ft 0in   **Private Site**

–		0-4-0DM	Dtz	47414	1951	
–		4wDM	Eclipse		c1956	
No.10		4wDM	Moës			
	(BECKY)	4wDM	MR	7215	1938	
2		4wDM	MR	7333	1938	
7494	ALNE	4wDM	MR	7494	1940	
		reb	York(BRE)		1991	
	IRTHING	4wDM	MR	8655	1941	

NORTH BAY RAILWAY CO LTD, NORTH BAY RAILWAY, NORTHSTEAD MANOR GARDENS, NORTH BAY, SCARBOROUGH

Gauge 1ft 8in   www.snbr.org.uk

YO12 6PF
TA 035898

1931	NEPTUNE		4-6-2DH	s/o	HC	D565	1931
No.570	ROBIN HOOD		4-6-4DH	s/o	HC	D570	1932
			reb		AK		1982
1932	TRITON		4-6-2DH	s/o	HC	D573	1932
1933	POSEIDON		4-6-2DM	s/o	HC	D582	1933
		rebuilt as	4-6-2DH	s/o	#		1991

    #   rebuilt by Lenwade Hydraulic Services, locomotive was probably DH from new and not as shown.

NORTH YORKSHIRE MOORS HISTORICAL RAILWAY TRUST LTD, NORTH YORKSHIRE MOORS RAILWAY

Locomotives are kept at :-

Grosmont	YO22 5QE	NZ 828049, 828053
Levisham	YO18 7NN	SE 818909
New Bridge	YO18 8JL	SE 803854
Pickering	YO18 7AJ	SE 797842

Gauge 4ft 8½in   www.nymr.co.uk

(30825)	825		4-6-0	OC	Elh		1927	
(30830)			4-6-0	OC	Elh		1930	
30841			4-6-0	OC	Elh		1936	Dsm
30926	REPTON		4-4-0	3C	Elh		1934	
34101	HARTLAND		4-6-2	3C	Elh		1950	
			reb		Elh		1960	
44806			4-6-0	OC	Derby		1944	
44871			4-6-0	OC	Crewe		1945	
(45428)	5428	ERIC TREACY	4-6-0	OC	AW	1483	1937	
(63395)	2238		0-8-0	OC	Dar		1918	
(65894)	2392		0-6-0	IC	Dar		1923	
75029	THE GREEN KNIGHT		4-6-0	OC	Sdn		1954	

76079		2-6-0	OC	Hor		1957	
80135		2-6-4T	OC	Bton		1956	
80136		2-6-4T	OC	Bton		1956	
92134		2-10-0	OC	Crewe		1957	
No.3672	DAME VERA LYNN	2-10-0	OC	NBH	25458	1944	
No.8	"LUCIE"	0-4-0VBT	OC	Cockerill	1625	1890	
			reb	Dorothea		1988	
68030		0-6-0ST	IC	HE	3777	1952	
No.29		0-6-2T	IC	K	4263	1904	
No.5		0-6-2T	IC	RS	3377	1909	
(D1661)	47077 (47613 47840) NORTH STAR						
		Co-CoDE		Crewe		1965	
D2207		0-6-0DM		_(VF	D208	1953	
				(DC	2482	1953	
(D3308	13308) 08238 CHARLIE	0-6-0DE		Dar		1956	
(D3723)	08556	0-6-0DE		Dar		1959	
(D4018)	08850	0-6-0DE		Hor		1961	
D5032	(24032) HELEN TURNER	Bo-BoDE		Crewe		1959	
D5061	(24061) (97201)						
	IAN JOHNSON	Bo-BoDE		Crewe		1960	
(D5533	31115) 31466	A1A-A1ADE		BT	132	1959	
(D6964)	37264	Co-CoDE		_(EE	3524	1965	
				(EEV	D953	1965	
D7628	(25278) SYBILLA	Bo-BoDE		BP	8038	1965	
12139	REDCAR	0-6-0DE		EEDK	1553	1948	
(50160)	53160	2-2w-2w-2DMR		MetCam		1956	
M50164	(53164) DAISY 1956-2003	2-2w-2w-2DMR		MetCam		1956	
E50204	(53204)	2-2w-2w-2DMR		MetCam		1957	
E51511		2-2w-2w-2DMR		MetCam		1959	
–		2w-2PMR		Wkm	417	1931	DsmT
–		2w-2PMR		Wkm	1305	1933	DsmT
–		2w-2PMR		Wkm			DsmT

POPPLETON COMMUNITY RAILWAY NURSERY LTD, STATION ROAD, POPPLETON, YORK

YO26 6QA

Gauge 2ft 0in   www.poppletonrailwaynursery.co.uk   **SE 558536**

No.1	LOWECO	4wDM		L	20449	1942	
			reb	ALR	No.1	1978	
	TERRY STANHOPE	4wDM		StanhopeT			

RAILWAY MUSEUM, LEEMAN ROAD, YORK   **YO26 4XJ**

(part of the Science Museum Group - formerly known as the National Railway Museum)
Some of the National collection is exhibited at 'Locomotion – at Shildon' and is rotated quite frequently.
Some locomotives may be used on 'Mainline Runs' and also exhibited at other locations.

Gauge 4ft 8½in   www.railwaymuseum.org.uk   **SE 593519**

	THE AGENORIA	0-4-0	VC	FosterRastrick		1829	
	ROCKET	0-2-2	OC	RS	4089	1934	a
No.1		4-2-2	OC	Don	50	1870	
No.3		0-4-0	IC	BuryC&K		1846	

82	BOXHILL	0-6-0T	IC	Bton		1880	
214	GLADSTONE	0-4-2	IC	Bton		1882	
673		4-2-2	IC	Derby		1897	
990	HENRY OAKLEY	4-4-2	OC	Don	769	1898	
No.1275		0-6-0	IC	D	708	1874	
4003	LODE STAR	4-6-0	4C	Sdn	2231	1907	
(30245)	No.245	0-4-4T	IC	9E		1897	
(31737)	No. 737	4-4-0	IC	Afd		1901	
(33001)	C1	0-6-0	IC	Bton		1942	
35029	ELLERMAN LINES	4-6-2	3C	Elh		1949	
		reb		Elh		1959	a
(42700	2700) 13000	2-6-0	OC	Hor		1926	
(46229)	6229 DUCHESS OF HAMILTON	4-6-2	4C	Crewe		1938	
(50621)	No.1008	2-4-2T	IC	Hor	1	1889	
(60022)	4468 MALLARD	4-6-2	3C	Don	1870	1938	
60103	(4472) FLYING SCOTSMAN	4-6-2	3C	Don	1564	1923	
		reb		Don		1947	
(68846)	1247	0-6-0ST	IC	SS	4492	1899	
92220	EVENING STAR	2-10-0	OC	Sdn		1960	
	BAUXITE No.2	0-4-0ST	OC	BH	305	1874	
KF7	(607)	4-8-4	OC	VF	4674	1935	
(D1656	47072, 47609,47834) 47798 PRINCE WILLIAM						
		Co-CoDE		Crewe		1965	
D2860		0-4-0DH		YE	2843	1961	
(D4105)	09017	0-6-0DE		Hor		1961	
D8000	(20050)	Bo-BoDE		_(EE	2347	1957	
				(VF	D375	1957	
D9002	(55002)	Co-CoDE		_(EE	2907	1960	
	THE KINGS OWN YORKSHIRE LIGHT INFANTRY			(VF	D559	1960	
W43002	253001 SIR KENNETH GRANGE						
		Bo-BoDE		Crewe		1976	
26020	(76020)	Bo-BoWE		Gorton	1027	1951	
(E3036)	84001	Bo-BoWE		NBH	27793	1960	
87001	STEPHENSON / ROYAL SCOT	Bo-BoWE		Crewe		1973	
3308		Bo-BoWE/RE		GEC-Alsthom		1995	
22-141		4w-4wWER				1976	
7050		0-4-0DM		_(EEDK	874	1934	
				(DC	2047	1934	
(W4W)	No.4	4w-2w+2DMR		_(AEC (852004?)		1934	
				(PRoyal B3550		1934	
M51562		2-2w-2w-2DMR		DerbyC&W		1959	
(51922)		2-2w-2w-2DMR		DerbyC&W		1960	
8143	1293	4w-4RER		_(MV		1925	
				(MetC&W		1925	
28249		4w-4wRER		Oerlikon		1915	

a   locomotive is sectioned

Gauge 900mm

RA 36		4wBE/WE		HE	9423	1990

Gauge 1ft 11½in

No.3	LIVINGSTON THOMPSON	0-4-4-0T	4C	FRCo		1885
		reb		FRCo		1905

Gauge 1ft 6in

PET	0-4-0ST	IC	Crewe	1865	

Gauge 1ft 4½in

–	2-4-0	OC	Young&Co	1856	

Gauge 1ft 2in

FIREFLY	2-2-2	IC	Sdn	1892	

Gauge 1ft 0⅛in

ALICE	2-2-2	IC	ClarkeE	c1845	Dsm

Gauge monorail

W.D. No.1	0-2-2-0BER	Brennan	1907	

RIPON & DISTRICT LIGHT RAILWAY, CANALSIDE, DALLAMIRES LANE, RIPON
Gauge 1ft 11½in   **SE 321705**

7		4wBE	GB	2848	1957
No.6		4wPM	L	7280A	1936
3		4wDM	L	7954	1936
1		4wDM	L	50191	1957
	THE WASP	2-2wPM	WilsonAJ		1969
		reb	WilsonAJ		1979

Gauge 1ft 3in

4	"FERRET"	4wDM	StanhopeT	c1977	

W.D. SMITH, AYSGARTH STATION, AYSGARTH   DL8 3TH
Gauge 4ft 8½in   **Private site**   **SE 013890**

(D2196 03196) 40 JOYCE / GLYNIS 0-6-0DM	Sdn		1961

WENSLEYDALE RAILWAY ASSOCIATION, LEEMING BAR STATION

Locomotives are kept at :-

Leeming Bar	DL7 9AR	SE 013889
Leyburn	DL8 5ET	SE 116902
Bedale	DL8 1AN	SE 268884

Gauge 4ft 8½in   www.wensleydale-railway.co.uk

	RICHARD III	0-6-0T	OC	RSHN	7537	1949
(D1909 47232 47665 47820) 47785	Co-CoDE			BT	671	1965 a
(D2144) 03144		0-6-0DM		Sdn		1961
(D6950) 37250		Co-CoDE		_(EE	3507	1964
				(EEV	D938	1964
(D9513) N.C.B. 38		0-6-0DH		Sdn		1964
D9523		0-6-0DH		Sdn		1964
WL 4		0-6-0DE		BT	804	1978
01526 (265)		4wDH		TH	307V	1983 a
E50256 (53256)		2-2w-2w-2DMR		MetCam		1957
(50746 53746)		2-2w-2w-2DMR		MetCam		1957

51210			2-2w-2w-2DMR	MetCam	1958	
51353	117301	LBRA COMMUNITY HUB				
			2-2w-2w-2DMR	PSteel	1960	b
W51400	117420		2-2w-2w-2DMR	PSteel	1960	
W55032	(121032)		2-2w-2w-2DMR	PSteel	1961	
55559	142018		2-2wDMR	_(BRE(D)	1986	
				(Leyland R5.42	1986	
		rebuilt as	2-2wDHR	RFSD	1989	
55609	142018		2w-2DMR	_(BRE(D)	1986	
				(Leyland R5.41	1986	
		rebuilt as	2w-2DHR	RFSD	1989	
55569	142028		2-2wDMR	_(BRE(D)	1986	
				(Leyland R5.04	1986	
		rebuilt as	2-2wDHR	AB	1990	
55619	142028		2w-2DMR	_(BRE(D)	1986	
				(Leyland R5.003	1986	
		rebuilt as	2w-2DHR	AB	1990	
55576	142035		2-2wDMR	_(BRE(D)	1986	
				(Leyland R5.68	1986	
		rebuilt as	2-2wDHR	AB	1990	
55626	142035		2w-2DMR	_(BRE(D)	1986	
				(Leyland R5.67	1986	
		rebuilt as	2w-2DHR	AB	1990	
55582	142041		2-2wDMR	_(BRE(D)	1986	
				(Leyland R5.86	1986	
		rebuilt as	2-2wDHR	AB	1990	
55632	142041		2w-2DMR	_(BRE(D)	1986	
				(Leyland R5.81	1986	
		rebuilt as	2w-2DHR	AB	1990	
55664	143623		2-2wDMR	_(AB 689	1986	
				(Alex 1784/45	1986	
		rebuilt as	2-2wDHR	AB	1990	
55689	143623		2w-2DMR	_(AB 690	1986	
				(Alex 1784/46	1986	
		rebuilt as	2w-2DHR	RFSD	1990	
55710	142060		2-2wDMR	_(BRE(D)	1986	
				(Leyland R5.168	1986	
		rebuilt as	2-2wDHR	RFSD	1989	
55756	142060		2w-2DMR	_(BRE(D)	1985	
				(Leyland R5.145	1985	
		rebuilt as	2w-2DHR	RFSD	1989	
55737	142087		2-2wDMR	_(BRE(D)	1987	
				(Leyland R5.110	1987	
		rebuilt as	2-2wDHR	RFSD	1989	
55783	142087		2w-2DMR	_(BRE(D)	1987	
				(Leyland R5.123	1987	
		rebuilt as	2w-2DHR	RFSD	1989	
55740	142090		2-2wDMR	_(BRE(D)	1987	
				(Leyland R5.120	1987	
		rebuilt as	2-2wDHR	RFSD	1989	
55786	142090		2w-2DMR	_(BRE(D)	1987	
				(Leyland R5.119	1987	
		rebuilt as	2w-2DHR	RFSD	1989	
55820	144020		2-2wDMR	_(BRE(D)	1987	
				(Alex 2785/39	1987	
		rebuilt as	2-2wDHR	AB	1990	
55843	144020		2w-2DMR	_(BRE(D)	1987	
				(Alex 2785/40	1987	
		rebuilt as	2w-2DHR	AB	1990	

55856	(144020)		2w-2DMR	_(BRE(D)		1988
				(Alex	1187/7	1988
	rebuilt as		2w-2DHR	AB		1990
LEV 1	RDB 975874		4wDMR	Leyland		1978
RTU 6898	99709 901038-8		2w-2PMR	Geismar ST/01/33		2001
WR 3012	MPP 0010		2w-2PMR	Wkm	10731	1974

a   property of Harry Needle Railroad Co Ltd, Derbyshire
b   engines removed; used as static community centre

Aysgarth Station, Aysgarth DL8 3TH
Gauge 4ft 8½in   Private site   SE 013890

| WENSLEY | 4wDM | RH | 476141 | 1963 |

YORKSHIRE DALES NATIONAL PARK AUTHORITY,
DALES COUNTRYSIDE MUSEUM, HAWES STATION, HAWES   DL8 3NT
Gauge 4ft 8½in   www.yorkshiredales.org.uk   SD 875899

| 67345 | 0-6-0T | OC | RSHN | 7845 | 1955 |

SOUTH YORKSHIRE

INDUSTRIAL SITES

C.F. BOOTH LTD, SCRAP MERCHANTS,
CLARENCE METAL WORKS, ARMER STREET, ROTHERHAM   S60 1AF
Locomotives for scrap are usually present at this location.
Gauge 4ft 8½in   www.cfbooth.com   SK 421924

506/1		0-4-0DH		AB	506/1	1969	OOU a
			reb	AB		1989	
506/2		0-4-0DH		AB	506/2	1969	OOU a
			reb	AB		1989	
	LAURA	0-6-0DH		AB	646	1979	
			reb	HAB	6767	1990	
			reb	HE		2004	
–	"LITTLE BLUE"	0-6-0DH		EEV	D1194	1967	
	MADDIE	0-6-0DH		HE	6662	1966	
01565		0-6-0DH		S	10144	1963	
426		4wDH		TH	170V	1966	OOU

a   rebuild of 0-8-0DH AB 506/1965

C.T.L. SEAL LTD,
BUTTERTHWAITE LANE, ECCLESFIELD, SHEFFIELD   S35 9WA
Gauge 4ft 8½in   www.b17steamloco.com // www.72010-hengist.co.uk   Private Site

| 61673 | SPIRIT OF SANDRINGHAM | 4-6-0 | OC | Llangollen | 2015 | u/c |
| 72010 | HENGIST | 4-6-2 | OC | CTL | 2019 | u/c |

Private site with visitors by appointment only

DB CARGO UK LTD

Doncaster Belmont Yard, Balby Carr Bank, Doncaster **DN4 8DE**

Gauge 4ft 8½in www.uk.dbcargo.com **SE 575013**

TWEETIE	4wBE	R/R	BEAZ	M001	2017	a

a property of MCL Rail Ltd, Selby, North Yorkshire

Doncaster Wood Yard, off Decoy Bank South, Doncaster **DN4 5PD**

Gauge 4ft 8½in www.uk.dbcargo.com **SE 577017**

99709 979113-6	4wDH	R/R	Zephir	2926	2021	

ELECTRO-MOTIVE DIESEL LTD, ROBERTS ROAD LOCOMOTIVE MAINTENANCE DEPOT, ROBERTS ROAD, DONCASTER **DN4 0JT**

(Subsidiary of Progress Rail; A Caterpillar Company)

Gauge 4ft 8½in www.progressrail.com **SE 566023**

–	2-2w-2BE	R/R	Harmill		c2018
–	4wBE		Niteq	B284	2009

EUROPEAN METAL RECYCLING LTD, EMR SHEFFIELD, EAST COAST ROAD, ATTERCLIFFE, SHEFFIELD **S9 3YD**

Locomotives for scrap or resale are occasionally present.

Gauge 4ft 8½in www.uk.emrgroup.com **SK 373888**

(D4040)	08872	0-6-0DE	Dar		1960	OOU
(D4111)	09023	0-6-0DE	Hor		1961	
1137		0-6-0DH	EEV	D1137	1966	
–		0-6-0DE	YE	2641	1957	
–		0-6-0DE	YE	2714	1958	

HITACHI RAIL EUROPE LTD, DONCASTER CARR DEPOT, TEN POUND WALK, DONCASTER **DN4 5HX**

Gauge 4ft 8½in www.hitachirail.com **SE 576016**

–	4wBE	R/R	Zephir	2589	2016
–	4wBE		Zephir	2590	2016

RON HULL Jnr LTD, RON HULL GROUP, MANGHAM WORKS, BARBOT HALL INDUSTRIAL ESTATE, MANGHAM ROAD, PARKGATE, ROTHERHAM **S62 6EF**

Locomotives for scrap or resale occasionally present. www.ronhull.co.uk **SK 433950**

LIBERTY STEELS, LIBERTY SPECIALITY STEELS
Aldwarke Works, Rotherham
(S65 3SR)

Locomotives may also be found at Roundwood Works SK 449961 (Closed) and Thrybergh Works
SK 456952 and SK 458953   www.libertysteelgroup.com

Gauge 4ft 8½in   SK 446951, 448955, 449946 and SK 451957

31		0-6-0DE	YE	2904	1964	
32		0-6-0DE	YE	2935	1964	
34		0-6-0DE	YE	2594	1956	OOU
35		0-6-0DE	YE	2635	1957	
37		0-6-0DE	YE	2736	1959	Dsm
93		0-6-0DE	YE	2889	1962	
94		0-6-0DE	YE	2890	1962	
96		0-6-0DE	YE	2905	1963	
No.1	714/37	4wDM	Robel 21.12RK3	1969		

Stocksbridge Works, Stocksbridge, Sheffield
S36 2JA

Gauge 4ft 8½in   www.libertysteelgroup.com   SK 260990, 267992

30	0-6-0DE	YE	2750	1959
33	0-6-0DE	YE	2740	1959
38	0-6-0DE	YE	2798	1961

MECHAN LTD,
DAVY INDUSTRIAL PARK, PRINCE OF WALES ROAD, SHEFFIELD
S9 4EX

UK suppliers of Zwiehoff rail shunters and depot equipment.   www.mechan.co.uk   SK 395875

STAGECOACH SUPERTRAM MAINTENANCE LTD,
NUNNERY SUPERTRAM DEPOT, WOODBOURN ROAD, SHEFFIELD
S9 3LS

(Subsidiary of Stagecoach Holdings)

Gauge 4ft 8½in   www.supertram.com   SK 374878

08	M992 NNB	4wDM	R/R	_(Multicar 000339	1995		
				(Perm	1995	Dsm	a
09	YM02 DJY	4wDM	R/R	_(Hako 000870	2002		
				(Harsco	2002		
	YS63 FMG	4wDM	R/R	_(Hako	2014		
				(APEL	2014		

a   rail wheels removed

WABTEC RAIL LTD,
DONCASTER WORKS, HEXTHORPE ROAD, DONCASTER
DN4 0BF

Locomotives and railcars under repair usually present.

Gauge 4ft 8½in   www.wabteccorp.com   SE 569031

(D1690	47514)	47703	Co-CoDE	BT	622	1967	a
(D1955	47511)	47714	Co-CoDE	BT	617	1966	a
(D3836)	08669	BOB MACHIN	0-6-0DE	Crewe		1960	
(D3892)	08724		0-6-0DE	Crewe		1960	

(D4021)	08853		0-6-0DE	Hor		1961
	PAMMY		4wDH	TH	166V	1966

a   property of Harry Needle Railroad Co Ltd, Derbyshire

PRESERVATION SITES

DARNALL LOCOMOTIVE and RAILWAY HERITAGE COLLECTION, SHEFFIELD
Gauge 2ft 0in   Private Site

740		0-6-0T	OC	OK	2343	1907
91016	RALPH	Bo-BoDE		GEU	38619	1973

DONCASTER COUNCIL, DANUM GALLERY, LIBRARY and MUSEUM,
junction of WATERDALE and CHEQUER ROAD, DONCASTER   DN1 2AA
Gauge 4ft 8½in   www.dglam.org.uk   SE 577030

No.251		4-4-2	OC	Don	991	1902
(60800)	4771 GREEN ARROW	2-6-2	3C	Don	1837	1936

ELSECAR STEAM RAILWAY,
ELSECAR HERITAGE CENTRE, WATH ROAD, ELSECAR, BARNSLEY   S74 8HJ
Gauge 4ft 8½in   (Closed)   SE 386998

	ELIZABETH		4wDH		TH	138C	1964
		a rebuild of	4wVBT	VCG	S	9584	1955
–			2w-2DM				a

a   dumper truck with fixed rail wheels

Gauge 2ft 2in

–		4wDM	RH	382808	1955

NATIONAL COLLEGE for ADVANCED TRANSPORT and INFRASTRUCTURE,
CAROLINA WAY, DONCASTER   DN4 5PN
Gauge 4ft 8½in   (Closed)   SE 593012

3101	#DONNYSTAR	Bo-BoWE/RE	GEC-Alsthom	1992	
93830 001006-4	AGV	4w-4wWER-artic	Alstom(L)	2007	a

a   power car from articulated Electric Multiple Unit

SHEFFIELD INDUSTRIAL MUSEUMS TRUST,
KELHAM ISLAND INDUSTRIAL MUSEUM, ALMA STREET, SHEFFIELD   S3 8RY
This locomotive is not on public display.

Gauge 4ft 8½in   www.sheffieldmuseums.org.uk   SK 352882

BSC 1	0-4-0DE	YE	2481	1950

SOUTH YORKSHIRE TRANSPORT MUSEUM,
UNIT 9, WADDINGTON WAY, ALDWARKE, ROTHERHAM   S65 3SH
Gauge 4ft 8½in   www.sytm.co.uk   SK 440944

(BROWN BAYLEY No.7)	0-4-0ST	OC	HC	1689	1937

WEST YORKSHIRE

INDUSTRIAL SITES

CROSSLEY EVANS LTD, METAL PROCESSORS,
STATION SIDINGS, OTLEY ROAD, SHIPLEY   BD18 2BH
Gauge 4ft 8½in   www.crossleyevans.co.uk   **Closed**   SE 148372

42 M	PRINCE OF WALES	0-4-0DH	HE	7159	1969	OOU
01507	425 VENOM	0-6-0DH	RH	459519	1961	OOU

W.H. DAVIS LTD, WAGON REPAIRERS, FERRYBRIDGE DEPOT,
c/o GMOS FERRYBRIDGE MAINTENANCE & ENGINEERS,
RWE GENERATION UK, OLD GREAT NORTH ROAD, KNOTTINGLEY   WF11 8NG
Gauge 4ft 8½in   www.whdavis.co.uk   SE 481252

01515	(304)	4wDH		TH	V321	1987	
			rep	LH Group	76629	2002	a
01520	(274)	4wDH		TH	V322	1987	a

a   property of Harry Needle Railroad Co Ltd, Worksop, Nottinghamshire

DB CARGO (UK) LTD, KNOTTINGLEY DEPOT,
SPAWD BONE LANE, KNOTTINGLEY, PONTEFRACT   WF11 OUG
Gauge 4ft 8½in   www.uk.dbcargo.com   SE 493235

99709 979114-4	4wDH	R/R	Zephir	2927	2021

Riviera Trains Ltd
Gauge 4ft 8½in   www.riviera-trains.co.uk   SE 493235

(D3871) 08704	0-6-0DE	Hor		1960

FREIGHTLINER MAINTENANCE LTD, MIDLAND ROAD, HUNSLET, LEEDS
(part of the America Genesee & Wyoming Railway Co)   LS10 2RJ
Gauge 4ft 8½in   www.freightliner.co.uk   SE 312311

–	4wBE	Express	ES407	2012

NORTHERN TRAINS LTD (owned by DfT OLR Holdings Ltd),
NEVILLE HILL TRAINCARE DEPOT, OSMONDTHORPE LANE, LEEDS　　**LS9 9BJ**
Locomotives and railcars from other railway operators may also be present

Gauge　4ft 8½in　　　www.northernrailway.co.uk　　　**SE 329331, 327329, 330330**

(D3520)	08405		0-6-0DE	Derby	1958	a
(D3687)	08525	DUNCAN BEDFORD	0-6-0DE	Dar	1959	**M** OOU
(D3857)	08690	DAVID THIRKILL	0-6-0DE	Hor	1959	**M** OOU
(D4138)	08908	IVAN STEPHENSON	0-6-0DE	Hor	1962	**M** OOU
(D4180)	08950	DAVID LIGHTFOOT	0-6-0DE	Dar	1962	**M**
	–		4wBE	Windhoff 101005675/30	2008	

a　　property of Railway Support Services Ltd, Wishaw, Warwickshire

PRESERVATION SITES

AIREDALE NHS FOUNDATION TRUST,
AIREDALE GENERAL HOSPITAL, SKIPTON ROAD, STEETON, KEIGHLEY BD20 6TD
Gauge　4ft 8½in　　　www.airdale-trust.nhs.uk　　　　　　　**SE 025446**

55801	144001		2-2wDMR	_(BRE(D)	1986
				(Alex　2785/01	1986
		rebuilt as	2-2wDHR	AB	1990

ALLAN BAMFORD, MODEL FARM,
TOFTSHAW FOLD, off TOFTSHAW LANE, EAST BIERLEY, BRADFORD　　**BD4 6QR**
Gauge　4ft 8½in　　　www.modelfarmshopbradford.co.uk　　　**SE 189296**

| – | 0-4-0F | OC | WB | 2473 | 1932 |

CITY OF BRADFORD METROPOLITAN COUNCIL ART GALLERIES & MUSEUMS,
BRADFORD INDUSTRIAL & HORSES AT WORK MUSEUM, MOORSIDE MILLS,
MOORSIDE ROAD, BRADFORD　　　　　　　　　　　　　　**BD2 3HP**
Gauge　4ft 8½in　　　www.bradfordmuseums.org　　　　　　**SE 184353**

| NELLIE | 0-4-0ST | OC | HC | 1435 | 1922 |

THE CHILDRENS MUSEUM LTD, EUREKA !,
THE NATIONAL CHILDRENS MUSEUM, DISCOVERY ROAD, HALIFAX　**HX1 2NE**
Gauge　4ft 8½in　　　www.eureka.org.uk　　　　　　　　　**SE 097247**

| 02641 | 0-4-0DM | HE | 2641 | 1941 |

FAGLEY PRIMARY SCHOOL, FALSGRAVE AVENUE, BRADFORD　　**BD2 3PU**
Gauge　4ft 8½in　　　www.fagley.bradford.sch.uk　　　　　**SE 187349**

55808	144008		2-2wDMR	_(BRE(D)	1986
				(Alex　2785/15	1986
		rebuilt as	2-2wDHR	AB	1991

KEIGHLEY & WORTH VALLEY LIGHT RAILWAY LTD

Locomotives are kept at :-

Haworth	BD22 8NJ	SE 034371
Ingrow	BD21 5AX	SE 058399
Oxenhope	BD22 9LB	SE 032355

Gauge 4ft 8½in   www.kwvr.co.uk

5775		0-6-0PT	IC	Sdn		1929
41241		2-6-2T	OC	Crewe		1949
43924		0-6-0	IC	Derby		1920
45212		4-6-0	OC	AW	1253	1935
45596	BAHAMAS	4-6-0	3C	NBQ	24154	1935
		reb		HE	5596	1968
47279		0-6-0T	IC	VF	3736	1924
48431		2-8-0	OC	Sdn		1944
51218	(68)	0-4-0ST	OC	Hor	811	1901
52044	(957)	0-6-0	IC	BP	2840	1887
(58926	7799) 1054	0-6-2T	IC	Crewe	2979	1888
75078		4-6-0	OC	Sdn		1956
78022		2-6-0	OC	Dar		1954
80002		2-6-4T	OC	Derby		1952
No.2258	TINY	0-4-0ST	OC	AB	2258	1949
	LORD MAYOR	0-4-0ST	OC	HC	402	1893
(31)	(HAMBURG)	0-6-0T	IC	HC	679	1903
No.1704	NUNLOW	0-6-0T	OC	HC	1704	1938
118	BRUSSELS	0-6-0ST	IC	HC	1782	1945
5820		2-8-0	OC	Lima	8758	1945
No.85		0-6-2T	IC	NR	5408	1899
–		0-4-0CT	OC	RSHN	7069	1942
90733		2-8-0	OC	VF	5200	1945
D2511		0-6-0DM		HC	D1202	1961
(D3336	13336) 08266	0-6-0DE		Dar		1957
(D3759	08592) 08993 ASHBURNHAM	0-6-0DE		Crewe		1959
		reb		Landore		1985
(D6775)	37075	Co-CoDE		_(EE	3067	1962
				(RSHD	8321	1962
(D8031)	20031	Bo-BoDE		_(EE	2753	1959
				(RSHD	8063	1959
D0226	VULCAN	0-6-0DE		_(EE	2345	1956
				(VF	D226	1956
23	MERLIN	0-6-0DM		HC	D761	1951
32	HUSKISSON	0-6-0DM		HE	2699	1944
	JAMES	0-4-0DE		RH	431763	1959
M50928		2-2w-2w-2DMR		DerbyC&W		1959
M51189		2-2w-2w-2DMR		MetCam		1958
M51565		2-2w-2w-2DMR		DerbyC&W		1959
Sc51803		2-2w-2w-2DMR		MetCam		1959
55666	143625	4wDMR		_(AB	693	1986
				(Alex	1784/49	1986
	rebuilt as	4wDHR		AB		1990
55691	143625	4wDMR		_(AB	694	1986
				(Alex	1784/50	1986
	rebuilt as	4wDHR		AB		1990

55811	144011		4wDMR	_(BRE(D)	1986
				(Alex 2785/21	1986
		rebuilt as	4wDHR	BRE(D)	1988
55834	144011		4wDMR	_(BRE(D)	1986
				(Alex 2785/22	1986
		rebuilt as	4wDHR	BRE(D)	1988
E79962			2w-2DMR	WMD 1267	1958
M79964			2w-2DMR	WMD 1269	1958

KEIGHLEY BUS MUSEUM TRUST LTD, KEIGHLEY BUS MUSEUM, UNIT 5, RIVER TECHNOLOGY PARK, RIVERSIDE, DALTON LANE, KEIGHLEY   BD21 4JP
Gauge 4ft 8½in   www.kbmt.org.uk   SE 071413

(9036)	2w-2PMR	Wkm	8196	1958

THE KINSLEY HOTEL, WAKEFIELD ROAD, KINSLEY, near PONTEFRACT   WF9 5EH
Gauge 2ft 0in   SE 419144

713009	4wBE	CE	B0182A	1974

LEEDS CITY COUNCIL, DEPARTMENT OF LEISURE SERVICES, LEEDS MUSEUMS & GALLERIES, LEEDS INDUSTRIAL MUSEUM, ARMLEY MILLS, CANAL ROAD, LEEDS   LS12 2QF
Gauge 4ft 8½in   www.museumsandgalleries.leeds.gov.uk   SE 275342

(GWR 252)		0-6-0	IC	EBW	1855	Dsm
	ELIZABETH	0-4-0ST	OC	HC	1888	1958
R.A.F.No.111	ALDWYTH	0-6-0ST	IC	MW	865	1882
–		4wBE		GB	1210	1930
			reb	HE	9146	1987
S 1986.0028		0-4-0WE		GB	2543	1955
	SOUTHAM 2	0-4-0DM		HC	D625	1942
B16	ND 3066	0-4-0DM		HE	2390	1941
		0-4-0DM		JF	22060	1937
S 1990.0013		0-4-0DM		JF	22893	1940

Gauge 3ft 6in

	PIONEER	0-6-0DMF	HC	DM634	1946
S 1985.0008		0-6-0DMF	HC	DM733	1950

Gauge 3ft 0in

S 1985.0014	4057	0-6-0DMF	HE	4057	1953
S 1985.0020		2-4-0DM	JF	20685	1935

Gauge 2ft 11in

–	4wDM	HC	D571	1932

Gauge 2ft 8in

–	0-4-0DMF	HE	3200	1945

Gauge 2ft 6in

S 1990.0015 "JUNIN"	2-6-2DM	HC	D557	1930	
S 1992.0009	4wBE	HT	9728	1985	

Gauge 2ft 1½in

No.5	0-4-0DMF	HE	4019	1949	

Gauge 2ft 0in

S 1992.0020 FAITH	0-4-0DMF	HC	DM664	1952	
–	0-4-0DMF	HC	DM749	1949	
1368	4wDMF	HC	DM1368	1965	
S.2000.0002	0-4-0DMF	HE	2008	1939	
8	4wDMF	HE	4756	1954	
–	0-4-0DMF	HE	5340	1957	
21294 LAYER	4wDM	JF	21294	1936	Dsm

Gauge 1ft 6in

–	4wBE	GB	1325	1933	Dsm

MIDDLETON RAILWAY TRUST,
MOOR ROAD STATION, TUNSTALL ROAD, HUNSLET, LEEDS   LS10 2JQ

Gauge 4ft 8½in   www.middletonrailway.org.uk   SE 305310

No.1310		0-4-0T	IC	Ghd	(38?)	1891
68153	59	4wVBT	VCG	S	8837	1933
Nr.385		0-4-0WT	OC	Hart	2110	1895
	HAWARDEN	0-4-0ST	OC	HC	526	1899
	HENRY DE LACY II	0-4-0ST	OC	HC	1309	1917
67		0-6-0T	IC	HC	1369	1919
	SLOUGH ESTATES No.3	0-6-0ST	OC	HC	1544	1924
	MIRVALE	0-4-0ST	OC	HC	1882	1955
–		0-4-0ST	OC	HE	1493	1925
	"PICTON"	2-6-2T	OC	HE	1540	1927
–		0-4-0T	OC	HE	1684	1931
	"BROOKES No.1"	0-6-0ST	IC	HE	2387	1941
	rebuilt as	0-6-0T	IC	Middleton		1999
	rebuilt as	0-6-0ST	IC	Middleton		2007
No.6	(SWANSCOMBE)	0-4-0ST	OC	HL	3860	1935
44	"CONWAY"	0-6-0ST	IC	K	5469	1933
	SIR BERKELEY	0-6-0ST	IC	MW	1210	1891
	MATTHEW MURRAY	0-6-0ST	IC	MW	1601	1903
D2999		0-4-0DE		_(BT	91	1958
				(BP	7856	1958
	MARY	0-4-0DM		HC	D577	1932
	CARROLL	0-4-0DM		HC	D631	1946
MDHB 45		0-6-0DH		HC	D1373	1965
7051	(WD 27 70027)	0-6-0DM		HE	1697	1932
	"JOHN ALCOCK"	reb		HE		1949
	"COURAGE" "SWEET PEA"	4wDM		HE	1786	1935
	CONOCO	0-4-0DH		HE	6981	1968
–		0-4-0DM		JF	3900002	1945
	HARRY	0-4-0DH		JF	4220033	1965

AUSTINS No.1	0-4-0DM		P	5003	1961	
(03-03-PO300)	4wWE		GB	420452	1979	
DB 998901 "OLIVE"	2w-2DMR		Bg/DC	2268	1950	

Gauge 3ft 0in

BEM 402	4wDHF	RACK	HE	8505	1981	

Gauge 2ft 2in

"FLYING SCOTSMAN"	4wDM	HE	7274	1973	

PLATFORM 1 : MENTAL HEALTH & CRISIS SUPPORT, PLATFORM 1, HUDDERSFIELD STATION, ST. GEORGES SQUARE, HUDDERSFIELD

HD1 1JF

Gauge 4ft 8½in   www.platform-1.co.uk

SE 143169

55824	144001		2w-2DMR	_(BRE(D)		1986
				(Alex	2785/02	1986
	rebuilt as		2w-2DHR	AB		1990	Dsm

THE SHIRES REMOVAL GROUP, THE DEPOSITORY, HOYLE MILL ROAD, HEMSWORTH WAY, KINSLEY, PONTEFRACT

WF9 5JB

Gauge 4ft 8½in   www.shiresremovals.com

SE 425146

(D6717	37017) 37503	Co-CoDE	_(EE	2880	1961	
			(VF	D596	1961	
60050	ROSEBERRY TOPPING	Co-CoDE	BT	952	1991	
60086	SCHIEHALLION	Co-CoDE	BT	988	1991	

    ·   private site with locomotives for storage

THE STEAM WORKSHOP, HECKMONDWIKE

Gauge 1ft 3in   www.steamworkshop.co.uk

JACK	0-6-0	OC	LemonB		c1956

    Locomotives for overhaul, restoration or resale occasionally present

Mr TAYLOR, CALDERDALE

Gauge 2ft 0in

Private Site

4470	4wPM	OK	4470	1931	

WHISTLESTOP VALLEY LTD, (Kirklees Light Railway), PARK MILL WAY, CLAYTON WEST, near HUDDERSFIELD

HD8 9XJ

Gauge 1ft 3in   www.whistlestopvalley.co.uk

SE 258112

KATIE	2-4-2	OC	Guest		1956	a
SIÂN	2-4-2	OC	Guest	18	1963	
FOX	2-6-2T	OC	TaylorB	No.9	1987	
BADGER	0-6-4ST	OC	TaylorB	No.10	1991	
HAWK	0-4-4-0T	4C	TaylorB		2007	b
OWL	4w-4wT	VC	TaylorB	No.12	2000	

(7)			2-2wPH	s/o	TaylorB	1991	Dsm
No.8	JAY		4wDH		TaylorB	1992	
7	"TOBY"		4wDM	s/o	KLR	c2019	

a   carries plate Guest 14/1954
b   rebuilt using parts from TaylorB No.11 1998

YORKSHIRE MINING MUSEUM TRUST,
NATIONAL COAL MINING MUSEUM FOR ENGLAND, CAPHOUSE COLLIERY,
NEW ROAD, OVERTON, WAKEFIELD   WF4 4RH

Gauge 4ft 8½in   www.ncm.org.uk   SE 248161, 249162, 253164

	PROGRESS	0-6-0ST	IC	RSHN	7298	1946	
(No.20)	MANTON	0-6-0DM		HC	D1121	1958	
NCB 44		0-6-0DH		HE	6684	1968	
–		0-6-0DH		HE	7307	1973	
(No.47)		0-6-0DH		TH	249V	1974	

Gauge 2ft 6in

No. 1	4471 COMPO	4w-4wBEF		CE	B3538	1989	
No.7	SICK NOTE	4w-4wBEF		CE			
	KIRSTIN	4w-4wDHF	RACK	GMT	0592	1981	
	(ANNA) YKSMM 2001.831	4w-4wDHF	RACK	GMT		1984	
20		4wDHF	RACK	HE	9271	1987	
(E682)		0-4-0DMF		HC	DM746	1951	
No.2	"DEBORAH"	0-4-0DMF		HC	DM1356	1965	
	"LARK"	0-6-0DMF		_(HC	DM1433	1978	
				(HE	8581	1978	
2	T198	4wDMF		RH	480679	1961	

Gauge 2ft 4in

–		4wBEF	Atlas	2463	1944	
	YKSMM 1997.857	4wDM	RH	375347	1954	

Gauge 2ft 3in

(8)		4wDH	HE	6273	1965	
	CAPHOUSE FLIER	4wDHF	HE	8832	1978	

Gauge 2ft 2in

	YKSMM 1992.156	4wDH	HE	7530	1977	

Gauge 2ft 1½in

T199		4wDM	RH	379659	1955	

Gauge 2ft 0in

	YKSMM 1986.54	0-4-0DMF	HC	DM655	1949	
713007		4wBE	CE	B0182B	1974	

SECTION 2 — SCOTLAND

Consequent upon the fragmentation of the County Geography of Scotland, due to the creation of a number of autonomous Unitary Authorities and similar bodies and, in view of the relatively few remaining locations hosting locomotives, this Section of this Handbook is now presented in two parts - Industrial sites and Preservation sites.

Defining Areas in accordance with the 1974-1996 "large counties" era, now provides a direct relationship between this volume and the "Historic Handbook" series which is also published by the Industrial Railway Society. The historic books contain the recorded details of all past and recent locomotives, at all known past and present locations, in the areas they cover, together with extensive texts describing the locations and the businesses they served.

INDUSTRIAL SITES

1ST BATTALION, THE ROYAL HIGHLAND FUSILIERS,
FORT GEORGE RANGE, INVERNESS (IV2 7TD)
Gauge 600mm **Highland NH 783571**

	BEN	2w-2PM	Wkm	11682	1990
	BRUCE	2w-2PM	Wkm	11683	1990

ALSTOM TRANSPORT, GLASGOW TRAINCARE DEPOT,
109 POLMADIE ROAD, POLMADIE, GLASGOW (part of Alstom Holdings S.A.) **G5 0BA**
Gauge 4ft 8½in www.alstom.com **City of Glasgow NS 598625**

(D3566)	08451	LOOPY LOU	0-6-0DE	Derby		1958
(D3932)	08764		0-6-0DE	Hor		1961
(D4184)	08954		0-6-0DE	Dar		1963

BRODIE ENGINEERING LTD, BONNYTON RAIL DEPOT,
1 BONNYTON INDUSTRIAL ESTATE, MUNRO PLACE, KILMARNOCK **KA1 2NP**
Gauge 4ft 8½in www.brodie-engineering.co.uk **East Ayrshire NS 421383**

18	TINY II	0-4-0DH	YE	2676	1959

Caledonia Works, West Langlands Street, Kilmarnock **KA1 2QD**
Gauge 4ft 8½in www.brodie-engineering.co.uk **East Ayrshire NS 425382**

01584		0-4-0DH	AB	482	1963
	GEORGE TOMS	0-4-0DH	_(WB	3209	1962
			(RSHD	8364	1962

R. & N. CESSFORD, WHANLAND FARM, FARNELL, BRECHIN **DD9 6UF**
Gauge 4ft 8½in www.cessfordgroup.co.uk **Angus NO 621543**

-		4wDM	R/R	S&H	7505	1967	OOU

stored off site. Visitors by appointment only.

CHEMRING ENERGETICS UK LTD, (Part of the Chemring Group plc), ARDEER, LUNDHOLM ROAD, STEVENSTON KA20 3NF

Gauge 2ft 6in   www.chemring.com   **North Ayrshire NS 290401, 290405**

05/582	4wDH		AK	21	1987
05/583	4wDH		AK	22	1987

DIRECT RAIL SERVICES LTD, MOTHERWELL DEPOT, PARKNEUK STREET, MOTHERWELL ML1 3ST

Gauge 4ft 8½in   www.directrailservices.com   **North Lanarkshire NS 749579**

–	4wBE	R/R	Zwiehoff		(2021)

DSM NUTRITIONAL PRODUCTS (UK) LTD, DRAKEMYRE, DALRY KA24 5JJ

Gauge 4ft 8½in   www.dsm.com   **North Ayrshire NS 295503**

50 GC 8	0-6-0DH		RR	10267	1967
		reb	RFSK		1990

EDINBURGH TRAMS LTD, GOGAR DEPOT, off GLASGOW ROAD, GOGAR, EDINBURGH EH12 9DH

Gauge 4ft 8½in   www.edinburghtrams.com   **City of Edinburgh NT 172726**

–	4wBE	R/R	Niteq	B300	2009
SN12 DWG	4wDM	R/R	Unimog	224283	2010

GIBSONS ENGINEERING LTD, SPRINGBURN RAIL DEPOT, 79 CHARLES STREET, off SPRINGBURN ROAD, GLASGOW G21 2PS

Gauge 4ft 8½in   **Glasgow NS 605665**

Locomotives and railcars for overhaul and repair usually present

INEOS, GRANGEMOUTH REFINERY, BO'NESS ROAD, GRANGEMOUTH

Gauge 4ft 8½in   www.ineos.com   **Falkirk NS 942817, 944814, 952822**

10	AVON	0-6-0DH		AB	600 1976	a
11	FORTH	0-6-0DH		AB	649 1980	
			reb	HE	750215 2006	a

a   property of Hunslet Ltd, Staffordshire, England

LITHGOWS LTD, NETHERTON, LANGBANK PA14 6YG

Gauge 2ft 0in   R.T.C.   **Renfrewshire NS 393722**

–	4wPM		MR	2097	1922 OOU
–	4wPM		MR	2171	1922 OOU
No.2	4wDM		MR	8700	1941 OOU

LONDON NORTH EASTERN RAILWAY LTD, t/a LNER,
Craigentinny Maintenance Depot, 167 Mountcastle Crescent, Edinburgh   **EH8 7TE**

Gauge 4ft 8½in   www.lner.co.uk   **City of Edinburgh**   **NT 298738**

(D3595)	08480	0-6-0DE		Hor	1958	a
–		4wBE	R/R	Niteq	2023	

    a   property of Railway Support Services Ltd, Wishaw, Warwickshire

Craigentinny Wheel Lathe Depot, Stanley Street, Edinburgh   **EH15 1JJ**

Gauge 4ft 8½in   **City of Edinburgh**   **NT 306732**

–	4wBE	Windhoff	101005675/10	2008

W & D McCULLOCH, CRAIGIE MAINS, MAIN STREET, BALLANTRAE   **KA26 0NB**

Gauge 4ft 8½in   www.mccullochrail.com   **Ayrshire**   **NX 083829**

FH06 LBG	99709 979073-4	4wDM	R/R	_(Unimog	209478	2006
				(LH Access		2006
YX67 AMV	99709 977049-4	4wDM	R/R	_(Unimog	249135	2017
				(Zagro	4453	2017
	99709 909319-4	4w-4wDHR		_(McCulloch		2016
	THE BIG GIRL			(McDowall		2016

    Other road/rail vehicles usually present at this location

MINISTRY OF DEFENCE, DEFENCE MUNITIONS, GLEN DOUGLAS

See Section 6 for details

NETWORK RAIL, HALKIRK BALLAST TIP, HALKIRK

Gauge 4ft 8½in   **Highland**   **ND 126580**

–	2w-2PMR		Wkm	Dsm

SCOTRAIL TRAINS LTD, t/a SCOTRAIL (subsidiary of Scottish Rail Holdings Ltd)
Haymarket Maintenance Depot, Russell Road, Edinburgh   **EH12 5NB**

Gauge 4ft 8½in   www.scotrail.co.uk   **City of Edinburgh**   **NT 229728**

–	4wBE	R/R	Zwiehoff	c2019

Inverness Rail Depot, Longman Road, Inverness   **IV1 1RY**

Gauge 4ft 8½in   www.scotrail.co.uk   **Highland**   **NH 668457**

(D3815)	08648	H 065	0-6-0DE	Hor	1959	a

    a   property of Rail Management Services Ltd, Chesterfield, Derbyshire

Shields Electric Traction Depot, 35 St. Andrews Drive, Pollokshields, Glasgow
G41 5SG

Gauge 4ft 8½in www.scotrail.co.uk **City of Glasgow** NS 371638, 367639

–		4wBE	Windhoff	101005675/20	2008	

THOS MUIR HAULAGE & METALS LTD, (part of the Thomas Muir Group), THOS MUIR METALS LTD, DEN ROAD, KIRKCALDY
KY1 2ER

Gauge 4ft 8½in www.thomasmuir.net **Fife** NT 282926

No.3		0-4-0ST	OC	AB	946	1902	OOU
No.22		0-4-0ST	OC	AB	1069	1906	OOU
–		0-4-0ST	OC	AB	1807	1923	OOU
No.7		0-4-0ST	OC	AB	2262	1949	OOU
No.12 H 662		0-4-0DH		NBQ	27732	1957	OOU

THOS MUIR ROSYTH LTD, (part of the Thomas Muir Group), CROMARTY CAMPUS, STABLES ROAD, ROSYTH
KY11 2YB

Gauge 4ft 8½in www.thomasmuir.net **Fife** NT 102823

51655	2w-2-2-2wDMR	DerbyC&W		1960	OOU

NUCLEAR DECOMMISSIONING AGENCY, CHAPELCROSS WORKS, ANNAN
DG12 6RF

(operated by Magnox Ltd) www.magnoxstakeholdergroups.com

Gauge 5ft 4in (Closed) **Dumfries & Galloway** NY 216695

No.1 PETER	4wDM	RH	411320	1958	
No.2 JIM	4wDM	RH	411321	1958	

QTS RAIL LTD, RENCH FARM, DRUMCLOG, near STRATHAVEN
ML10 6QJ

(part of the QTS Group) Plant depot with various road-rail and rail plant present between contracts. Permission is required before visiting this location.

Gauge 4ft 8½in www.qtsgroup.com **Lanarkshire** NS 630386

RRU 09 AD02 FKU 99709 979060-9	4wDM	R/R	_(Unimog	197737	2002	
			(Zagro		2002	

RAIL SIDINGS LTD, EASTRIGGS

Gauge 4ft 8½in www.railsidings.com **Dumfries and Galloway** NY 246656

Railway storage facility with railway vehicles usually present

JOHN G. RUSSELL (TRANSPORT) LTD, t/a RUSSELL LOGISTICS, DEANSIDE ROAD, HILLINGTON, GLASGOW   G52 4XB
(a member of The Russell Group)

Gauge 4ft 8½in   www.johngrussell.co.uk   **City of Glasgow NS 522659**

(D3562)	08447	0-6-0DE	Derby	1958	

SCOTTISH WATER, STORNOWAY WATERWORKS, ISLE OF LEWIS
Gauge 2ft 0in   www.scottishwater.co.uk   **Western Isles NB 410375**

–	4wPM	(MR ?)		Dsm

E.G. STEELE & CO LTD, WINTON WAGON WORKS, 25 DALZELL STREET, HAMILTON   ML3 9AU
The following is a FLEET LIST of the locomotives owned by this contractor.

Gauge 4ft 8½in   www.egsteele.com   **North Lanarkshire NS 708561**

–		4wDH	R/R	NNM	75511	1979	OOU	
–		4wDH	R/R	NNM	81504	1983		
T4	05/273	4wDH	R/R	NNM	83503	1984		
–		4wDH	R/R	NNM			a	

a   one of NNM 82503/1983, NNM 83501/1983 or NNM 83504/1984

Assenta Rail Ltd
Gauge 4ft 8½in   www.assentarail.co.uk   **North Lanarkshire NS 708561**

D1388	6 "CLAIRE"	0-4-0DH		HC	D1388	1970	
3		0-6-0DH		Jung	12842	1958	
			reb	Newag	273	2015	

STRATHCLYDE PARTNERSHIP FOR TRANSPORT, BROOMLOAN DEPOT, ROBERT STREET, GOVAN   G51 3HB
Glasgow underground railway maintenance locomotives.

Gauge 4ft 0in   www.spt.co.uk   **City of Glasgow NS 555655**

L2	LOBEY DOSSER	4wBE		CE	B0965B	1977	
L3	RANK BAJIN	4wBE		CE	B0965A	1977	
W5		4wBE		CE	B0186	1974	Dsm a
L6		4wBE		CE	B4477A	2010	
L7		4wBE		CE	B4477B	2010	
–		4wBE	R/R	NNM	78101E	1979	
–		4wBE	R/R	Zephir	2766	2018	
–		4wBE	R/R	Zephir	2922	2020	
–		2w-2BER		Bance	ERV2 276	2015	
–		2w-2BER		Bance	ERV2 278	2015	
–		2w-2BER		Consillia		20xx	

a   rebuilt into non-powered permanent way vehicle

Stadler Rail Service UK Ltd,
Glasgow Metro Contract, off Broomloan Road, Ibrox   (G51 2XW)
Gauge 4ft 0in   www.stadlerrail.com   City of Glasgow NS 551645

–		4wBE		CE	B4624A	2017
–		4wBE		CE	B4624B	2017

TARMAC plc – A CRH Company,
OXWELLMAINS CEMENT WORKS, DUNBAR   EH42 1SL
Gauge 4ft 8½in   www.tarmac.com   East Lothian NT 708768

DOON HILL	0-6-0DH		HE	7304	1972
		reb	HE	9374	2009
BLACK AGNES	0-6-0DH		HE	8979	1979
		reb	HE	9373	2009
DOVEDALE	0-6-0DH		RR	10284	1969
		reb	TH		1974

UPM-KYMMENE (UK) LTD, CALEDONIAN PAPER MILL,
MEADOWHEAD ROAD, SHEWALTON, IRVINE   KA11 5AT
Gauge 4ft 8½in   www.upmpaper.com   North Ayrshire NS 335354

–	0-6-0DH		HE	9092	1988
CHRISTIAN	0-6-0DH		RR	10217	1965
		reb	HE	9371	2006

WILSONS AUCTIONS LTD, 6 KILWINNING ROAD, DALRY   KA24 4LG
Gauge 3ft 0in   www.wilsonsauctions.com   North Ayrshire NS 297478

LM 205	A10	0-4-0DM		HE	6238	1963

PRESERVATION SITES

ABERDEEN CITY COUNCIL, SEATON PARK, DON STREET, ABERDEEN AB24 1XQ
Gauge 4ft 8½in   www.aberdeencity.gov.uk   Aberdeenshire NJ 943092

MR THERM	0-4-0ST	OC	AB	2239	1947

ALFORD VALLEY COMMUNITY RAILWAY LTD,
ALFORD STATION, MURRAY PARK, ALFORD   AB33 8DG
Gauge 2ft 0in   www.avcr.org.uk   Aberdeenshire NJ 579159

	JAMES GORDON	0-4-0DH	s/o	AK	63	2001	
AVR No.1	HAMEWITH	4wDM		L	3198	c1930	Pvd
	THE BRA'LASS	4wDM	s/o	MR	9381	1948	
87022	THE WEE GORDON HIGHLANDER / BYDAND						
		4wDM		MR	22221	1964	

ALMOND VALLEY HERITAGE TRUST,
ALMOND VALLEY HERITAGE CENTRE, MILLFIELD, LIVINGSTON   **EH54 7AR**

Gauge **2ft 6in**   www.almondvalley.co.uk   **West Lothian** **NT 034667**

–		4wDH	AB	557	1970
–		4wDH	BD	3752	1980
"OAKBANK No.2"		4wWE	BLW	20587	1902
20		4wBE	BV	612	1972
38		4wBE	BV	698	1974
42		4wBE	BV	700	1974
7A	TAM	4wBE	BV	1143	1976
–		4wBEF	GB	1698	1940
YARD No.B10		0-4-0DM	HE	2270	1940
–		4wDM	HE	7330	1973
–		4wDM	SMH	40SPF522	1981

AYRSHIRE RAILWAY PRESERVATION GROUP, DOON VALLEY RAILWAY,
SCOTTISH INDUSTRIAL RAILWAY CENTRE, DUNASKIN HERITAGE CENTRE,
DALMELLINGTON ROAD, WATERSIDE, PATNA   **KA6 7JF**

Gauge **4ft 8½in**   www.doonvalleyrailway.co.uk   **East Ayrshire** **NS 443083**

No.16		0-4-0ST	OC	AB	1116	1910
(19)		0-4-0ST	OC	AB	1614	1918
–		0-4-0F	OC	AB	1952	1928
No.10		0-4-0ST	OC	AB	2244	1947
NCB No.23		0-4-0ST	OC	AB	2260	1949
–		0-6-0ST	OC	AB	2358	1954
–		0-4-0ST	OC	AB	2368	1955
No.1		0-4-0DM		AB	347	1941
–		0-4-0DM		AB	399	1956
–		4wDMR		Donelli	163	1979
–		0-4-0DM		HE	3132	1944
–		0-4-0DM		JF	22888	1939
–	(ARMY 409)	0-4-0DH		NBQ	27644	1959
–		4wDM		RH	224352	1943
M/C 324	"BLINKIN BESS"	4wDM		RH	284839	1950
–		4wDM		RH	417890	1959
–		0-4-0DM		RH	421697	1959
–		4wDH		S	10012	1959
L482 LNU		4wDM	R/R	_(Multicar		1993
				(Perm		1993

Gauge **3ft 0in**

–	"KILLOCH"	4wDH	HE	8816	1981

Gauge 2ft 6in

05/579		4wDH	AB	561	1971	
3		4wBE	BV	307	1968	
43	CC 4	4wBE	BV	701	1974	
7329		4wDM	HE	7329	1973	
2		4wDM	RH	183749	1937	Dsm
No.3		4wDM	RH	210959	1941	
No.1		4wDM	RH	211681	1942	
(1)		4wDM	RH	422569	1959	

THE KILN GROUP,
BARCLAY HOUSE, CALEDONIA WORKS OFFICE BLOCK SITE,
WEST LANGLANDS STREET, KILMARNOCK   KA1 2PR
Gauge 4ft 8½in   East Ayrshire NS 424381

DRAKE	0-4-0ST	OC	AB	2086	1940	Pvd

CALEDONIAN RAILWAY (BRECHIN) LTD, BRECHIN, near MONTROSE

Locomotives are kept at :	Brechin	DD9 7AF	Angus	NO 603603
	Bridge of Dun	DD10 9LH	Angus	NO 663587

Gauge 4ft 8½in   www.caledonianrailway.co.uk

46464		2-6-0	OC	Crewe		1950	
–		0-4-0ST	OC	AB	1863	1926	
1		0-6-0T	OC	AB	2107	1941	
45		0-4-0ST	OC	AB	2352	1954	Dsm
–		0-6-0ST	IC	HE	2879	1943	
6		0-4-0ST	OC	P	1376	1915	
MENELAUS		0-6-0ST	OC	P	1889	1934	
–		0-6-0ST	OC	P	2153	1954	
–		0-6-0ST	IC	WB	2749	1944	
–		0-6-0ST	IC	WB	2759	1944	
(D)3059 (08046 13059) BRECHIN CITY	0-6-0DE		Derby		1954	
(D5222) 25072		Bo-BoDE		Derby		1963	
(D5233) 25083		Bo-BoDE		Derby		1963	
D5301 (26001)		Bo-BoDE		BRCW	DEL46	1958	
(D5302 26002)		Bo-BoDE		BRCW	DEL47	1958	a
D5314 (26014)		Bo-BoDE		BRCW	DEL59	1959	
(D5325 26025)		Bo-BoDE		BRCW	DEL70	1959	a
(D5335) 26035		Bo-BoDE		BRCW	DEL80	1959	
(D5353 27007) 27015 UNIVERSITY OF WIBBLEFROTH						
		Bo-BoDE		BRCW	DEL196	1961	
D5370 (27024)		Bo-BoDE		BRCW	DEL213	1962	
(D6797) 37097 OLD FETTERCAIRN	Co-CoDE		_(EE	3226	1962	
				(VF	D751	1962	
(D8016) 20016		Bo-BoDE		_(EE	2363	1957	
				(VF	D391	1957	

(D8081)	20081	Bo-BoDE	_(EE	2987	1961
			(RSHD	8239	1961
(D8088	20088) 2017 37	Bo-BoDE	_(EE	2994	1961
			(RSHD	8246	1961
(D8166)	20166	Bo-BoDE	_(EE	3637	1966
			(EEV	D1036	1966
D9553	54	0-6-0DH	Sdn		1965
12052		0-6-0DE	Derby		1949
12093		0-6-0DE	Derby		1951
–		4wDM	FH	3747	1955
	DEWAR HIGHLANDER	4wDM	RH	458957	1961
211	ROLLS	0-4-0DE	YE	2628	1956
	"DONCASTER"	0-4-0DE	YE	2654	1957
(212)	MAVIS	0-4-0DE	YE	2684	1958
205028	60146	4-4wDER	Afd/Elh		1962
1132	S60150 (205032)	4-4wDER	Afd/Elh		1962

a   stored off site at Hydrus Group, Brechin Business Park, Brechin

CAMERON RAILWAY TRUST, BALBUTHIE OPEN FARM VISITOR CENTRE, BALBUTHIE ROAD, near KILCONQUHAR, ST. MONANS   KY9 1EX

Gauge 4ft 8½in   Fife NO 502020

60009	UNION OF SOUTH AFRICA	4-6-2	3C	Don	1853	1937
61994	THE GREAT MARQUESS	2-6-0	3C	Dar	(1761?)	1938

CARNEGIE DUNFERMLINE TRUST / FIFE COUNCIL, PITTENCREIFF PARK, DUNFERMLINE   KY12 8QH

Gauge 4ft 8½in   www.fife.gov.uk   Fife NT 086872

No.29		0-4-0ST	OC	AB	1996	1934

CLYDE VALLEY FAMILY PARK, CROSSFORD, CARLUKE, near LANARK ML8 5NJ

Gauge 600mm   www.clydevalleyfamilypark.co.uk   South Lanarkshire NS 831461

1863		4w-4wDH	s/o	Chance	
				64-5031-24	1964
–		4-4wBE	s/o	Schwingel	
	rebuild of	4wDH		Schöma	

COUNTESS OF SUTHERLAND, DUNROBIN STATION, BRORA, near GOLSPIE   KW10 6SF

Gauge 2ft 0in   www.dunrobincastle.co.uk   Highland NC 849013

BRORA		0-4-0PM	s/o	Bg	1797	1930
			reb	Bg	2083	1934

DB CARGO (UK) LTD,
FORT WILLIAM FREIGHT CENTRE, TOM-NA-FAIRE, FORT WILLIAM   PH33 6QT

Gauge 4ft 8½in   Private site with no public access   **Highland**   NN 119752

Seasonal base for "The Jacobite" steam services between Fort William and Mallaig

JOHN DEWAR & SONS LTD, DEWAR'S WORLD OF WHISKY,
ABERFELDY DISTILLERY, ABERFELDY   PH15 2EB

Gauge 4ft 8½in   www.dewars.com   **Perth & Kinross**   NN 865496

–		0-4-0ST	OC	AB	2073	1939

DUNDEE MUSEUM OF TRANSPORT,
MARKET MEWS, MARKET STREET, DUNDEE   DD1 3LA

Gauge 4ft 8½in   www.dmoft.co.uk   **City of Dundee**   NO 417309

–		4wWE	BTH		1908	a	Dsm

a   currently at Ribble Steam Railway, Preston, Lancashire

EAST LINKS FAMILY PARK,
JOHN MUIR PARK, BELTONFORD, near DUNBAR   EH42 1XF

Gauge 2ft 0in   www.eastlinks.co.uk   **East Lothian**   NT 648786

No.49	SANDY	4wDH		AK	49	1994
	THE EXPRESS	0-6-0DM	s/o	Bg	3014	1938
	rebuilt as	0-6-0DH	s/o	BES		1993

EAST LOTHIAN COUNCIL, MUSEUM SERVICES,
PRESTONGRANGE MUSEUM, COAST ROAD, PRESTONGRANGE   EH32 9RX

Gauge 4ft 8½in   www.eastlothian.gov.uk   **East Lothian**   NT 374737

17		0-4-0ST	OC	AB	2219	1946	
No.7	PRESTONGRANGE	0-4-2ST	OC	GR	536	1914	
–		4wDH		EEV	D908	1964	
No.33		4wDM		RH	221647	1943	
No.2	IVOR GEORGE EDWARDS	4wDM		RH	398613	1956	a
	SAM	4wDM		RH	458960	1962	

a   carries plate 398163 in error

EASTRIGGS & GRETNA HERITAGE GROUP, DEVIL'S PORRIDGE MUSEUM,
STANFIELD, ANNAN ROAD, EASTRIGGS, ANNAN   DG12 6TF

Gauge 4ft 8½   www.devilsporridge.org.uk   **Dumfries & Galloway**   NY 252662

	"SIR JAMES"	0-6-0F	OC	AB	1550	1917

FALKIRK DISTRICT COUNCIL, DEPARTMENT OF LIBRARIES & MUSEUMS, MUSEUM WORKSHOP, ABBOTSINCH COURT, 7–11 ABBOTSINCH ROAD, ABBOTSINCH INDUSTRIAL ESTATE, GRANGEMOUTH FK3 9UX

Gauge 4ft 8½in   www.falkirk.gov.uk   Falkirk NS 936814

| No.1 | | 0-4-0DM | | JF | 22902 | 1943 | |

FERRYHILL RAILWAY HERITAGE TRUST, FERRYHILL RAILWAY HERITAGE CENTRE, off POLMUIR AVENUE, ABERDEEN AB11 7TH

Gauge 4ft 8½in   www.frht.org.uk   Aberdeenshire NJ 941045

	NORTH DOWNS No.3	0-6-0T	OC	RSHN	7846	1955	
No.6	144-6	0-4-0DM		RH	421700	1959	
(900338	LNER 338)	2w-2PMR		Wkm	626	1934	a
A37W		2w-2PMR		Wkm	8502	1960	
DX 68002	DB 965330	2w-2PMR		Wkm	10180	1968	

a   rebuilt using parts from Wkm 1583

Gauge 600mm

| RTT/767162 | | 2w-2PM | | Wkm | 3235 | 1943 | |

FIFE REGIONAL COUNCIL, LOCHORE MEADOWS COUNTRY PARK, LOCHORE KY5 8AL

Gauge 4ft 8½in   www.lochoremeadows.org   Fife NT 172963

| – | | 0-4-0ST | OC | AB | 2259 | 1949 | a |

a   currently at Smeaton Engineering Ltd, Kirkcaldy, Fife

FRASERBURGH HERITAGE SOCIETY LTD, FRASERBURGH HERITAGE CENTRE, QUARRY ROAD, FRASERBURGH AB43 9DT

Gauge 2ft 0in   www.fraserburghheritage.com   Aberdeenshire NJ 997675

| 677 | KESSOCK KNIGHT | 4wDM | s/o | LB | 53541 | 1963 | Pvd |

FRIENDS OF CRAIGTOUN COUNTRY PARK RAILWAY, RIO GRANDE RAILWAY, CRAIGTOUN COUNTRY PARK, near ST.ANDREWS KY16 8NX

Gauge 1ft 3in   www.friendsofcraigtoun.org.uk   Fife NO 482150

| 278 | | 2-8-0DH | s/o | SL | R8 | 1976 | |

THE GARDEN OF COSMIC SPECULATION, PORTRACK SCOTTI GARDEN OF RAILS 2004, PORTRACK HOUSE, HOLYWOOD, near DUMFRIES DG2 0RW

Gauge 4ft 8½in   www.charlesjencks.com   Dumfries & Galloway NX 939830

| – | | 0-4-0DM | | RH | 418790 | 1958 | |

GLASGOW CITY COUNCIL, CULTURAL AND SPORT GLASGOW
Glasgow Museums Resource Centre,
200 Woodhead Road, South Nitshill Industrial Estate, Nitshill, Glasgow   G53 7NN

Gauge 4ft 8½in   www.glasgowlife.org.uk   **City of Glasgow**   **NS 518601**

–	0-4-0VBT	VCG	Chaplin	2368	1885
No.1	0-6-0F	OC	AB	1571	1917
–	2w-2PM		Albion		c1916

Gauge 4ft 0in

–	4wBE		_(JF	16559	1925
			(WR	583	1927
SUBWAY CAR No.1	4w-4wRER		OldburyC&W		1896

Riverside Museum, 100 Pointhouse Road, Glasgow   G3 8RS

Gauge 4ft 8½in   www.glasgowlife.org.uk   **City of Glasgow**   **NS 557659**

103	4-6-0	OC	SS	4022	1894
123	4-2-2	IC	N	3553	1886
(62469) No.256 GLEN DOUGLAS	4-4-0	IC	Cowlairs		1913
9	0-6-0T	OC	NBH	21521	1917

Gauge 3ft 6in

3007	4-8-2	OC	NBQ	25546	1945

GRAMPIAN TRANSPORT MUSEUM, MONTGARRIE ROAD, ALFORD   AB33 8AE

Gauge 4ft 8½in   www.gtm.org.uk   **Aberdeenshire**   **NJ 577161**

No.3	0-4-0ST	OC	AB	1889	1926

D. HERBERT, GLASGOW AREA
Locomotives stored at a private location.

Gauge 2ft 6in   **City of Glasgow**   **Private Site**

2209	4wDM		HE	2209	1941
No.4 ND 10394	0-4-0DM		HE	2243	1941
R9 ND 6473	4wDM		RH	235727	1944

THE HIGHLAND LIGHT RAILWAY, THE RAILWAY FARM,
MILL OF LOGIERAIT FARM, BALLINLUIG, near PITLOCHRY   PH9 0LH

Gauge 2ft 0in   www.railwayfarm.co.uk   **Perth & Kinross**   **NN 975516**

–	4wDM		AK	No.5	1979

I. HUGHES, PRIVATE LOCATION
Gauge 2ft 0in   Perth & Kinross   Private Site

T7	DOLLY		4wDH		AB	560	1971
1	5		4wBE		CE	B2905	1981
		reb		CE	B3550A	1988	a
–			4wDM		MR	9846	1952
–			4wDM		MR	40S383	1971
–			4wDM		RH	283513	1949
	rebuilt as		4wDH		AK	20R	1986

a   plate reads CE B2903

THE INVERGARRY & FORT AUGUSTUS RAILWAY MUSEUM LTD,
INVERGARRY STATION, SOUTH LAGGAN, near SPEAN BRIDGE   PH34 4EA
Gauge 4ft 8½in   www.invergarrystation.org.uk   Highland NN 303983

–	4wDM	RH	236364	1946
99709 901002-4	2w-2PMR	Lesmac LMS012	2006	

KEITH & DUFFTOWN RAILWAY ASSOCIATION, DUFFTOWN   AB55 4BA
Gauge 4ft 8½in   www.keith-dufftown-railway.co.uk   Moray NJ 323414

–		0-4-0DH	AB		1979	
	a rebuild of	0-4-0DM	AB	415	1957	
KDR 40	THE WEE MAC	4wDH	CE	B1844	1979	
(KDR 41)		0-6-0DH	EEV	D1193	1967	
51568	(KDR 6) SPIRIT OF BANFFSHIRE	2-2w-2w-2DMR	DerbyC&W		1959	
Sc 52008		2-2w-2w-2DMR	DerbyC&W		1961	
(52030 960 932 977831)		2-2w-2w-2DMR	DerbyC&W		1961	
52053	(KDR 7) SPIRIT OF BANFFSHIRE	2-2w-2w-2DMR	DerbyC&W		1960	
(50628)	53628 (KDR 4) SPIRIT OF SPEYSIDE					
		2-2w-2w-2DMR	DerbyC&W		1958	
(55500)	140.001	4wDMR	_(DerbyC&W		1981	
			(Leyland (R2.001)	1981	
(55501)	140.001	4wDMR	_(DerbyC&W		1981	
			(Leyland (R2.002)	1981	
55822	144022	2w-2DMR	_(BRE(D)		1987	
			(Alex 2785/43	1987	
	rebuilt as	2w-2DHR	AB		1991	
55845	144022	2-2wDMR	_(BRE(D)		1987	
			(Alex 2785/44	1987	
	rebuilt as	2-2wDHR	AB		1991	
55858	(144022)	2w-2DMR	_(BRE(D)		1988	
			(Alex 1187/9	1988	
	rebuilt as	2w-2DHR	AB		1991	
CAR 1	C N RAIL 144-5	2-2wPMR	Fairmont 244095			
(CAR 2	KDR 43) 144-55	2-2wPMR	Fairmont			
CAR 3		2-2wPMR	Fairmont 252180			
PQ 364		4wDH R/R	NNM	80505	1980	

KINGDOM OF FIFE PRESERVATION SOCIETY, FIFE HERITAGE RAILWAY, KIRKLAND YARD, BURNMILLS INDUSTRIAL ESTATE, LEVEN KY8 4RB

Gauge 4ft 8½in   www.fifeheritagerailway.co.uk   Fife   NO 373007

No.10	"LOCHGELLY"	0-4-0ST	OC	AB	1890	1926		
	(B.A.CO.LTD No.3)	0-4-0ST	OC	AB	2046	1937		
No.17		0-4-0ST	OC	AB	2292	1951		
400	RIVER EDEN	0-4-0DH		NBQ	27421	1955		
	N.C.B.No.10	0-6-0DH		NBQ	27591	1957		
No.7	YARD No.DP35	0-4-0DM		RH	313390	1952		
No.4	"NORTH BRITISH"	4wDM		RH	421415	1958		
2	THE GARVIE FLYER	0-4-0DE		RH	431764	1960		
1	"LARGO LAW"	0-4-0DE		RH	449753	1961		
Sc52029		2-2w-2w-2DMR		DerbyC&W		1961		
–		2w-2PMR		Fife		2018		
	rebuilt as	2w-2DMR		Fife		2020		

LOWTHERS RAILWAY SOCIETY LTD, LEADHILLS & WANLOCKHEAD RAILWAY, THE STATION, STATION ROAD, LEADHILLS ML12 6XS

Gauge 2ft 6in   www.leadhillsrailway.co.uk   South Lanarkshire   NS 888145

| – | | 4wBE | | WR | (1614 | 1940)? | | Dsm |

Gauge 2ft 0in

(9)	"CHARLOTTE"	0-4-0WT	OC	OK	6335	1913		
(12)		4wDH		CE	B1819D	1978		
(8)	NITH	0-4-0DMF		HC	DM1002	1956		
(6)	CLYDE	4wDH		HE	6347	1975		
(10)	(MENNOCK)	4wDH		HE	9348	1994		
20		4wDM		MR	5880	1935		Dsm
250	8 8564	4wDM		MR	8564	1940		Dsm
251		4wDM		MR	8863	1944	a	Dsm
(253)	53	4wDM		MR	8884	1944	b	Dsm
(2)	ELVAN	4wDM		MR	9792	1955		
(5)	LITTLE CLYDE	4wDM		RH	7002/0467/2	1966		
(4)	LUCE	4wDM		RH	7002/0467/6	1966		

a   converted into a coach.
b   converted into a brake van

NATIONAL MINING MUSEUM SCOTLAND, RAIL & MINING HERITAGE CENTRE, LADY VICTORIA COLLIERY, NEWTONGRANGE EH22 4QN

Gauge 4ft 8½in   www.nationalminingmuseum.com   Mid Lothian   NT 332638

| – | | 0-6-0ST | OC | AB | 1458 | 1916 | |
| No.21 | | 0-4-0ST | OC | AB | 2284 | 1949 | |

Gauge 3ft 6in

| | TRAINING LOCO No.2 | 0-6-0DMF | | HE | 4074 | 1955 | a |

Gauge 2ft 6in

–	4wBE	CE	5871A	1971	a
–	4wBEF	CE	B3325A	1986	a

a   These locomotives are in storage and can only be viewed from public tours

NATIONAL MUSEUM OF SCOTLAND, CHAMBERS STREET, EDINBURGH EH1 1JF
Gauge 5ft 0in   www.nms.ac.uk   City of Edinburgh   NT 258734

"WYLAM DILLY"	4wG	VC	Hedley	1827-1832	a

a   incorporates parts of locomotive of same name built c1814 to c1815

Gauge 4ft 8½in

"ELLESMERE"	0-4-0WT	OC	H(L)	244	1861

NATIONAL MUSEUMS SCOTLAND, NATIONAL MUSEUMS COLLECTION CENTRE, GRANTON STORE, 242 WEST GRANTON ROAD, EDINBURGH   EH5 1JA
Gauge 2ft 0in   www.nms.ac.uk   City of Edinburgh   NT 229769

5	0-4-0T	OC	AB	988	1903

Gauge 1ft 7in

WYLAM DILLY	4wG	VC	RSM		1885

ORKNEY ISLANDS COUNCIL, ORKNEY ARTS, MUSEUMS and HERITAGE, SCAPA FLOW VISITOR CENTRE and MUSEUM, LYNESS, HOY   KW16 3NU
Gauge 600mm   www.orkneymuseums.co.uk   Orkney Islands   ND 310947

–	2w-2PM	Wkm	3030	1941	

PRIVATE OWNER, PORT ELPHINSTONE SIDINGS, INVERURIE MILLS, INVERURIE   AB51 5NR
Gauge 4ft 8½in   Private Site   Aberdeenshire   NJ 780192

D9500	(9312/92 No.1)	0-6-0DH	Sdn	1964
61287	4311	4w-4wRER	Afd/Elh	1959

Private site with no public access

D. RITCHIE & FRIENDS
Gauge 3ft 0in   City of Edinburgh   Private Site

–	4wDM	RH	466591	1961

Gauge 2ft 6in

–	4wDM	RH	189992	1938
–	4wDM	RH	242916	1946
–	4wDM	RH	273843	1949
10553	0-4-0DM	RH	338429	1955
P 9303	YARD No.1018	4wBE	VE	7667	

Gauge 2ft 0in

		4wDM		HE	2654	1942
–	TERRAS	4wDM		MR	7189	1937
CCC 51		4wDM		MR	7330	1938
–		4wDM		MR	9982	1954
–		4wDM		RH	179005	1936
–		4wDM		RH	249530	1947

F. ROACHE, SLEEPERZZZ,
ROGART STATION, PITTENTRAIL, near GOLSPIE   IV28 3XA

Gauge 4ft 8½in   www.sleeperzzz.com   **Closed**   **Highland NC 725020**

8016	ERIC	4wDM	RH	294263	1950

THE ROYAL DEESIDE RAILWAY, MILTON OF CRATHES, BANCHORY   AB31 5QH

Gauge 4ft 8½in   www.deeside-railway.co.uk   **Aberdeenshire NO 740962, 724964**

	BON ACCORD	0-4-0ST	OC	AB	807	1897
	SALMON	0-6-0ST	OC	AB	2139	1942
RRM 10	8 "WELBECK No.6"	0-4-0ST	OC	P	2110	1950
(D2037	03037)	0-6-0DM		Sdn		1959
D2094	(03094)	0-6-0DM		Don		1960
D2134	(03134)	0-6-0DM		Sdn		1960
	rebuilt as	0-6-0DH				
(No.1)		4wDM		MR	5763	1957
(Sc 79998	RDB 975003)	4w-4wBER		DerbyC&W		1956
	reb	2w-2BER		Cowlairs		1958
–		2w-2BER		TS&S	N/P1023	1985
	(PWM 2830)	2w-2PMR		Wkm	5008	1949
A34W		2w-2PMR		Wkm	8501	1960

SAUGHTREE STATION, NEWCASTLETON   TD9 0SP

Gauge 4ft 8½in   **Borders NT 565981**

275882	MEG OF SAUGHTREE	4wDM	RH	275882	1949
99709	909135-4	2w-2PMR	Geismar	ST/04/03	2004

SCOTTISH RAILWAY PRESERVATION SOCIETY, BO'NESS & KINNEIL RAILWAY,
BO'NESS STATION, UNION STREET, BO'NESS   EH51 9AQ

Gauge 4ft 8½in   www.bkrailway.co.uk   **West Lothian NT 003817**

(55189)	419	0-4-4T	IC	StRollox		1907	
(62277)	No.49 GORDON HIGHLANDER	4-4-0	IC	NBH	22563	1920	
62712	(246) MORAYSHIRE	4-4-0	3C	Dar	(1391?)	1928	a

65243	MAUDE	0-6-0	IC	N	4392	1891	
68095	No.42	0-4-0ST	OC	Cowlairs		1887	
80105		2-6-4T	OC	Bton		1955	b
(45170	WD 554)	2-8-0	OC	NBH	24755	1942	
No.3		0-4-0ST	OC	AB	1937	1928	
	(LORD ASHFIELD)	0-6-0F	OC	AB	1989	1930	
No.6		0-4-0ST	OC	AB	2043	1937	
THE WEMYSS COAL CO LTD No.20		0-6-0T	IC	AB	2068	1939	
GLENGARNOCK WORKS No.6		0-4-0CT	OC	AB	2127	1942	
No.24		0-6-0T	OC	AB	2335	1953	
	CITY OF ABERDEEN	0-4-0ST	OC	BH	912	1887	
No.19		0-6-0ST	IC	HE	3818	1954	
No.5		0-6-0ST	IC	HE	3837	1955	
6		0-4-0ST	OC	HL	3640	1926	
No.1		0-4-0WT	OC	N	1561	1870	
No.13	N.C.B.13	0-4-0ST	OC	N	2203	1876	
(No.1	LORD ROBERTS)	0-6-0T	IC	NR	5710	1902	
	(JOHN)	4wVBT	VCG	S	9561	1953	b
	(RANALD)	4wVBT	VCG	S	9627	1957	
	(DENIS)	4wVBT	VCG	S	9631	1958	b
68007	(WD 75254)	0-6-0ST	IC	WB	2777	1945	
(D1970)	47643	Co-CoDE		Crewe		1965	
D2767		0-4-0DH		NBQ	28020	1960	
			reb	AB		1968	
D3558	(08443)	0-6-0DE		Derby		1958	
(D5324)	26024	Bo-BoDE		BRCW	DEL69	1959	
(D5338)	26038 TOM CLIFT 1954 - 2012	Bo-BoDE		BRCW	DEL83	1959	
(D5347)	27001	Bo-BoDE		BRCW	DEL190	1961	
(D5351)	27005	Bo-BoDE		BRCW	DEL194	1961	
(D6607)	(37307) 37403	Co-CoDE		_(EE	3567	1965	
	ISLE OF MULL			(EEV	D996	1965	
(D6725)	37025 INVERNESS TMD	Co-CoDE		_(EE	2888	1961	
				(VF	D604	1961	
(D6914)	37214	Co-CoDE		_(EE	3392	1963	
				(EEV	D858	1963	
(D6961)	37261	Co-CoDE		_(EE	3521	1965	
				(EEV	D950	1965	
(D7585)	25235	Bo-BoDE		Dar		1964	
(D8020)	20020	Bo-BoDE		_(EE	2742	1959	
				(RSHD	8052	1959	
(91031)	91131	Bo-BoWF		Crewe		1991	
19001	(82113)	4w-4wDH		Artemis		2018	
	a rebuild of non powered bogie driving van trailer			BRE(D)		1988	
No.1		0-6-0DM		AB	343	1941	
	F.G.F.	0-4-0DH		AB	552	1968	
F.82		4wWE/BE		EEDK	1131	1940	
	(TIGER)	0-4-0DH		NBQ	27415	1954	
	(KILBAGIE)	4wDM		RH	262998	1949	
–		4wDM		RH	275883	1949	
P6687		0-4-0DE		RH	312984	1951	

–		4wDM		RH	321733	1952
"521"		0-4-0DH		RH	457299	1962
Sc51017		2-2w-2w-2DMR		Sdn		1959
Sc51043		2-2w-2w-2DMR		Sdn		1959
61503	(303 032)	4w-4wRER		PSteel		1960
	99709 909131-3	2w-2PMR		Geismar	ST/03/57	2003
SV138	99709 909264-2	2w-2PMR		Geismar	ST/09/04	2009
SV142	99709 909305-3	2w-2PMR		Geismar	ST/06/03C	2006
	(HCT 022)	4wDH		Perm	022	1988	c
–		2w-2PMR		Wkm	10482	1970

a   currently off site for restoration
b   property of the Locomotive Owners Group (Scotland) Ltd
c   property of Northumbria Rail Ltd, Bedlington, Northumberland

Gauge 4ft 0in

55		4w-4wRER		OldburyC&W		1901
			reb	Govan		1935

Gauge 3ft 0in

	(FAIR MAID OF FOYERS)	0-4-0T	OC	AB	840	1899
–		4wDH		MR	110U082	1970

SHED 47 RAILWAY RESTORATION GROUP, LATHALMOND RAILWAY MUSEUM, SCOTTISH VINTAGE BUS MUSEUM, M90 COMMERCE PARK, LATHALMOND

KY12 OSJ

Gauge 4ft 8½in   www.shed47.org   **Fife NT 093922**

	N.C.B. No.29	0-4-0ST	OC	AB	1142	1908
No.17		0-4-0ST	OC	AB	2296	1950
236		0-4-0DM		AB	372	1945
			reb	YEC	L123	1994
B.R.I.L. No.001	"JINTY"	0-6-0DM		AB	385	1952
D2650		0-4-0DH		HE	9045	1980
			reb	YEC	L135	1994
(251)		0-4-0DH		HE	9046	1980
	TEXACO	0-4-0DM		JF	4210140	1958
–		4wDM		MR	9925	1963
–		4wDM		RH	265617	1948
A315885	(A241059)	2w-2PMR		Geismar	ST/02/02	2002	a
(970213)		2w-2PMR		Wkm	6049	1952

a   rebuilt using parts from Geismar ST/03/52

Gauge 2ft 0in – The West Fife Munitions Railway

–		0-4-0WTT	OC	Dec	917	1917	Dsm
	BIG DAVE	0-4-0ST	OC	NBRES	(014)	2021
(T9	T4)	4wDM		AK	28	1989
L1	SANDRA	4wDH		AK	47	1994
	N.C.B. No.10	4wDM		HE	4440	1952
(T11)	"ELOUISE"	4wDM		MR	21505	1955
L3	SYLVIA	4wDM		MR	22128	1961

STRATHSPEY RAILWAY CO LTD

Locomotives are kept at :–

Aviemore	PH22 1PY	NH 898131
Boat of Garten	PH24 3BQ	NH 943189
Broomhill	(PH26 3LU)	NH 997226

Gauge 4ft 8½in www.strathspeyrailway.co.uk **Highland**

(45025)	5025		4-6-0	OC	VF	4570	1934
46512	E.V.COOPER ENGINEER	2-6-0	OC	Sdn		1952
92219		2-10-0	OC	Sdn		1960
No.17	BRAERIACH	0-6-0T	IC	AB	2017	1935
6		0-4-0ST	OC	AB	2020	1936
No.9	CAIRNGORM	0-6-0ST	IC	RSHN	7097	1943
	BIRKENHEAD	0-4-0ST	OC	RSHN	7386	1948
D2774		0-4-0DH		NBQ	28027	1960
			reb	AB		1968
D3605	(08490)	0-6-0DE		Hor		1958
D5394	(27050)	Bo-BoDE		BRCW	DEL237	1962
D5862	31327	A1A-A1ADE		BT	398	1962
(D6869	37169) 37674	Co-CoDE		_(EE	3347	1963
				(EEV	D833	1963
–		0-4-0DH		AB	517	1966
14		0-4-0DH		NBQ	27549	1956
–		0-4-0DM		RH	260756	1950
	QUEEN ANNE	4wDM		RH	265618	1948
–		4wDH		TH	277V	1977
Sc51367		2-2w-2w-2DMR		PSteel		1960
Sc51402		2-2w-2w-2DMR		PSteel		1960
(51990	960 932) 977830	2-2w-2w-2DMR		DerbyC&W		1960
99709 909005-9 T111719 S45373	2w-2PMR		Geismar ST/02/03	2002
99709 909153-7 T114925 65058	2w-2PMR		Geismar ST/02/21	2002
99709 909157-8 T113892 65095	2w-2PMR		Geismar ST/03/36	2003
99709 909158-6 T111720 S45401	2w-2PMR		Geismar ST/03/37	2002
99709 909162-8 T111722	2w-2PMR		Geismar ST/03/60	2003
99709 909265-9 T117259	2w-2PMR		Geismar ST/04/17	2004
99709 901047-9	2w-2PMR		Lesmac LMS010	2006
(DE 940099) 813 CALE	2w-2PMR		Wkm	1288	1933

SUMMERLEE – THE MUSEUM OF SCOTTISH INDUSTRIAL LIFE, HERITAGE WAY, COATBRIDGE

ML5 1QD

Gauge 4ft 8½in www.culturenl.co.uk/summerlee **North Lanarkshire NS 728655**

No.11		0-4-0ST	OC	GH		1898
No.9		0-6-0T	IC	HC	895	1909
	ROBIN	4wVBT	VCG	S	9628	1957
(62174)	(936 103) 977845	4w-4wWER		Cravens		1967
W280		0-4-0DH		AB		1966
	a rebuild of	0-4-0DM		AB	472	1961

Gauge 3ft 6in

4112	(SPRINGBOK)	4-8-2+2-8-4T 4C		_(BP	7827	1957
				(NBH	27770	1957

TWEEDDALE HERITAGE RAILWAY SOCIETY,
TWEEDDALE HERITAGE RAILWAY, BROUGHTON, near BIGGAR

Gauge 2ft 6in   **Peeblesshire**   **Private site**

	CHRISTINE	4wDH	Byers		1998	a
	RACHEL	4wDH	Byers		1998	b
9	YARD No.24	0-4-0DM	HE	2017	1939	

a   incorporates frame of MR 115U094 / 1970
b   incorporates frame of AB 562 / 1971

WAVERLEY ROUTE HERITAGE ASSOCIATION,
WHITROPE HERITAGE CENTRE, WHITROPE SUMMIT, near HAWICK   **TD9 9TY**

Gauge 4ft 8½in   www.wrha.org.uk   **Roxburghshire**   **NT 525001**

(D5340)	26040		Bo-BoDE	BRCW	DEL85	1959	
	ARMY 110 A2 EQ		4wDM	RH	411319	1958	
55560	142019		2w-2DMR	_(BRE(D)		1985	
				(Leyland	R5.44	1985	
		rebuilt as	2w-2DHR	RFSD		1988	
55610	142019		2-2wDMR	_(BRE(D)		1986	
				(Leyland	R5.43	1986	
		rebuilt as	2-2wDHR	RFSD		1988	
55561	142020		2w-2DMR	_(BRE(D)		1985	
				(Leyland	R5.45	1985	
		rebuilt as	2w-2DHR	RFSD		1989	
55611	142020		2-2wDMR	_(BRE(D)		1985	
				(Leyland	R5.46	1985	
		rebuilt as	2-2wDHR	RFSD		1989	
RB004			4wDMR	_(DerbyC&W		1984	
				(Leyland	RB004	1984	a

a   property of Northumbria Rail Ltd, Bedlington, Northumberland

WEST LOTHIAN COUNCIL, POLKEMMET COUNTRY PARK, WHITBURN   **EH47 0AD**

Gauge 4ft 8½in   www.visitwestlothian.co.uk   **West Lothian**   **NS 924649**

No.8	DARDANELLES	0-6-0ST	OC	AB	1175	1909

SECTION 3 — WALES

Consequent upon the fragmentation of the County Geography of Wales, due to the creation of a large number of autonomous Unitary Authorities and similar bodies and, in view of the relatively few remaining locations hosting locomotives, this section of this handbook is now presented in two areas of coverage.

Defining Areas in accordance with the 1974-1996 "large counties" era, now provides a direct relationship between this volume and the "Historic Handbook" series which is also published by the Industrial Railway Society. The historic books contain the recorded details of all past and recent locomotives, at all known past and present locations, in the areas they cover, together with extensive texts describing the locations and the businesses they served.

The Areas can, in the main, be summarised as follows:

NORTH & MID WALES
Breconshire (apart from a narrow strip at its southern boundary which was transferred to Gwent and Mid-Glamorgan – note that Breconshire was also known as Brecknock, and Brecknockshire), Caernarfonshire, Ceredigion, Clwyd, Conwy, Denbighshire, Dyfed (north; i.e. Ceredigion), Flintshire, Flintshire Detatched, Gwynedd, Isle of Anglesey (Ynys Môn), Merionethshire, Montgomeryshire, Powys, Radnorshire and Wrexham.

SOUTH WALES
Blaenau Gwent, Bridgend, Caerphilly, Cardiff, Carmarthenshire, Dyfed (excluding Ceredigion), Glamorgan (prior to 1974), Gwent, Merthyr Tydfil, Mid Glamorgan, Monmouthshire, Neath Port Talbot, Newport, Pembrokeshire, Rhondda Cynon Taff, Swansea, Torfaen, South Glamorgan, West Glamorgan and Vale of Glamorgan.

The relevant Historic Series Handbooks are as follows:

NORTH & MID WALES
Handbook NW – Industrial Locomotives of North Wales   (1992)
Handbook DP – Industrial Locomotives of Dyfed & Powys   (1994)

SOUTH WALES
Handbook DP – Industrial Locomotives of Dyfed & Powys   (1994)
Handbook WG – Industrial Locomotives of West Glamorgan   (1996)
Handbook GT – Industrial Locomotives of Gwent   (1999)
Handbook GL – Industrial Locomotives of Glamorgan (Mid and South) (2007)

Update Bulletins for the published Historic Handbooks are available. For information see the Society website at   www.irsociety.co.uk or write to the address on page 2.

NORTH AND MID WALES

INDUSTRIAL LOCATIONS

ORTHIOS ECO PARKS (ANGLESEY) LTD,
PENRHOS WORKS, LONDON ROAD, HOLYHEAD

Gauge 4ft 8½in	R.T.C. (in administration)			Isle of Anglesey	SH 264807		
–	0-4-0DH	HE		7183	1970	a

a   privately owned

HANSON QUARRY PRODUCTS EUROPE LTD, PENMAENMAWR QUARRY, BANGOR ROAD, PENMAENMWAR (part of the Heidelberg Cement Group) LL34 6NA

Locomotive abandoned in a remote location on "Level 2", officially known as "Bottom Bank East Quarry".

Gauge 3ft 0in   www.heidelbergmaterials.com   **R.T.C.**   **Conwy SH 701758**

1878	PENMAEN		0-4-0VBT	VC	DeW	1878	Dsm

TATA STEEL EUROPE, TATA STEEL COLORS, SHOTTON WORKS   CH5 2NH

(part of the Tata Group)

Gauge 4ft 8½in   www.tatasteeleurope.com   **Flintshire SJ 302704, 305705**

(D3782)	08615	UNCLE DAI	0-6-0DE		Derby	1959	a
(D3991)	08823	KEVLA	0-6-0DE		Derby	1960	a

a   property of Hunslet Ltd, Barton-under-Needwood, Staffordshire

UNITED UTILITIES GROUP plc, MILWR TUNNEL, HALKYN MINE, RHYDYMWYN

Locomotives underground – no public access

Gauge 1ft 10½in   www.unitedutilities.com   **Flintshire SJ 296536**

–		4wDM	RH	182138	1936	OOU
–		4wDM	RH	226309	1943	Dsm a
–		4wDM	RH	354029	1953	OOU
774		4wBE	WR	744	1929	OOU
3		4wBE	WR	773	1930	OOU
–		2-2wDM	EdwardsE&J		2004	b

a   dumped underground in workshops adjacent to Pen-y-Bryn shaft
b   owned and operated by the Grosvenor Caving Club

PRESERVATION LOCATIONS

ANGLESEY CENTRAL RAILWAY // LEIN AMLWCH, c/o LLANERCH-Y-MEDD STATION COMMUNITY GARDEN LTD, LLANERCHYMEDD STATION, off BRIDGE STREET, LLANERCHYMEDD LL71 8EU

Gauge 4ft 8½in   www.leinamlwch.co.uk   **Anglesey SH 416840**

	ELISEG	0-4-0DM	JF	22753	1939	
–		0-4-0DH	HE	7460	1977	a

a   currently stored off site at an unknown location

ANGLESEY TRANSPORT & AGRICULTURE MUSEUM // TACLA TAID, TYDDYN PWRPAS, NEWBOROUGH   LL61 6TN

Gauge 4ft 8½in   www.angleseytransportmuseum.co.uk   **Anglesey SH 431666**

–	4wDM	RH	321727	1952	

BALA LAKE RAILWAY LTD,
THE STATION, STATION ROAD, LLANUWCHLLYN, Near BALA   LL23 7DD
Gauge 1ft 11½in   www.bala-lake-railway.co.uk   Gwynedd SH 881300

–		0-4-0T	OC	AE	1909	1922	
	WINIFRED	0-4-0ST	OC	HE	364	1885	
1	GEORGE B	0-4-0ST	OC	HE	680	1898	
No.3	HOLY WAR	0-4-0ST	OC	HE	779	1902	
	ALICE	0-4-0ST	OC	HE	780	1902	
No.5	MAID MARIAN	0-4-0ST	OC	HE	822	1903	
	TRIASSIC	0-6-0ST	OC	P	1270	1911	
	DOROTHY	0-4-0ST	OC	WB	1568	1899	
3776	BOB DAVIES	4wDH		BD	3776	1983	
			reb	YEC	L125	1994	
–		4wDM		HE	2024	1940	
	CHILMARK	4wDM		RH	194771	1939	
	LADY MADCAP	4wDM		RH	283512	1949	
LM 30	"MURPHY"	4wDH		Schöma	5697	2001	
D1087	MEIRIONNYDD	4w-4wDH		SL	22	1973	
–		2w-2PMR		Bala		2011	
–		2w-2PMR		Wkm	1548	1934	DsmT

CAERNARFON AIR MUSEUM // AMGUEDDFA AWYRENNOL,
CAERNARFON AIRPORT AIRWORLD, DINAS DINLLE, CAERNARFON   LL54 5TP
Gauge 2ft 0in   www.airworldmuseum.com   Gwynedd SH 435584

RTT 767150	2w-2PM	Wkm	(3152	1943)?	

CAERNARFONSHIRE SLATE RAILWAY, PORTHMADOG
Gauge 2ft 0in   Gwynedd Private Site

4	4wDM	RH	177638	1936	a
22	4wDM	RH	226302	1944	
–	0-4-0BE	WR	(3867	1948?)	
–	4wBE	WR			
–	4wBE	WR			

a   carries plate RH 177642/1938

Gauge 1ft 11½in

E2 No.7179 GWYNFYNYDD	0-4-0BE	WR	G7179	1967	

CONWY VALLEY RAILWAY MUSEUM, BETWS-Y-COED   LL24 0AL
Gauge 1ft 3in   www.conwyrailwaymuseum.co.uk   Conwy SH 796565

5 BATTISON	2-6-4DE	s/o	MossDW	2012	
	rebuild of	2-6-4DH	s/o	Battison	1958

CORRIS RAILWAY COMPANY LTD, MAESPOETH, near CORRIS   SY20 9RD

Gauge 2ft 6in   www.corris.co.uk   Gwynedd   SH 753069

(SIR NEVILLE LUBBOCK)	0-4-2ST	OC	KS	857	1904	a

Gauge 2ft 3in

No.10		0-4-2ST	OC	AK	91	2022	
No.7		0-4-2ST	OC	Winson	17	2005	
No.5	ALAN MEADEN	4wDM		MR	22258	1965	
(11)	VLAD	0-4-0DH		OK	25721	1957	
No.6	MALCOLM	4wDH		RH	518493	1966	
(9)	ABERLLEFENNI	4wBE		CE	B0457	1974	b

a   not on public display
b   currently at Talyllyn Railway Co, Tywyn, Gwynedd

DOLGARROG RAILWAY SOCIETY LTD,
ALUMINIUM WORKS SIDING, CLARK STREET, DOLGARROG   LL32 8QE

Gauge 4ft 8½in   www.dolgarrograilway.co.uk   Conwy   SH 774674

2	TAURUS	0-4-0DM	_(VF	D139	1951	
			(DC	2273	1951	

ERWOOD STATION GALLERY,
LLANDEILO GRABAN, near BUILTH WELLS   LD2 3SJ

Gauge 4ft 8½in   www.erwoodstation.com   Powys   SO 089439

A.W.M. No.169	ALAN	0-4-0DM	JF	22878	1939	

FAIRBOURNE RAILWAY PRESERVATION SOCIETY, FAIRBOURNE RAILWAY,
BEACH ROAD, FAIRBOURNE   LL38 2EX

Gauge 1ft 3in   www.fairbournerailway.com   Gwynedd   SH 615128

(DINGO)	4w-4wPM	Fairbourne		1951	a Dsm
WHIPPIT QUICK	4wPM	L	6502	1935	
rebuilt as	4w-4PM	Fairbourne		1955	
rebuilt as	4w-4PMR	Fairbourne		1962	
rebuilt as	4w-4DMR	Moss AJ			
GWRIL	4wPM	L	20886	1943	

a   frames utilised in sector turnout

Gauge 1ft 0¼in

759	YEO	2-6-2T	OC	CurwenD	1978	
	BEDDGELERT	0-6-4STT	OC	CurwenD	1979	
	SHERPA	0-4-0STT	OC	Milner	1978	
	RUSSELL	2-6-4T	OC	Milner	1979	
	TONY	4w-4wDM		Guest	1961	
	GWRIL	4wDH		HE	9354	1994	a
–		4wDM		MR	8937	1944	
–		4wDM		RH	435398	1959	

a   carries works number 9332

FFESTINIOG & WELSH HIGHLAND RAILWAYS
The Ffestiniog Railway Company, Porthmadog

Locomotives may be transferred to/from the Welsh Highland Railway on occasion, as required.

Locomotives are kept at :-   Boston Lodge Shed & Works, Gwynedd   LL48 6HT   SH 584379, 585378
           Glan-y-Pwll Depot, Blaenau Ffestiniog, Gwynedd   LL41 3PF   SH 693461
           Minffordd P.W. Depot, Gwynedd   (LL48 6HP)   SH 599386

Gauge 750mm   www.festrail.co.uk

B4539/03		4wBE		CE	B4539.3	2012		
B4539/05		4wBE		CE	B4539.5	2012		
			reb	CE	B4632	2018		

Gauge 2ft 3in

–		4wDH		HE	6292	1967	a	Dsm

     a   frames utilised in a hydraulic press

Gauge 2ft 0in

608		4-6-0T	OC	BLW	45190	1917	
–		4wDM		RH	174536	1936	

Gauge 1ft 11½in

NOTE : Most of the locomotives have been "rebuilt" by the FRCo many times during their working lives. In this list we detail only those rebuilds which made significant alteration to the loco's appearance.

	MOUNTAINEER	2-6-2T	OC	Alco(C)	57156	1916		
			reb	FRCo		c1968		
No.10	MERDDIN EMRYS	0-4-4-0T	4C	FRCo		1879		
			reb	FRCo		1988		
	LIVINGSTON THOMPSON	0-4-4-0T	4C	FRCo		1885	a	
			reb	FRCo		1905		
	EARL OF MERIONETH / IARLL MEIRIONNYDD	0-4-4-0T	4C	FRCo		1979		
12	DAFYDD LLOYD GEORGE / DAVID LLOYD GEORGE	0-4-4-0T	4C	FRCo		1992		
	TALIESIN	0-4-4T	OC	FRCo		1999		
	LYD 190	2-6-2T	OC	FRCo	14	2010		
	WELSH PONY	0-4-0STT	OC	FRCo		2015	b	
	JAMES SPOONER II	0-4-4-0T	4C	FRCo		2021		
No.1	PRINCESS	0-4-0TT	OC	GE	(200?)	1863		
	rebuilt as	0-4-0STT		FRCo		1895		
No.2	PRINCE	0-4-0TT	OC	GE	(199?)	1863		
	rebuilt as	0-4-0STT		FRCo		1891		
			reb	FRCo		1955		
			reb	FRCo		1979		
No.4	PALMERSTON	0-4-0TT	OC	GE		1864		
	rebuilt as	0-4-0STT		FRCo		1888		
			reb	FRCo		1910		
			reb	FRCo		1933		
(No.5)	WELSH PONY	0-4-0STT	OC	GE	234	1867		
			reb	FRCo		1891		Dsm
4		0-8-0	OC	Harbin	221	1988		
	VELINHELI	0-4-0ST	OC	HE	409	1886		
	LILLA	0-4-0ST	OC	HE	554	1891		
	BLANCHE	0-4-0ST	OC	HE	589	1893		
	rebuilt as	2-4-0STT		FRCo		1972		
	LINDA	0-4-0ST	OC	HE	590	1893		
	rebuilt as	2-4-0STT		FRCo		1970		

1	BRITOMART	0-4-0ST	OC	HE	707	1899		
	HUGH NAPIER	0-4-0ST	OC	HE	855	1904	c	
	"PANAD"	0-4-0VBT	VCG	FRCo				u/c
CASTELL HARLECH / HARLECH CASTLE	0-6-0DH		BD	3767	1983			
7011	MOELWYN	0-4-0PM		BLW	49604	1918		
	rebuilt as	0-4-0DM		FRCo		1956		
	rebuilt as	2-4-0DM		FRCo		1957		
CRICCIETH CASTLE / CASTELL CRICIETH	0-6-0DH		FRCo		1995			
	VALE OF FFESTINIOG	4w-4wDH		Funkey		1968		
			reb	FRCo				
	MONSTER / AFANC	4wDM		FRSociety		1974		
	MOEL HEBOG	0-4-0DM		HE	4113	1955		
P13350	MOEL Y GEST	4wDH		HE	6659	1965		
	HAROLD	4wDM		HE	7195	1974		
–		4wDH		HE	9349	1994		
			reb	AK		2004		
4415		6wDM		KS	4415	1928		
(MPU 8)		4wDH		Matisa		1956		
	MARY ANN	4wPM		MR		1917		
	rebuilt as	4wDM		FRCo		c1957		
	rebuilt as	4wPM		FRCo		c2016	d	
–		4wDM		MR	435	1917		
			reb	MR	3663	1924		
	THE COLONEL	4wDM		MR	8788	1943	Dsm	
	ANDY	4wDM		MR	21579	1957		
			reb	FRCo		2010		
	DOLGARROG / INNOGY	4wDM		MR	22154	1962		
	"BUSTA"	2w-2PMR		FRCo		2006		
1543		2w-2PMR		Wkm	1543	1934		

a   currently on display at National Railway Museum, York
b   worksplate dated 2020
c   property of The National Trust; based here, but visits other locations
d   carries incorrect plate 507/1917

Gauge 1ft 10¾in

NESTA	0-4-0ST	OC	HE	704	1899	

Welsh Highland Railway // Rheilffordd Eryri, Dinas, near Caernarfon   LL55 2YD

Locomotives may be transferred to/from the Festiniog Railway on occasion, as required.
Public railway operating from Caernarfon to Porthmadog.
Locomotives are kept at :-   Dinas shed and workshops, Caernarfon LL54 5UP   SH 476585, 478589

Gauge 3ft 0in — Static display at Dinas   Gwynedd SH 477587

LLANFAIR	0-4-0VBT	VC	DeW		1895	Pvd

Gauge 3ft 0in — Static display at Caernarfon   Gwynedd SH 480625

WATKIN	0-4-0VBT	VC	DeW		1893	Pvd

Gauge 1ft 11½in   www.festrail.co.uk

133	2-8-2	OC	AFB	2683	1952	OOU
134	2-8-2	OC	AFB	2684	1952	
130	2-6-2+2-6-2T 4C		BP	7431	1951	
138	2-6-2+2-6-2T 4C		BP	7863	1958	
(140)	2-6-2+2-6-2T 4C		BP	7865	1958	
143	2-6-2+2-6-2T 4C		BP	7868	1958	
87	2-6-2+2-6-2T 4C		Cockerill	3267	1937	
9	0-6-0DM		Bg/DC	2395	1952	
CONWAY CASTLE / CASTELL CONWY	4wDM		FH	3831	1958	
UPNOR CASTLE	4wDM		FH	3687	1954	
CASTELL CAERNARFON	4w-4wDH		Funkey		1968	
No.1 BILL	4wDHF		HE	9248	1985	
No.2 BEN	4wDHF		HE	9262	1985	

FLINTSHIRE MUSEUM SERVICE,
SHOTTON STORE, UNIT 3, ROWLEY PARK, EVANS WAY, SHOTTON   CH5 1QJ
(operated by Aura Leisure and Libraries)
Gauge 1ft 10½in   www.aura.wales   Flintshire SJ 310687

–	4wDM	RH	331250	1952	
–	4wBE	WR	898	1935	

GLYN VALLEY TRAMWAY and INDUSTRIAL HERITAGE TRUST,
INDUSTRIAL MUSEUM, THE OLD TRAMWAY ENGINE SHED,
NEW ROAD, GLYN CERIOG, near LLANGOLLEN   LL20 7HE
Gauge 700mm   www.glynvalleytramway.org.uk   Clywd SJ 202378

–	4wDM	RH	371932	1954	

Gauge 600mm

2	4wPM	Fordson		a

a   locomotive stored at private site with no public access

GREENFIELD VALLEY TRUST LTD,
GREENFIELD VALLEY HERITAGE PARK, HOLYWELL   CH8 7GH
Gauge 1ft 10½in   www.greenfieldvalley.com   Flintshire SJ 193773

– "1987/67"	0-4-0BE	WR	(1080 1937?)	

GWYNEDD COUNCIL,
DIFFWYS CAR PARK, HIGH STREET, BLAENAU FFESTINIOG   LL41 3ES
Gauge 2ft 0in   www.gwynedd.llyw.cymru   Gwynedd   SH 702459

2207	5	4wDM	HE	2207	1941	Pvd

I.B. JOLLY, MOLD
Locomotives are stored at two private locations.

Gauge 4ft 8½in   Flintshire   Private Site

MRTC 1944	4wDM	MR	1944	1919	

Gauge 2ft 7in

–	4wDM	MR	5025	1929	Dsm

Gauge 600mm

–	0-2-2GasE	HopleyCP		c1982	
(LR 2718) MRTC 997	4wPM	MR	997	1918	Dsm
–	4wDM	MR	4803	1934	Dsm
–	4wDM	MR	5852	1933	Dsm
–	4wDM	MR	8723	1941	Dsm
No.9	4wDM	MR	9547	1950	

Gauge 1ft 11½in

–	4wDM	L	30233	1946	
–	4wPM	MR	6013	1931	Dsm

INIGO JONES & CO LTD, TUDOR SLATE WORKS,
Y GROESLON, near PENYGROES (subsidiary of Wincilate Ltd)   LL54 7UE
Gauge 2ft 3in   www.inigojones.co.uk   Gwynedd   SH 471551

–	4wBE	LMM		Pvd

LLANGOLLEN RAILWAY TRUST LTD
Locomotives are kept at :–   Llangollen Shed   LL20 8SW   SJ 212422
                                               Carrog   LL21 9BD   SJ 118435
                                 Glyndyfrdwy   LL21 9HF   SJ 150428
                 Pentrefelin C & W Shed   LL20 8EE   SJ 207434
                           Pentrefelin Sidings   SJ 209432

Gauge 4ft 8½in   www.llangollen-railway.co.uk   Denbighshire

3802		2-8-0	OC	Sdn		1938
5532		2-6-2T	OC	Sdn		1928
6430		0-6-0PT	IC	Sdn		1937
7754		0-6-0PT	IC	NBQ	24042	1930
80072		2-6-4T	OC	Bton		1953
68067	ROBERT	0-6-0ST	IC	HC	1752	1943
	AUSTIN No.1	0-6-0ST	IC	K	5459	1932
(D1566)	47449 ORION	Co-CoDE		Crewe		1964

(D2162)	03162	0-6-0DM	Sdn			1960
(D3265	08195) 13265	0-6-0DE	Derby			1956
(D3272)	08202	0-6-0DE	Derby			1956
(D)5310	(26010)	Bo-BoDE	BRCW	DEL55		1960
(D5801)	31271 STRATFORD 1840 – 2001					
		A1A-A1ADE	BT	302		1961
(2145)		0-4-0DM	HE	2145		1940
	PILKINGTON	0-4-0DE	YE	2782		1960
–		0-4-0DE	YE	2854		1961
E50416		2-2w-2w-2DMR	Wkm	7346		1957
M50447	M53447 610	2-2w-2w-2DMR	BRCW			1957
M50454		2-2w-2w-2DMR	BRCW			1957
M50528		2-2w-2w-2DMR	BRCW			1958
M51618		2-2w-2w-2DHR	DerbyC&W			1959
M51907	LO262	2-2w-2w-2DMR	DerbyC&W		1960	a
M51933		2-2w-2w-2DMR	DerbyC&W			1960
DR 90005	"NEPTUNE"	4wDMR	Matisa	PV6 627		1967
	Q317 GRN	4wDM	R/R	Unimog	008983	1992
	99709 909266-7	2w-2PMR	Geismar	ST/00/24	2000
(68012	DB 965065)	2w-2PMR	Wkm	7580		1956

a   currently at the Midland Railway - Butterley, Swanwick, Derbyshire

DAVID MITCHELL, LLANNERCHYMEDD, ANGLESEY
Gauge 1ft 3in   **Anglesey**   **Private Site**

–	2w-2PMR	Wkm	4816	1948	Dsm

NARROW GAUGE RAILWAY MUSEUM TRUST LTD, WHARF STATION, TYWYN
LL36 9EY

Gauge 3ft 2¼in   www.narrowgaugerailwaymuseum.org.uk   **Gwynedd SH 586004**

No.5	WILLIAM FINLAY	0-4-0T	OC	FJ	173L	1880

Gauge 2ft 0in

(OAKLEY)	0-4-0PM		BgC	774	1919

Gauge 1ft 10¾in

GEORGE HENRY	0-4-0VBT	VC	DeW		1877
ROUGH PUP	0-4-0ST	OC	HE	541	1891

Gauge 1ft 10in

13	0-4-0T	IC	Spence	13L	1895

Gauge 1ft 6in

DOT	0-4-0WT	OC	BP	2817	1887

NATIONAL MUSEUM OF WALES,
WELSH SLATE MUSEUM, GILFACHDDU, LLANBERIS
LL55 4TY

Gauge 1ft 11½in   www.museum.wales/slate/   **Gwynedd**   **SH 586603**

	UNA	0-4-0ST	OC	HE	873	1905	
	–	4wBE		BE		1917	
"ALC/118"		4wPMR		WilliamsWJ			a
"1999/315"	CILGWYN	4wDM		RH	175414	1936	

a   vintage motorcycle converted for rail use

NATIONAL TRUST, INDUSTRIAL RAILWAY MUSEUM,
PENRHYN CASTLE, LLANDYGÁI, near BANGOR
LL57 4HN

Gauge 2ft 0in   www.nationaltrust.org.uk/penrhyn-castle   **Gwynedd** **SH 603720**

HUGH NAPIER	0-4-0ST	OC	HE	855	1904	a

a   normally based at the Ffestiniog Railway, but occasionally visits other locations

Gauge 1ft 11½in

ACORN	4wDM	RH	327904	1951

Gauge 1ft 10¾in

CHARLES	0-4-0ST	OC	HE	283	1882

NORTH WALES MINERS ASSOCIATION TRUST LTD,
BERSHAM COLLIERY, RHOSTYLLEN, WREXHAM
LL14 4EG

Gauge 1ft 10½in   www.colinw1248.wixsite.com/nwmat   **Clwyd** **SJ 314481**

–	"HAWYS"	4wDM	RH	183727	1937	a

a   currently stored off site

EXECUTORS of A. PILBEAM, LLANRHOS

Gauge 1ft 3in   **Powys** **Private Site**

–	2-2wBER	MossAJ		2000	
–	4wBE	Riordan	T6664	1967	

PRIVATE OWNER, CRIGGION, near OSWESTRY

Gauge 4ft 8½in   **Powys** **SJ 284150**

12	ALEXANDRA	0-4-0ST	OC	AB	929	1902	a

a   carries plate AB 979/1903

QUARRY TOURS LTD,
LLECHWEDD SLATE MINE, BLAENAU FFESTINIOG
LL41 3NB

Gauge 2ft 0in   www.zipworld.co.uk   **Gwynedd** **SH 699468**

–	4wBE	BEV	308	1921	Pvd

RHEILFFORDD LLYN PADARN CYFYNGEDIG,
LLANBERIS LAKE RAILWAY, GILFACHDDU, LLANBERIS

LL55 4TY

Gauge 1ft 11½in   www.lake-railway.co.uk   **Gwynedd**   SH 586603

No.1	ELIDIR	0-4-0ST	OC	HE	493	1889	
No.2	THOMAS BACH	0-4-0ST	OC	HE	849	1904	
No.3	DOLBADARN	0-4-0ST	OC	HE	1430	1922	
–		4wDM		MR	7927	1941	Dsm
No.9	GARRET	4wDM		RH	198286	1940	OOU
No.8	YD No.AD689 TWLL COED	4wDM		RH	268878	1952	
–		4wDM		RH	425796	1958	Dsm
No.7	COED GORAU // TOPSY	4wDM		RH	441427	1961	
No.19	LLANELLI	4wDM		RH	451901	1961	

RHIW VALLEY LIGHT RAILWAY,
LOWER HOUSE, MANAFON, near WELSHPOOL

SY21 8BJ

Gauge 1ft 3in   www.rvlr.co.uk   **Closed**   **Powys**   SJ 143028

JACK	0-4-0	OC	_(Rhiw		2003	
			(TMA	25657	2003	a
POWYS	0-6-2T	OC	SL	20	1973	
MONTY	4wPM		_(Jaco		1989	
			(Brunning		1989	

a   construction commenced as an 0-4-2T in 1985, but completed 2003 as listed

RHYL STEAM PRESERVATION TRUST, RHYL MINIATURE RAILWAY // RHEILFFORDD
FACH Y RHYL, MARINE LAKE, WELLINGTON ROAD, RHYL

LL18 1AQ

Gauge 1ft 3in   www.rhylminiaturerailway.co.uk   **Denbighshire**   SJ 003805

	JOAN	4-4-2	OC	Barnes	101	1920		
	RAILWAY QUEEN	4-4-2	OC	Barnes	102	1921		
	(BILLIE)	4-4-2	OC	Barnes	104	c1927		
105	MICHAEL	4-4-2	OC	Barnes	105	c1928		
106	BILLY	4-4-2	OC	Barnes	106	c1934		
44	CAGNEY	4-4-0	OC	McGarigle		c1910		
99	No.15	4-4-2	OC	_(WalkerG		1991		
	PRINCE EDWARD OF WALES			(MossAJ		1991		
	incorporates parts of			BL	15	1909		
		reb		Rhyl		2017		
	CLARA	0-4-2DM	s/o	Guest		1961		
	rebuilt as	0-4-2DH	s/o	Rhyl		2012		
	LISTER	4wDM		L	10498	1938		
–		2w-2-4BER		Hayne		1983	DsmT	a

a   converted to unpowered coaching stock

SNOWDON MOUNTAIN RAILWAY, LLANBERIS   LL55 4TY
(Operated by Heritage Attractions Ltd)
All locomotives drive through the rack gear, with the "driving wheels" free to rotate on their axles.

Gauge 800mm   www.snowdonrailway.co.uk   **Gwynedd   SH 582597**

2	ENID	0-4-2T OC	RACK	SLM	924	1895	
3	WYDDFA	0-4-2T OC	RACK	SLM	925	1895	
4	SNOWDON	0-4-2T OC	RACK	SLM	988	1896	
5	MOEL SIABOD	0-4-2T OC	RACK	SLM	989	1896	OOU
6	PADARN	0-4-2T OC	RACK	SLM	2838	1922	
7	RALPH	0-4-2T OC	RACK	SLM	2869	1923	OOU
8	ERYRI	0-4-2T OC	RACK	SLM	2870	1923	OOU
9	NINIAN	0-4-0DH	RACK	HE	9249	1986	
10	YETI	0-4-0DH	RACK	HE	9250	1986	
11	PERIS	0-4-0DH	RACK	_(HAB	775	1991	
				(HE	9305	1991	
12	GEORGE	0-4-0DH	RACK	HE	9312	1992	

TALYLLYN RAILWAY CO, TYWYN

Locomotives are kept at :-			Tywyn Pendre Station	LL36 9EN	SH 590008
			Tywyn Wharf	LL36 9EY	SH 585005

Gauge 2ft 6in   www.talyllyn.co.uk   **Gwynedd**

7	T 0009 00 NZ 35	4wDH		BD	3781	1984	a	OOU

Gauge 2ft 3in

(1)	TAL-Y-LLYN	0-4-2ST	OC	FJ	42	1865		
(No.2)	DOLGOCH	0-4-0WT	OC	FJ	63	1866		
No.3	SIR HAYDN	0-4-2ST	OC	HLT	323	1878		
(No.4)	EDWARD THOMAS	0-4-2ST	OC	KS	4047	1921	b	
(No.6)	DOUGLAS // AIR SERVICE CONSTRUCTIONAL CORPS No.1							
		0-4-0WT	OC	AB	1431	1918		
(No.7)	TOM ROLT	0-4-2T	OC	Pendre		1991		
	built using parts of	0-4-0WT	OC	AB	2263	1949		
No.11	TRECWN	4wDH		BD	3764	1983		
No.12	ST. CADFAN	4wDH		BD	3779	1983		
–		0-4-0DM		HE	4135	1950	c	Dsm
(No.9)	ALF	0-4-0DM		HE	4136	1950		
5	MIDLANDER	4wDM		RH	200792	1940		
	"THE FLAIL MOWER"	2-2-0DH		RH	476109	1964	d	Dsm
(19)		2w-2PM		CurwenD		1952	e	Dsm
	TOBY	2w-2PMR		BateJ		1954		
	rebuilt as	2w-2DMR		Pendre		2001		
(9)	(ABERLLEFENNI)	4wBE		CE	B0457	1974		

a	located at Quarry Shed, Dolgoch (SH 654050)
b	currently at Vale of Rheidol Railway, Aberystwyth
c	stored at Brynglas Station (SH 628031)
d	frames utilised in self-propelled flail mower
e	currently off site for restoration

Gauge 2ft 0in

JMLM19	"6502"	4wBE			WR	6502	1962	
			reb		WR	10102	1983	f
JMLM22	"MURPHY"	4wBE			WR	6504	1962	
			reb		WR	10106	1983	f

f   currently at Alan Keef Ltd, Lea Line, Ross-on-Wye, Herefordshire

TEIFI VALLEY RAILWAY, VALE OF TEIFI NARROW GAUGE RAILWAY, STATION YARD, HENLLAN, near LLANDYSUL   SA44 5TD

Gauge 2ft 0in   www.teifivalleyrailway.wales   Ceredigion   SN 357406

	ALAN GEORGE	0-4-0ST	OC	HE	606	1894	a
	SGT MURPHY	0-6-0T	OC	KS	3117	1918	
	rebuilt as	0-6-2T	OC	Winson	9	1991	
	SINEMBE	4-4-0T	OC	WB	2287	1926	
–		4wDM		HE	2433	1941	
004	SAMMY	4wDM		MR	11111	1959	
	JOHN HENRY	4wDM		RH	(433390	1959?)	

a   currently off site for restoration

JOHN TENNENT & P. SMITH, BRECON

These locomotives are in store at a private location.

Gauge 1ft 3in   Powys   Private Site

–		4wDM		AK	6	1981
–		4wDM		LB	52579	1961

UNKNOWN OWNER, BETTISFIELD, near WHITCHURCH   SY13 2LB

Gauge 4ft 8½in   Flintshire   SJ 460358

2107		4wDM		Bg/DC	2107	1937

VALE OF RHEIDOL LIGHT RAILWAY, PARK AVENUE, ABERYSTWYTH

Locomotives are kept at :-			
	Aberystwyth	SY23 1PG	SN 587812
	Capel Bangor	SY23 4EL	SN 647798
	Devils Bridge	SY23 3JL	SN 737769

Gauge 4ft 8½in   www.rheidolrailway.co.uk   Ceredigion

(3217)	9017 EARL OF BERKELEY	4-4-0	IC	Sdn		1938

Gauge 4ft 0in

	FIRE QUEEN	0-4-0	OC	AH	1848

Gauge 1000mm

–		4wPM		RP	51168	1916

Gauge 2ft 3in

(No.4)	(EDWARD THOMAS)	0-4-2ST	OC	KS	4047	1921	

Gauge 2ft 0in [nominal] (Static display)

–		0-4-0T	OC	Dec	1027	1926	a
	MARGARET	0-4-0ST	OC	HE	605	1894	
SPSM 6		0-6-0TT	OC	JF	10249	1905	a
21		0-4-2T	OC	JF	11938	1909	a
23		0-6-2T	OC	JF	15515	1920	a
	WREN	0-4-0ST	OC	KS	3114	1918	
No.31		0-8-0T	OC	Maffei	4766	1917	
	JUBILEE 1897	0-4-0ST	OC	MW	1382	1897	
18BG		0-4-4T	OC	WB	2228	1924	a
762		4-6-2	OC	WB	2457	1932	
765		4-6-2	OC	WB	2460	1932	
No.18		4wPE		DK		c1918	
	rebuilt as	4wDE					
D564		4wDM		HC	D564	1930	a
–		2-2wPMR		HE	9901	2008	

Gauge 1ft 11¾in (Operational)

7	(OWAIN GLYNDWR)	2-6-2T	OC	Sdn		1923	
8	(LLYWELYN)	2-6-2T	OC	Sdn		1923	
(9)	1213 (PRINCE OF WALES)	2-6-2T	OC	DM	2	1902	b
60	DRAKENSBERG	2-6-2+2-6-2T	4C	Hano	10551	1927	
	GRAF SCHWERIN-LÖWITZ	0-6-2WT	OC	Jung	1261	1908	
10		0-6-0DH		BMR	002	1987	c
11		0-6-0DMF		HC	DM1366	1965	
				reb	STRPS	1997	
				reb	Bredgar	2009	
–		4wDH		HE	7495	1977	d
(68804)	THUNDERBIRD 4 No.5	4wDHR		Perm	005	1985	

Gauge 600mm

–		0-4-0WT	OC	Borsig	5913	1908
1	SABERO	0-6-0T	OC	Couillet	1140	1895

Gauge 1ft 10¾in (Static display)

106	1877 KATHLEEN	0-4-0VBT	VC	DeW		1877	a

a	stored at Capel Bangor Station
b	possibly a new locomotive built by Sdn in 1924
c	built with parts supplied by BD
d	converted for use as a flail-mower

Gauge 1ft 3in

7	TYPHOON	4-6-2	OC	DP	22073	1926

R.WATSON-JONES,
WOODLANDS, THE DINGLE, CONSTITUTION HILL, PENMAENMAWR

Gauge 3ft 0in   Conwy   SH 720765

–		4wDM	RH	202987	1941	a

a   currently at Coleg Llandrillo, Rhos-on-Sea, Colwyn Bay

WELSH HIGHLAND RAILWAY // RHEILFFORDD ERYRI,
DINAS, near CAERNARFON   LL54 5UP

(Operated by the Ffestiniog Railway Company)   Gwynedd   SH 477587

Public railway operating from Caernarfon to Porthmadog.

For details of locomotives, see entry under Ffestiniog & Welsh Highland Railways

WELSH HIGHLAND HERITAGE RAILWAY, WELSH HIGHLAND RAILWAY LTD,
TREMADOG ROAD, PORTHMADOG   LL49 9DY

Public railway operating northwards from terminus in Porthmadog.
Locomotive shed and workshops at Gelert's Farm, Porthmadog.

Gauge 2ft 0in   www.whr.co.uk   Gwynedd   SH 571393

"No.10"		0-4-0PM		BgC	736	1917
–		0-4-0PM		BgC	760	1918
No.5		4wDM		HE	6285	1968
3	ODIN	4wDM		MR	5859	1934
			reb	ALR	No.3	1989
–		4wDM		MR	22237	1965

Gauge 1ft 11½in

590		4-6-0T	OC	BLW	44699	1917	
	RUSSELL	2-6-2T	OC	HE	901	1906	
	KAREN	0-4-2T	OC	P	2024	1942	
	GELERT	0-4-2T	OC	WB	3050	1953	
			reb	Winson		1991	
	(LADY MADCAP)	0-4-0ST	OC	WHR(GF)		2006	a
"2"		4wDH		AB	554	1970	
			reb	AB	6613	1987	
1		4wDH		AB	555	1970	
No.58		0-6-0DH		U23A	23387	1979	
No.60		0-6-0DH		U23A	23389	1977	
No.08		0-6-0DH		U23A	24051	1980	
59105		0-4-0DMF		HE	3510	1947	
	WEIGHITIN	4wDH		HE	7535	1977	
	EMMA	4wDH		HE	9346	1994	
2	"KATHY"	4wDH		HE	9350	1994	
264		4wPM		MR	264	c1916	
36	CNICHT	4wDM		MR	8703	1941	
6		4wDM		MR	11102	1959	
4		4wDM		MR	60S333	1966	
9		4wDM		MR	60S363	1968	

–		4wDM		RH	203031	1941	
1	GLASLYN	4wDM		RH	297030	1952	
			reb	WHR(GF)		1981	
	KINNERLEY	4wDM		RH	354068	1953	
L5	HOVERINGHAM	4wDM		RH	370555	1953	
No.2	BERWYN	4wDMF		RH	481552	1962	

a   under construction. Incorporates some components from HE 652

WELSHPOOL & LLANFAIR LIGHT RAILWAY PRESERVATION CO LTD

Locomotives are kept at :-   Llanfair Caereinion SY21 0SF   SJ 106068
                                   Welshpool (Raven Square) SY21 7LT   SJ 215074

Gauge 2ft 6in   www.wllr.org.uk   **Powys**

822	THE EARL	0-6-0T	OC	BP	3496	1902	
823	COUNTESS	0-6-0T	OC	BP	3497	1902	
G.C.G.D. PROVAN WORKS No.1							
8	DOUGAL	0-4-0T	OC	AB	2207	1946	
10	699.01 SIR DREFALDWYN	0-8-0T	OC	AFB	2855	1944	
No.14	SLR 85	2-6-2T	OC	HE	3815	1954	
No.2	ZILLERTAL	0-6-2T	OC	KraussL	4506	1900	
12	JOAN	0-6-2T	OC	KS	4404	1927	
5	LBC 1 NUTTY	4wVBT	ICG	S	7701	1929	
No.6	MONARCH	0-4-4-0T	4C	WB	3024	1953	Pvd
No.7	CHATTENDEN YD No.AD 690	0-6-0DM		Bg/DC	2263	1949	
17	175	6wDM		Diema	4270	1979	
11	FERRET YARD No.86	0-4-0DM		HE	2251	1940	OOU
"No.16"	ND 3082 SCOOBY / SCWBI	0-4-0DM		HE	2400	1941	
			reb	W&LLR		1992	
DL-34		0-4-4-0DH		Mitsubishi	1871	1972	
9150		4wDHR		BD	3746	1976	
–		2w-2PMR		Wkm	2904	1940	

SOUTH WALES

INDUSTRIAL LOCATIONS

CAF ROLLING STOCK UK LTD,
1 MONKS DITCH DRIVE, CELTIC BUSINESS PARK, NEWPORT   **NP19 4RH**
Gauge 4ft 8½in   www.caf.net   **Newport ST 370871**

–	4wBE	R/R	Zephir	2828	2018

CARDIFF CAPITAL REGION,
ABERTHAW POWER STATION, THE LEYS, ABERTHAW   **CF62 4ZW**
Gauge 4ft 8½in   www.cardiffcapitalregion.wales   **Vale of Glamorgan ST 026664**

–	2w-2PMR	Bance	169/06	2006

CELSA MANUFACTURING (UK) LTD, CARDIFF
Locomotives are kept at :-   Castle Works   CF24 5NN   ST 195755
                                                             Tremorfa Works CF24 2YE   ST 209758
Gauge 4ft 8½in   www.celsauk.com   **Cardiff**

(D3504) 08389		0-6-0DE	Derby	1958	a
(D3797 08630) 3		0-6-0DE	Derby	1959	a
(D3927 08759 09106) 6		0-6-0DE	Hor	1961	a
(D4154 08924) 2		0-6-0DE	Hor	1962	a

a   property of Harry Needle Railroad Co Ltd, Derbyshire

Gauge 2ft 0in

–	2w-2CE	AS&W	1990	OOU

CHRYSALIS RAIL SERVICES LTD,
LANDORE DEPOT, NEATH ROAD, LANDORE, SWANSEA   **SA1 2BD**
Gauge 4ft 8½in   www.chrysalisrail.com   **Swansea SS 658952**

(D3963) 08795	0-6-0DE	Derby	1960

DOW SILICONES UK LTD, BARRY PLANT, CARDIFF ROAD, BARRY   **CF63 2YL**
Gauge 4ft 8½in   www.gb.dow.com   **Vale of Glamorgan ST 142685**

DC 27584	4wDM	R/R	Trackmobile LGN963990492	1992

FITZGERALD PLANT SERVICES LTD,
AVONDALE WAY, AVONDALE INDUSTRIAL ESTATE, CWMBRAN   NP44 1TS

www.fitzgeraldplant.co.uk   **Torfaen**   ST 295966

Road-Rail plant under conversion, overhaul or repair usually present

MAYPHIL (UK) LTD, GOAT MILL ROAD, DOWLAIS, MERTHYR TYDFIL   CF48 3TF

Gauge 5ft 3in	www.mayphil.co.uk			**Merthyr Tydfil**	SO 062073	
1		0-6-0DH	EEV	_(D1266	1969	
				(3954	1969	OOU

ORB STEEL TERMINAL, ORB, NEWPORT   NP19 0RB

Gauge 4ft 8½in	**(Closed)**			**Newport**	ST 326863	
JAMO		4wDH	RR	10198	1965	OOU

PULLMAN RAIL LTD, (CANTON TRACTION & ROLLING STOCK DEPOT),
TRAIN MAINTENANCE DEPOT, LECKWITH ROAD, CARDIFF   CF11 8HP

(part of Colas Rail Group, which is part of the Bouygues Group of companies)

Gauge 4ft 8½in		www.pullmanrail.co.uk			**Cardiff**	ST 171758
(D3654) 08499	REDLIGHT	0-6-0DE	Don		1958	

PUMA ENERGY (UK) LTD, HERBRANDSTON, MILFORD HAVEN

Gauge 4ft 8½in		www.pumaenergy.com			**Pembrokeshire**	SM 888085	
077	53 MR 77	0-6-0DH		EEV	D1198	1967	
			reb	YEC	L122	1993	
078	53 MR 78	0-6-0DH		HE	6663	1969	
			reb	HAB	6586	1999	
	EDWARD	4wDH		TH	267V	1976	a

a   property of Hunslet Ltd, Barton-under-Needwood, Staffordshire

SIMEC ATLANTIS ENERGY LTD, (part of the SIMEC Group),
USKMOUTH POWER STATION, WEST NASH ROAD, NEWPORT   NP18 2BZ

Gauge 4ft 8½in	www.saerenewables.com	**(Closed)**		**Newport**	ST 327837	
–		0-6-0DH	RH	468046	1963	
			reb YEC	L106	1992	OOU

SIMS METAL MANAGEMENT LTD, (division of SIMS Ltd),
NORTHSIDE, SOUTH DOCK, ALEXANDRA DOCK, NEWPORT   NP20 2WF

Gauge 4ft 8½in   www.simsmm.co.uk   **Newport**   ST 317850

Rolling stock for disposal usually present

SOFIDEL UK LTD, BRUNEL WAY,
BAGLAN ENERGY PARK, BAGLAN BAY, PORT TALBOT   SA11 2FP
Gauge 4ft 8½in   www.sofidel.com   Neath Port Talbot SS 736928

ZZ 44	0-6-0DH	EEV-AEI	D3989	1970	OOU

SOUTH WALES FIRE & RESCUE SERVICE,
MAINDEE FIRE STATION, ARCHIBALD STREET, MAINDEE   NP19 8EP
Gauge 4ft 8½in   www.southwales-fire.gov.uk   Newport ST 325881

CN70 UWW		4wDH	R/R	_(Mitsubishi C00092	2019
	99709 978027-9			(GOS RRC440	2020

SOUTH WALES POLICE, LEARNING & DEVELOPMENT CENTRE,
3 WATERTON ROAD, BRIDGEND   CF31 3YY
Gauge 4ft 8½in   www.south-wales.police.uk   Bridgend SS 919785

55574	142033		2w-2DMR	_(BRE(D)	1985
				(Leyland R5.64	1985
		rebuilt as	2w-2DHR	HAB	1990
55624	142033		2-2wDMR	_(BRE(D)	1985
				(Leyland R5.63	1985
		rebuilt as	2-2wDHR	HAB	1990

TALYWAIN SALVAGE, UNIT 23, STAR TRADING ESTATE, PONTHIR   NP18 1PQ
Gauge 1000mm   Torfaen ST 333922

–	4wDH	MR	121UA117	1974	OOU

TATA STEEL EUROPE, TATA STEEL PACKAGING,
TROSTRE WORKS, LLANELLI (part of the Tata Group)   SA14 9SD
Gauge 4ft 8½in   www.tatasteeleurope.com   Carmarthenshire SS 531994

DH50-2	TAZ	0-6-0DHF		TH	246V	1973	
			reb	HE	9377	2011	a
DH50-1	SAMMY	0-6-0DH		TH	278V	1978	
			reb	HE	9376	2011	a

a   property of Hunslet Ltd, Barton-under-Needwood, Staffordshire

TATA STEEL EUROPE, TATA STEEL STRIP PRODUCTS UK,
LLANWERN WORKS, NEWPORT (part of the Tata Group)   NP19 4QZ
Gauge 4ft 8½in   www.tatasteeleurope.com   Newport ST 385863

302	0-6-0DH	GECT	5379	1972	
304	0-6-0DH	EEV	_(D1248	1968	
			(3946	1968	
	reb	LlanwernBSC		1996	OOU

DE1		6wDE	GECT	5409	1976	
DE2		6wDE	GECT	5410	1976	
DE3		6wDE	GECT	5411	1976	
DE4		6wDE	GECT	5412	1976	
DE5		6wDE	GECT	5413	1976	
EM 1		6wDE	GECT	5431	1977	a
903		Bo-BoDE	BBT	3065	1955	

a　　property of Ed Murray & Sons Ltd, Hartlepool, Teesside

TATA STEEL EUROPE, TATA STEEL STRIP PRODUCTS UK,
PORT TALBOT WORKS, PORT TALBOT (Part of the Tata Group)　　　SA13 2NG
Gauge 4ft 8½in　www.tatasteeleurope.com　**Neath Port Talbot SS 771859, 775861, 775876, 775878**

1		0-4-0WE	_(BD	3748	1979	
			(GECT	5476	1979	
2		0-4-0WE	_(BD	3749	1979	
			(GECT	5477	1979	
501		0-4-0DE	BBT	3066	1954	OOU
503	"No.1"	0-4-0DE	BBT	3068	1954	a
504		0-4-0DE	BBT	3069	1954	OOU
(505)	BT-2	0-4-0DE	BBT	3070	1954	b
506		0-4-0DE	BBT	3071	1954	OOU
509		0-4-0DE	BBT	3099	1956	OOU
(512)	BT-4	0-4-0DE	BBT	3102	1956	b
(513)	BT-3	0-4-0DE	BBT	3103	1957	
		reb	PortTalbotBSC	c1972		b
(514)	BT-1	0-4-0DE	BBT	3120	1957	
		reb	PortTalbotBSC	c1972		b
901		Bo-BoDE	BBT	3063	1955	
902		Bo-BoDE	BBT	3064	1955	OOU
904		Bo-BoDE	_(BT	92	1957	
			(WB	3137	1957	
905		Bo-BoDE	_(BT	93	1957	
			(WB	3138	1957	
906		Bo-BoDE	_(BT	94	1957	
			(WB	3139	1957	
07		Bo-BoDE	_(BT	95	1957	
			(WB	3140	1957	
		reb	HAB		1993	
08		Bo-BoDE	_(BT	96	1957	
			(WB	3141	1957	
		reb	HAB		1993	
09		Bo-BoDE	_(BT	97	1957	
			(WB	3142	1957	
		reb	HAB	6063	1994	
910		Bo-BoDE	_(BT	98	1957	
			(WB	3143	1957	OOU
920	BRANWEN	4w-4wDH	CNES	0001	2009	
		refurbished	HE	9381	2012	
921	RHIANNON	4w-4wDH	CNES	0002	2010	
		refurbished	HE	9380	2012	c

922	GUINEVERE	4w-4wDH	CNES	0003	2010		
		refurbished	HE	9379	2012	c	d
923	AURORA	4w-4wDH	CNES	0004	2011		
	'	refurbished	HE	9378	2012	c	
930	FIONA	4w-4wBE	CE	B4646	2019		
931		4w-4wBE	CE	B4649/1	2019		
932	BETSY	4w-4wBE	CE	B4649/2	2019		
933		4w-4wBE	CE	B4662.1	2021		
934	ABIGAIL	4w-4wBE	CE	B4662.2	2021		
935		4w-4wBE	CE	B4666/1	2023		
936		4w-4wBE	CE	B4666/2	2023		
953		Bo-BoDE	BBT	3113	1957		
305		0-6-0DH	GECT	5382	1973		

a   operates at Grange Coke Ovens
b   converted to a brake tender runner
c   worksplate dated 2009
d   currently at Loram UK Ltd, Derby, Derbyshire

TRANSPORT for WALES, SOUTH WALES METRO, TAFF'S WELL RAIL DEPOT, MOY ROAD INDUSTRIAL ESTATE, TAFF'S WELL, PONTYPRIDD   CF15 7QR

Gauge 4ft 8½in   www.tfw.wales   **Rhondda Cynon Taf ST 125833**

–		4wBE	R/R	Niteq	c2023

PRESERVATION LOCATIONS

BRECON MOUNTAIN RAILWAY CO LTD, PANT STATION, MERTHYR TYDFIL   CF48 2DD

Pontsticill Station, and the majority of the route of this railway, lies in Powys. The terminus, and principal locomotive shed and workshops are, however, at Pant, Merthyr Tydfil, and thus within our "South Wales Area".

Gauge 1ft 11¾in   www.bmr.wales   **Merthyr Tydfil SO 060098**

1	SAINT THERESA	2-6-2	OC	BMR		2014	
	a rebuild of	2-6-0	OC	BLW	15511	1897	
2		4-6-2	OC	BLW	61269	1930	
–		0-6-0DH		BMR	001	1987	a
Tu46 001		4w-4wDH		Kambarka		1981	
Tu7 1698		4w-4wDH		Kambarka		1981	

a   built from parts supplied by BD

Brecon Mountain Railway, Steam Museum, Pontsticill
Gauge 1ft 11¾in   **Powys SO 063120**

"PENDYFFRYN"	0-4-0VBT	VC	DeW	1894	
"REDSTONE"	0-4-0VBT	VC	Redstone	1905	

BRIDGEND VALLEYS RAILWAY RAILWAY COMPANY LTD, t/a GARW VALLEY RAILWAY // RHEILFFORD CWM GARW, PONTYCYMER LOCOMOTIVE WORKS, OLD STATION YARD, PONTYCYMER, BRIDGEND CF32 8AZ

Gauge 4ft 8½in   www.garwvalleyrailway.co.uk   Bridgend   SS 904914

68070	(PAMELA)	0-6-0ST	IC	HE	3840	1956
–		0-4-0ST	OC	RSHN	7705	1952
–		4wDH		FH	3890	1959
–		4wDH		FH	4006	1963
(W51919)		2-2w-2w-2DMR		DerbyC&W		1960
W52048		2-2w-2w-2DMR		DerbyC&W		1960

CAMBRIAN TRANSPORT (CAMBRIANCO) LTD, BARRY TOURIST RAILWAY & RAIL CENTRE, BARRY ISLAND STATION, BARRY ISLAND, VALE OF GLAMORGAN CF62 5TH

Gauge 4ft 8½in   (Closed)   Vale of Glamorgan   ST 117667

5539		2-6-2T	OC	Sdn		1928	
	SUSAN	4wVBT	VCG	S	9537	1952	
(D3658 | 08503)	0-6-0DE		Don		1958	a
(E6024)	73118	Bo-BoDE/RE		_(EE	3586	1966	
				(EEV	E356	1966	
7043		4w-4wRER		MetCam		1980	
7501		4w-4wRER		MetCam		1979	
–		4wDM	R/R	Unilok	2183	1982	
DR 98308		4w-4wDHR		Geismar	826	1998	
F621 PWP	14201	4wDM	R/R	_(Renault		c1988	
				(Bruff		c1988	
F591 RWP	68906	4wDM	R/R	_(Mercedes		1988	
				(Bruff		1988	

a   property of Railway Support Services Ltd, Wishaw, Warwickshire

CARMARTHENSHIRE COUNTY COUNCIL, MUSEUM SERVICES, CofGAR, KIDWELLY INDUSTRIAL MUSEUM TRUST, KIDWELLY INDUSTRIAL MUSEUM, BROADFORD, KIDWELLY SA17 4LW

Gauge 4ft 8½in   www.kidwellyindustrialmuseum.org.uk   Carmarthenshire   SN 422078

–		0-4-0ST	OC	AB	1081	1909
–		0-6-0ST	OC	P	2114	1951
No.2		0-4-0DM		AB	393	1954

Gauge 2ft 0in

–		4wDM		RH	398063	1956
			reb	ESCA		1986

The museum is temporarily closed

CEFN MABLY FARM PARK, BEGAN ROAD, MICHAELSTON-Y-FEDW   CF3 6XL
Gauge 1ft 0¼in   www.cefnmablyfarmpark.com   **Cardiff ST 231842**

–	2-4-2BE	s/o	WhalleyB	2020	

DARE VALLEY COUNTRY PARK, ABERDARE   CF44 7PS
Gauge 3ft 0in   www.darevalleycountrypark.co.uk   **Rhondda Cynon Taf SN 983026**

–	0-4-0DMF	HC	DM1314	1963	

GWENDRAETH VALLEY RAILWAY, near KIDWELLY
Gauge 4ft 8½in   **Carmarthenshire Private Site**

(D2141 03141)	0-6-0DM	Sdn		1960
(D3577 08462) 08994	0-6-0DE	Crewe		1958
(D3854 08687) 08995	0-6-0DE	Hor		1959

GWILI RAILWAY CO LTD, BRONWYDD ARMS
Locomotives are kept at :-

Abergwili Junction, Carmarthenshire	(SA31 2AQ)	SN 430213
Bronwydd Arms, Carmarthenshire	SA33 6HT	SN 417239
Llwyfan Cerrig, Carmarthenshire		SN 405258

Gauge 4ft 8½in   www.gwili-railway.co.uk

(No.28)		0-6-2T	IC	Cdf	306	1897
1369		0-6-0PT	OC	Sdn		1934
–		0-6-0ST	IC	HE	3829	1955
–		0-4-0ST	OC	P	1345	1914
47	MOORBARROW	0-6-0ST	OC	RSHN	7849	1955
	HAULWEN	0-6-0ST	IC	VF	5272	1945
			reb	HE	3879	1961
D2178		0-6-0DM		Sdn		1962
43056		Bo-BoDE		Crewe		1977
–		0-6-0DH		GECT	5391	1973
			reb	RMS(D)LWO	2108	2002
–		0-6-0DM		P	5014	1959
	ABIGAIL	4wDM		RH	312433	1951
W 51347		2-2w-2w-2DMR		PSteel		1960
W 51401		2-2w-2w-2DMR		PSteel		1960
RTU 6910	99709 901037-0	2w-2PMR		Geismar	ST/01/23	2001

LLANELLI & MYNYDD MAWR RAILWAY CO LTD, CYNHEIDRE   SA15 5YG
Gauge 4ft 8½in   www.llanellirailway.co.uk   **Carmarthenshire SN 495071**

(D3963) 08795	0-6-0DE	Derby		1960	a
10222	4wDH	RR	10222	1965	
(M55019) 975042 960015	2-2w-2w-2DMR	GRC&W		1958	

55547	142006		2w-2DMR	_(BRE(D)		1985	
				(Leyland R5.16		1985	
		rebuilt as	2w-2DHR	AB		1991	
55597	142006		2-2wDMR	_(BRE(D)		1985	
				(Leyland R5.15		1985	
		rebuilt as	2-2wDHR	AB		1991	
55647	143606		2w-2DMR	_(AB	695	1986	
				(Alex	1784/11	1986	
		rebuilt as	2w-2DHR	RFSD		1988	
55672	143606		2-2wDMR	_(AB	696	1986	
				(Alex	1784/12	1986	
		rebuilt as	2-2wDHR	RFSD		1988	
55648	143607		2w-2DMR	_(AB	687	1986	
				(Alex	1784/13	1986	
		rebuilt as	2w-2DHR	RFSD		1988	
55673	143607		2-2wDMR	_(AB	688	1986	
				(Alex	1784/14	1986	
		rebuilt as	2-2wDHR	RFSD		1988	
55653	143612		2w-2DMR	_(AB	681	1986	
				(Alex	1784/23	1986	
		rebuilt as	2w-2DHR	RFSD		1988	
55678	143612		2-2wDMR	_(AB	682	1986	
				(Alex	1784/24	1986	
		rebuilt as	2-2wDHR	RFSD		1988	
57374	153374 (57324)		4w-4DHR	Leyland R6.047		1988	
		reb		AB		1992	Dsm
(10297)	126 483006		2-2w-2w-2RER	MetCam		1938	
(11297)	226 483006		2-2w-2w-2RER	MetCam		1938	
(10255)	128 483008		2-2w-2w-2RER	MetCam		1938	
(11255)	228 483008		2-2w-2w-2RER	MetCam		1938	
64571	315856		4w-4wWER	York(BRE)		1981	
64572	315856		4w-4wWER	York(BRE)		1981	
–			4wDHR	BD	3706	1975	
–			2w-2PMR	DC	1895	1950	DsmT
RTU 50097	99709 901042-0		2w-2PMR	Geismar ST/03/38		2003	
MPP 13736	99709 901024-8		2w-2BER	Geismar 5E/0008		2006	
A13W	PWM 2801 (TR 3)		2w-2PMR	Wkm	6884	1954	

a   currently at Chrysalis Rail Services Ltd, Landore Depot, Swansea

LLANELLI RAILWAY GOODS SHED TRUST, MARSH STREET, LLANELLI SA15 1BG

Gauge 4ft 8½in   www.llanellirailwaygoodsshedtrust.org.uk **Carmarthenshire SN 509993**

55663	143622		2w-2DMR	_(AB	717	1986	
				(Alex	1784/43	1986	
		rebuilt as	2w-2DHR	AB		1990	
55688	143622		2-2wDMR	_(AB	718	1986	
				(Alex	1784/44	1986	
		rebuilt as	2-2wDHR	AB		1990	

DAVID MATTHEW, PEN-Y-BRYN LIGHT RAILWAY, CARDIFF
Gauge 2ft 0in   **Private Site**

–		4wDM		HE	6298	1964
33	DRUID	4wDM		MR	8644	1941
			reb	ALR	6	1999
8	ND 6455 DADDADONKEY	4wDM		RH	221625	1943
–		4wDM		RH	487963	1963

NATIONAL MUSEUM of WALES,
Big Pit National Coal Museum, Big Pit Colliery Site, Blaenavon   **NP4 9RL**
Gauge 4ft 8½in   www.museum.wales/bigpit   **Torfaen SO 236088**

NORA No.5	THE BLAENAVON CO. LTD	0-4-0ST	OC	AB	1680	1920
	P.D. 10	0-6-0ST	IC	HC	544	1900

Gauge 2ft 0in

–	0-4-0DMF	HE	6049	1961	
15/10	4wBEF	_(EE	3147	1961	
		(RSHD	8289	1961	Dsm

Collection Centre, Heol Crochendy, Parc Nantgarw, Nantgarw   **CF15 7QT**
Storage depot, visitors by prior appointment only.
Gauge 4ft 8½in   www.museum.wales/collections-centre   **Rhondda Cynon Taf ST 114861**

52/001	0-4-0F	OC	AB	1966	1929
–	0-4-0F	OC	AB	2238	1948

Gauge 3ft 0in

"OGGY"	4wDM	RH	187100	1937

Gauge 2ft 0in

4	4wBE	GB	

Waterfront Museum, Oystermouth Road, Maritime Quarter, Swansea   **SA1 3RD**
Gauge 4ft 4in   www.museum.wales/swansea   **Swansea SS 659927**

–	4wG	IC	NMW	1981

NATIONAL TRUST,
DOLAUCOTHI GOLD MINES, PUMSAINT, LLANWRDA   **SA19 8US**
Gauge 1ft 10½in   www.nationaltrust.org.uk   **Carmarthenshire SN 666404**

–	4wBE	CE	B2944I	1982	
–	4wBE	WR	899	1935	OOU
–	0-4-0BE	WR	5311	1955	

NEATH PORT TALBOT COUNTY BOROUGH COUNCIL,
Cefn Coed Colliery Museum, Neath Road, Crynant, Neath   **SA10 8SN**

Gauge 4ft 8½in   www.beta.npt.gov.uk   **(Closed)**   Neath Port Talbot   SN 786034

2758	CEFN COED	0-6-0ST	IC	WB	2758	1944

Gauge 2ft 0in

–		4wDH	HE	8812	1978

Margam Country Park // Parc Gwledig, off Margam Road, Groes, Margam   **SA13 2AG**

Gauge 2ft 0in   www.margamcountrypark.co.uk   Neath Port Talbot   SS 806863

	MARGAM CASTLE	0-4-0DH	s/o	AK	65	2001

JOE NEMETH ENGINEERING LTD,
BAY 1, SPACEMAN BUSINESS PARK, SEVERN BRIDGE BUSINESS PARK,
SYMONDSCLIFFE WAY, CALDICOT   **NP26 5PW**

www.joenemethengineering.co.uk   **Newport**   ST 492880

Locomotives for sale / under construction / overhaul / repair occasionally present

OAKWOOD LEISURE LTD,
OAKWOOD THEME PARK, CANASTON BRIDGE, NARBERTH   **SA67 8DE**

(part of Aspro Parks Group S.L.)   www.oakwoodthemepark.co.uk

Gauge 600mm — Brer Rabbit's Burrow   **Pembrokeshire**   SN 072124

–	4w-4wRE	s/o	SL	275/1.5.90	1990	
–	4w-4wRE	s/o	SL	275/2.5.90	1990	

Gauge 1ft 3in

LORNA	4-4wDHR		Goold		1989	
LINDY-LOU	0-8-0DH	s/o	SL	7218	1972	OOU
LENKA	4-4wDHR		SL	7322	1973	
		reb			1998	
–	0-8-0PH	s/o	SL	R9	1976	OOU

D. PARFITT, PORTHCAWL
Gauge 1ft 3in   **Bridgend**   **Private Site**

1935	SILVER JUBILEE	4-6-4PE	s/o	SmithP	1935

PEMBROKESHIRE COUNTY COUNCIL, SCOLTON MANOR MUSEUM
& COUNTRY PARK, BETHLEHEM, near HAVERFORDWEST   **SA62 5QL**

Gauge 4ft 8½in   www.scoltonmanor.co.uk   **Pembrokeshire**   SM 991222

(No.1378) GWENDREATH RAILWAY No.2 // MARGARET

		0-6-0ST	IC	FW	410	1878
(A 123 W)		2w-2PMR		Wkm	3361	1942	a

    a   not on public display

PONTYPOOL & BLAENAVON RAILWAY COMPANY (1983) LTD,
t/a BLAENAVON'S HERITAGE RAILWAY, BLAENAVON   NP4 9SF
Gauge 4ft 8½in   www.bhrailway.co.uk   Torfaen   SO 237093, 234093

	ROSYTH No.1	0-4-0ST	OC	AB	1385	1914	
2	"PONTYBEREM"	0-6-0ST	OC	AE	1421	1900	a
	(SIR JOHN)	0-6-0ST	OC	AE	1680	1914	
1857		0-6-0T	OC	HC	1857	1952	
18	JESSIE	0-6-0ST	IC	HE	1873	1937	
	rebuilt as	0-6-0T	IC	Llangollen		2007	
	rebuilt as	0-6-0ST	IC	Llangollen		2010	
	rebuilt as	0-6-0T	IC	Llangollen		2012	
	rebuilt as	0-6-0ST	IC	Barry		2019	
71515		0-6-0ST	IC	RSHN	7169	1944	
	EMPRESS	0-6-0ST	OC	WB	3061	1954	
D5627	(31203) STEVE ORGAN G.M.	A1A-A1A DE		BT	227	1960	
(D6723	37023)	Co-CoDE		_(EE	2886	1961	
				(VF	D601	1961	
D6916	(37216)	Co-CoDE		_(EE	3394	1963	
				(EEV	D860	1963	
	BLAENAVON No.14	0-4-0DH		HC	D1344	1965	
	JOHN RODEN	0-6-0DH		HE	5511	1960	
170		0-8-0DH		HE	7063	1971	
RT1		0-6-0DM		JF	22497	1938	
–		4wDM		RH	394014	1956	
51351	(117418)	2-2w-2w-2DMR		PSteel		1960	
51397	(117418)	2-2w-2w-2DMR		PSteel		1960	
RTU 9218		2w-2PMR		Geismar	97/17	1997	
	99709 901009-9	2w-2PMR		Lesmac	LMS019	2006	
(PWM 3962)		2w-2PMR		Wkm	6947	1955	

a   currently off site for restoration

G. REES, UNKNOWN LOCATION
Gauge 750mm   Private Site

13		4wDM		Moës		

Gauge 2ft 0in

263 051		4wBE		CE	B0119B	1973
11	BARGEE	4wDM		MR	8540	1940
–		4wDM		MR	9932	1972
			reb	AK		1988

RHONDDA CYNON TAFF COUNTY BOROUGH COUNCIL,
RHONDDA HERITAGE PARK, LEWIS MERTHYR COLLIERY,
COED CAE ROAD, TREHAFOD   CF37 2NP
Gauge 4ft 8½in   www.rctcbc.gov.uk   Rhondda Cynon Taff ST 040912

–	4wDM	RH	441936	1960

Gauge 3ft 0in

–		0-4-0DMF		HE	6696	1966

R. STAPLE, ABERDARE MOTOR REPAIRS, OLD TRAMWAY, ROBERTSTOWN INDUSTRIAL ESTATE, ABERDARE   CF44 8HD
Gauge 2ft 0in   Rhondda Cynon Taff   SO 005028

ROBERT		4wDM	s/o	RH	393327	1956

SOUTH WALES LOCO CAB PRESERVATION GROUP, "THE CAB YARD", off HORSEFAIR ROAD, BRIDGEND   CF31 3YN
Gauge 4ft 8½in   www.traincabs.co.uk   **Bridgend**   SS 940787

DX 50002		4wDHR		Matisa PV5 570	1957

SWANSEA MUSEUM, COLLECTIONS CENTRE, CROSS VALLEY LINK ROAD, LANDORE STORE, SWANSEA   SA1 2JT
Gauge 4ft 8½in   www.swanseamuseum.co.uk   **Swansea**   SS 662953

	SIR CHARLES	0-4-0F	OC	AB	1473	1916
1426		0-6-0ST	OC	P	1426	1916
	incorporates parts of			P	1187	
–		0-4-0DM		_(RSHD	7910	1963
				(WB		1963

SECTION 4 IRELAND

NORTHERN IRELAND

REPUBLIC OF IRELAND

+   No known locomotives exist in this county

RAILWAY CONTRACTORS

Listed below are the Railway Engineering Contractors / Plant Hire specialists who own locomotives for use on railway contracts etc. The locomotives are to be found in all parts of Ireland but the details of the locomotive fleets are listed under the firm's main depot in the County shown below.

TITLE OF FIRM	COUNTY
Breffni Hire Plant Ltd	Dublin
Carra Plant Services	Galway
Dixon Bros. (Agriculture & Plant) Ltd	Meath
John Dixon Plant Hire Ltd	Westmeath
Oliver Dixon (Hedgecutting & Plant Hire) Ltd	Meath
P.F. Dixon Plant Hire	Meath
Doyle Agricultural & Railplant Services Ltd	Wexford
S. Duffy Plant Hire Ltd	Dublin
Linmag Rail GmbH	Laois
Mike Lynch Excavations Ltd	Tipperary
Tom Lynes Plant Hire Ltd	Cork
Declan Maher Plant Ltd	Galway
McCormick Bros. Rail & Machinery Hire Ltd	Longford
Moloney Agricultural Services	Tipperary
Northern Excavators Ltd	Down (NI)
Railway Plant Services Ltd	Armagh (NI)
Sanline Systems Ltd	Dublin

NORTHERN IRELAND

LISTINGS OF LOCATIONS IN NORTHERN IRELAND ARE PRESENTED ON A SINGLE "LAND AREA" BASIS, RATHER THAN A COUNTY BASIS.

INDUSTRIAL SITES

BULRUSH PEAT CO LTD (Subsidiary of Pindstrup Mosebrug A/S)
Locomotivess are kept at :-
B - Newferry Road, Bellaghy, Magherafelt, Co. Derry   BT45 8ND   H 986985
G - Grove Bog, Carrickmore, Co.Tyrone   BT79 9NR   H 600707
GW - Glenconway Workshops, near Dungiven, Co. Derry   BT47 4NS
R - Randalstown, Co.Antrim   BT41 2PD   J 090935

Gauge 750mm   www.bulrush.co.uk

2		4wDM		MR	40S307	1967		OOU
	reb			Bulrush(B)		2009	R	a
–		4wDH		Schöma	4978	1988	GW	Dsm
L5		4wDH		Schöma	4979	1988		
	rebuilt as	4wDM		Bulrush(B)		2018	B	
–		4wDH		Schöma	4980	1988	B	b
–		4wDH		Schöma	4992	1989	R	b c
L4		4wDH		Schöma	5601	1999	R	
L6		4wDH		Schöma	5602	1999	R	
–		4wDH		Schöma	5603	1999	R	b

a   rebuilt using parts from MR 22220 and MR 40S307
b   slave unit for use with a master unit.
c   worksplate reads Schöma 4902

NORTHERN EXCAVATORS LTD, t/a NE RAIL, 103 CULCAVEY ROAD, HILLSBOROUGH   BT26 6HH

Gauge 5ft 3in   www.northernexcavators.com   Co. Down

RR14		4wPMR	R/R	Kawasaki KAF950B/ 00729C/0359	2011

NORTHERN IRELAND RAILWAYS / TRANSLINK LTD
YORK ROAD DEPOT, BELFAST   BT15 3RP

Gauge 5ft 3in   www.translink.co.uk   Co.Antrim

7015	(K-355 APT)	4w-4wDMR		P&T	46	1985
			reb	P&T		2002
	AFZ 7054	4wDM	R/R	_(Mitsubishi		2009
				(LH Group		2009
11	99709 428011-9	4w-4wDHR		Windhoff		2016
L2		4wDH	R/R	Windhoff		2022
–		4wBE		Niteq	B177	2001

RAILWAY PLANT SERVICES LTD,
19 DRUMARAN ROAD, GILFORD, CRAIGAVON   BT63 6DB

Gauge 5ft 3in   Co.Armagh

SV 062	S948 DWL	4wDM	R/R	_(Unimog	191802	1998
	99709 979081-5			(Zweiweg	1812	1998

plant yard with road/rail vehicles usually present

JAMES STEVENSON (QUARRIES) LTD,
CLINTY QUARRY, 215 DOURY ROAD, BALLYMENA   BT43 6SS

Gauge 2ft 0in   Co. Antrim   D 102073

–		4wDM	MR	8684	1941	OOU

SUNSHINE PEAT CO, 33 DERRYHUBBERT ROAD, DUNGANNON   BT71 6NW

Gauge 2ft 6in   Closed   Co.Armagh

–		0-4-0DM	HE	(2242	1941?)	Dsm
L3	HENRY ABRAHAM	0-4-0DM	HE	2250	1940	OOU
L4	VICTORIA	0-4-0DM	HE	2252	1940	OOU
L1	ND3053 GAVIN	0-4-0DM	HE	2264	1940	OOU
L2	ND3061 W.P.O'KANE	0-4-0DM	HE	2399	1941	OOU

SEYMOUR SWEENEY CO LTD, COLERAINE

Gauge 3ft 0in   Co. Antrim

–	2w-2PMR	Wkm	7441	1956	OOU

PRESERVATION SITES

DEPARTMENT OF THE ENVIRONMENT (N.I.), COUNTRYSIDE AND WILDLIFE BRANCH, BIRCHES PEATLANDS PARK, DERRYHUBBERT ROAD, DUNGANNON — BT71 6NW

Gauge 3ft 0in   R.T.C.   Co.Armagh

–		4wDH	AK	44	1993	
1		4wDM	FH	3719	1954	OOU
2	HENRY HEAL	4wDM	Schöma	1727	1955	OOU

Locomotives viewable by appointment

DESTINED GROUP plc, FOYLE VALLEY RAILWAY MUSEUM, FOYLE VALLEY RAILWAY COMPANY, 1 FOYLE ROAD, LONDONDERRY BT48 6SQ

Gauge 3ft 0in   www.destined.ie   R.T.C.   Co.Derry

(No.4)	(MEENGLAS)	2-6-4T	OC	NW	828	1907
6	COLUMBKILLE	2-6-4T	OC	NW	830	1907
–		4wDM		MR	11039	1956
12		0-4-0+4DMR		_(WkB	M44719	1934
				(Dundalk		1934

DOWNPATRICK & COUNTY DOWN RAILWAY, DOWNPATRICK RAILWAY MUSEUM, DOWNPATRICK STATION, MARKET STREET, DOWNPATRICK — BT30 6LZ

Gauge 5ft 3in   www.downrail.co.uk   Co.Down

90		0-6-0T	IC	Inchicore		1875	
1		0-4-0T	OC	OK	12475	1934	
3		0-4-0T	OC	OK	12662	1935	
146		Bo-BoDE		GM	27472	1962	
A39R		Co-CoDE		MV	925	1955	
C231		Bo-BoDE		MV	977	1956	
G611		4wDH		Dtz	57225	1962	
G613		4wDH		Dtz	57226	1962	
G617		4wDH		Dtz	57229	1962	
E421	W F GILLESPIE OBE	6wDH		Inchicore		1961	OOU
E432		6wDH		Inchicore		1962	OOU
2509		0-4-0+4DMR		_(WkB	1747D	1947	
				(ELC		1947	OOU
6111	(2624)	4w-4wDMR		_(AEC	8031 046	1953	
				(PRoyal	B35171	1953	OOU
(RB 3)	(RDB 977020)	4wDMR		_(Leyland		1981	
				(BRE(D)		1981	OOU
72	No.2	4+0-4-0VBTR		MetAmal		1905	DsmT a

8069			4w-4wDER		DerbyC&W		1977	
8090			4w-4wDER		DerbyC&W		1977	
8458	(ANTRIM CASTLE)		4w-4wDER		DerbyC&W		1986	
(7017)	VMT 850GR		4wDHR		_(Donelli	558	1998	
					(Geismar		1998	
(713)	ROSIE		2w-2DMR		Wkm	8916	1962	
416A	(712)		2w-2PMR		Wkm	8919	1962	
		rebuilt as	2w-2DMR		Inchicore		1986	OOU
(7020)	JXI 2860		4wDM	R/R	_(Bedford		1986	
					(Bruff	531	1986	

a   former railmotor, converted to hauled coaching stock

GIANT'S CAUSEWAY & BUSHMILLS RAILWAY, CAUSEWAY STATION, RUNKERRY ROAD, near BUSHMILLS   BT57 8SZ

Gauge 3ft 0in   **Co.Antrim**

No.3	SHANE	0-4-0WT	OC	AB	2265	1949	OOU
No.1	TYRONE	0-4-0T	OC	P	1026	1904	OOU
–		4w-4wDH		SL	8165	2010	
	RORY	4wDH		SMH	102T016	1976	

RAILWAY PRESERVATION SOCIETY OF IRELAND, WHITEHEAD DEPOT, CASTLEVIEW ROAD, WHITEHEAD   BT38 9NA

Gauge 5ft 3in   www.steamtrainsireland.com   **Co.Antrim J 474922**

4			2-6-4T	OC	Derby		1947	
27			0-6-4T	IC	BP	7242	1949	OOU
171	SLIEVE GULLION		4-4-0	IC	BP	5629	1913	
				reb	Dundalk	42	1938	
No.184			0-6-0	IC	Inchicore		1880	OOU
186			0-6-0	IC	SS	2838	1879	
461			2-6-0	IC	BP	6112	1922	
(No.3)			0-6-0ST	OC	AE	2021	1928	OOU
3	GUINNESS		0-4-0ST	OC	HC	1152	1919	a
(B142)			Bo-BoDE		GM	27468	1962	
(23)			4wDM		FH	3509	1951	OOU
(1)			4w-4wPMR		NCC		1933	
		rebuilt as	4w-4wDHR		NCC		1947	OOU
1			4wDM		RH	382827	1955	
–			4wPM	R/R	_(Unilok		1965	
					(Jung	A114	1965	

a   ex works 30/9/1915 but worksplate dated 1919

PETER SCOTT, FINNAGHY PARK CENTRAL, BELFAST
Private site. No casual visitors
Gauge 1ft 7¼in.   **Co.Antrim   Private Site**

–		2-4-0	OC	ScottP	c1969	
	rebuilt as	2-4-0T	OC	ScottP	1970	

ULSTER FOLK & TRANSPORT MUSEUM,
153 BANGOR ROAD, HOLYWOOD   **BT18 0EU**
Gauge 5ft 3in   www.nationalmuseumsni.org   **Co.Down   J 419806**

No.30		4-4-2T	IC	BP	4231	1901	
74	DUNLUCE CASTLE	4-4-0	IC	NBQ	23096	1924	
93		2-4-2T	IC	Dundalk	16	1895	
800	MAEDHBH	4-6-0	3C	Inchicore		1939	
No.1		0-6-0ST	OC	RS	2738	1891	
(102)	FALCON	Bo-BoDE		_(Don		1970	
				(HE	7198	1970	
B113		Bo-BoDE		Inchicore		1950	
1	(8178)	2w-2DMR		ADC		1928	
			reb	Dundalk		1933	
–		4wPM	R/R	_(Zagro	80159	2010	
				(Zwiehoff		2010	

Gauge 3ft 0in

2		0-4-0Tram	OC	K	T84	1883	a
2	BLANCHE	2-6-4T	OC	NW	956	1912	
2		0-4-0T	OC	P	1097	1906	
–		4-4-0T	OC	RS	2613	1887	
(1)		2-2-0PMR		A&O		1905	
11	PHOENIX	4wVBT		AtW	114	1928	
	rebuilt as	4wDM		Dundalk		1932	
3		2-4w-2PMR		_(DC	1519	1926	
				(D&BST		1926	Dsm b
10		0-4-0+4DMR		_(WkB	M42830	1932	
				(KMB		1932	
No.2		4+0-4-0RE/WER		_(M&P		1885	
				(Ashbury		1885	

a	worksplate reads 84
b	now unmotorised

Gauge 2ft 0in

–	4wDM		FH	3449	1950
–	4wDM		HE	3127	1944
246	4wPM		MR	246	1916
–	4wDM		MR	9202	1946

Gauge 1ft 10in

20	0-4-0T	IC	Spence		1905

REPUBLIC OF IRELAND

LISTINGS OF LOCATIONS IN THE REPUBLIC OF IRELAND ARE PRESENTED
ON A COUNTY BASIS, WITH COUNTIES LISTED ALPHABETICALLY.

BORD NA MÓNA LOCATIONS ARE DESCRIBED UNDER THEIR COUNTY HEADINGS

A combined fleet list of Bord na Móna locomotives follows the county-based listings.

CARLOW

PRESERVATION SITES

GLYNN BUSES & COACHES LTD, BALLYKEALY, near BALLON
Also known as Celtic steamers, with additional yard at Wexford Road, Carlow
Gauge 1ft 0in

TITAN	0-4-0ST	OC	Manean	1973

PRIVATE OWNER, CARLOW AREA
Gauge 3ft 0in

–	4wDM	RH	

CAVAN

PRESERVATION SITES

**BELTURBET RAILWAY AND VINTAGE RESTORATION ASSOCIATION,
BELTURBET STATION, RAILWAY ROAD, BELTURBET**

Gauge 5ft 3in	www.belturbetheritagerailway.com				H 363167
–	4wDM	RH	312425	1951	
Gauge 3ft 0in					
LM 137	4wDM	RH	382817	1955	

CLARE

PRESERVATION SITES

THE WEST CLARE RAILWAY CO LTD, MOYASTA JUNCTION, KILRUSH
Gauge 3ft 0in   www.westclarerailway.ie

No.5	SLIEVE CALLAN	0-6-2T	OC	D	2890	1892	
			reb	AK		2009	
–		0-6-0DMF		HC	DM719	1950	OOU
–		4wDH		RFSD	L101	1989	OOU
LM 60		4wDM		RH	259738	1948	
LM 64		4wDM		RH	259745	1948	
LM 125	O	4wDM		RH	379084	1954	OOU
LM 117	MÍCHAÉL	4wDM		RH	379910	1954	
LM 189		0-4-0DM		Dtz	57125	1960	OOU
(LM 257)		0-4-0DM		Dtz	57836	1965	OOU
LM 356	SEAN	4wDH		DunEW			

WEST OF IRELAND TRANSPORT MUSEUM, MOYASTA
Gauge 5ft 3in

A3R	Co-CoDE	MV	889	1955	
A15	Co-CoDE	MV	901	1955	
124	Bo-BoDE	GM	26274	1960	
152	Bo-BoDE	GM	27478	1962	
190	Bo-BoDE	GM	31257	1966	

J. WHELAN LTD, WHELAN TRACTOR SALES & AGRICULTURAL ENGINEERS, ENNIS ROAD, KILRUSH
Gauge 3ft 0in   www.whelansgarage.com

LM 133	4wDM	RH	379090	1955	Dsm
LM 168	4wDM	RH	402980	1956	Dsm

J.WHELAN & ASSOCIATES, TULLYGOWER QUARRY, near KILRUSH
Gauge 5ft 3in   R 039603

(C227 106) C202	Bo-BoDE	MV	973	1957	a	
(710)	2w-2DMR	Wkm	8918	1962	OOU	b
(714)	2w-2DMR	Wkm	8920	1962	OOU	b

Gauge 3ft 0in

LM 25	4wDM	RH	244871	1946	
LM 59	4wDM	RH	259737	1948	
LM 98	4wDM	RH	375335	1954	

LM 146		4wDM	RH	394024	1956	
LM 147		4wDM	RH	394025	1956	
LM 314		0-4-0DM	HE	8548	1977	c
C 53		4wPMR	_(BnM	9	1958	
			(SM		1958	c
C 78		4wPMR	BnM		1972	c

a   property of Irish Traction Group
b   property of Irish Traction Group, Wickham Awareness Group
c   stored for West Clare Railway Co Ltd, Moyasta Junction, Kilrush

CORK

INDUSTRIAL SITES

COURTMACSHERRY MACHINERY LTD and BENNETT TRACTOR SALES LTD, CULLENAGH, COURTMACSHERRY, BANDON
Gauge 3ft 0in   www.cmstractorparts.com

LM 268	0-4-0DM	HE	7234	1971
LM 325	0-4-0DM	HE	8942	1981
LM 329	0-4-0DM	HE	8936	1980
LM 333	0-4-0DM	HE	8940	1980
LM 338	0-4-0DM	HE	8945	1981
F 836	4wDM	BnM(BI)		2008
F 878	4wDM	BnM		1987

IARNRÓD ÉIREANN, MAINTENANCE DEPOT CORK, WATER STREET, CORK
Gauge 5ft 3in   www.irishrail.ie

–	4wBH	R/R	AVB	AVB5741/1	2010	OOU

TOM LYNES PLANT HIRE LTD, ORCHARD GROVE, LOHORT, CECILSTOWN, MALLOW
Gauge 4ft 8½in

(XLR 8101) 97 C 40013 (R624 JFE)	4wDMR	R/R	_(Landrover	1997
			(Raynesway	1998

Other road/rail plant usually present

THE RAILROAD FACTORY LTD,
UNIT E3, SOUTHLINK PARK, FRANKFIELD, CORK
Gauge 4ft 8½in   www.trrf.ie

–		4wDH	R/R	_(TRRF	2015	
				(Mecalac 08801	2015	a

a   demonstrator locomotive

new TRRF locomotives may be present

CHRIS SHILLING, t/a OIM LTD, OIBREACHA IARNROD NA MUMHAM,
MUNSTER RAILWAY WORKS, KEALKILL, near BANTRY
Gauge 1ft 3in

–		4wPE	OIM	010	2015

Locomotives under construction may be present

PRESERVATION SITES

J. & R. BRADLEY, BALLYNAKILLWEST, NEWTOWNSHANDRUM, near CHARLEVILLE
Gauge 2ft 6in   R 461244, 482223

SO 35 S 35	4wDM	MR	9709	1952

DENNIS COLLINS, NEWMARKET ROAD, KANTURK
Gauge 3ft 0in

LM 360	4wDH	DunEW			
–	2w-2PM	Ferguson		c1938	a

a   former road vehicle, converted to rail use 1985

Gauge 2ft 0in

BR 1949	(WANDERING WILLIE)	0-4-0VBT	VCG		(1949?)	
	rebuilt as	0-4-0PM				
	rebuilt as	4-2wPM		CollinsD	c2003	OOU
–		2-2-0PM		CollinsD	2011	OOU

TIM CROWLEY LTD, T/A CROWLEY ENGINEERING,
UPPER GLANMIRE BRIDGE, GLANMIRE
Gauge 3ft 0in   www.crowley.ie

LM 371	4wDH	DunEW	

TIM CROWLEY, YOUGHAL CASTLE, YOUGHAL
Gauge 3ft 0in

LM 140		4wDM	RH	392139	1955

MR FITZGERALD, RATHCORMAC
Gauge 5ft 3in

–		4wDM	RH	252843	1948	a

a   currently at Iarnród Éireann, Inchicore Works, Dublin

IARNRÓD ÉIREANN, CORK (CEANNT / KENT) STATION
Gauge 5ft 3in   www.irishrail.ie

36	2-2-2	IC	BuryC&K	1847

THE JAMESON EXPERIENCE, OLD DISTILLERY WALK, MIDLETON
Gauge 2ft 6in   www.jamesonwhiskey.com

–	2-2-2	VC	Midleton	2012

Non-working replica locomotive

TIM NAGLE (deceased),
GREY GABLES, FEATHERBED LANE, off EASTERN ROAD, KINSALE
Gauge 600mm

–	4wDM		Strüver	60385	c1955	Dsm

JOHN TOUIG, LISSAGROOM, UPTON
Gauge 2ft 0in   R 522612

–	4wDM	RH	264244	1949	OOU

TIM and SANDRA ROWE, MARKET GARDENERS, near BANTRY
Gauge 1ft 3in   Private Site

TORNADO	2w-2CA	OC	_(OIM	011	2022
			(Rowe		2022

WEST CORK MODEL RAILWAY VILLAGE,
CLONAKILTY STATION, INCHYDONEY ROAD, CLONAKILTY
Gauge 5ft 3in   www.modelvillage.ie

–	4wDM	RH	305322	1951

P. WILSON, ANTIQUES DEALER, CORK
Gauge 3ft 0in

LM 149	X	4wDM	RH	394028	1956
	rebuilt as	4wDH	BnM(Dg)		
LM 306		0-4-0DM	HE	8541	1977

DONEGAL

INDUSTRIAL SITES

FOREST ENTERPRISES, BELLANAMORE, near FINTOWN
Gauge 2ft 0in   (Closed)

–	4wDM	MR	7944	1943	Dsm

PRESERVATION SITES

COMHLACHT TRAENACH na GAELTACHTA LAIR (2000) TEO //
CENTRAL GAELTACHT TRAIN COMPANY,
FINTOWN RAILWAY, FINTOWN STATION, FINTOWN
Gauge 3ft 0in   www.antraen.com

–	4wDH	MR	102T007	1974
	reb	AK	78R	2007
18	0-4-0+4DMR	_(WkB	1361D	1940
		(Dundalk		1940

DIFFLIN LAKE RAILWAY, OAKFIELD PARK, RAPHOE
Open June to September inclusive.
Gauge 1ft 3in   www.oakfieldpark.com

No.1	THE DUCHESS OF DIFFLIN	0-4-2T	OC	ESR	317	2003
No.2	THE EARL OF OAKFIELD	0-4-0DH	s/o	AK	66	2002
	BISHOP TWYSDEN	4wDH		DLR		2018
	rebuild of			CCLR		2016
	rebuild of	4wVBT	VCG	StanhopeT		c1987

DONEGAL RAILWAY RESTORATION CLG,
THE DONEGAL RAILWAY HERITAGE CENTRE, THE OLD STATION HOUSE,
TYRCONNEL STREET, DONEGAL TOWN

Gauge 3ft 0in   www.donegalrailway.com

5	DRUMBOE	2-6-4T	OC	NW	829	1907	
(C 18)		2w-2PMR		RPSI(W)		2006	a

     a   incorporates parts from Wkm 4808 1948; kept in Goods Shed; not on public display

MICHAEL GALLAGHER, ARDRASOOL, BALLINDRAIT, near LIFFORD
Gauge 2ft 0in

_	4wDM	RH	371544	1954

ST. CONNELL'S MUSEUM & HERITAGE CENTRE SOCIETY LTD,
ST. CONNELL'S MUSEUM & GLENTIES HERITAGE CENTRE, MILL ROAD, GLENTIES
Gauge 2ft 0in   (Closed)

LM 263	4wDM	RH	7002/0600-1	1968

FREDDIE SYMONDS, MULLANBOYS HALT, near INVER
Gauge 3ft 0in

Visiting locomotives and railcars occasionally present

DUBLIN / BAILE ÁTHA CLIATH

INDUSTRIAL SITES

BREFFNI HIRE PLANT LTD, BALLYHACK FARM, KILSALLAGHAN
Gauge 5ft 3in   www.breffnigroup.ie

RR02	07 D 53644		4wDM	R/R	_(Mecalac	030485	1999	
					(Rexquote	1268	1999	a
RR22			4wDHR	R/R	Dieci	1530204		a OOU
L73	11 D 56041	99609 949001-2	4wDM	R/R	Mercedes	416326	2011	b
RRJ 01	131 LH 2432	99609 976017-4	4wDM	R/R	Landrover	433396	2014	
L36	06 D 124163	99609 977003-3	4wDM	R/R	_(Mitsubishi	40516	2006	
					(Aquarius		2006	

     a   Teleporter/bridge inspection unit
     b   rail vactor unit

Gauge 4ft 8½in

E1235	4wBER	Bance	133/04	2004	DsmT

(R473 JGG)	99709 976018-0	4wDMR	R/R	_(Landrover (Permaquip		1998 1998
GK02 UKV	99709 917115-6	4wDM	R/R	_(DAF (SRS	231200	2002 2002
GK02 UKW	99709 919079-2	4wDM	R/R	_(DAF (SRS	231279	2002 2002
GN51 MBU	99709 919051-1	4wDM	R/R	_(DAF (SRS	231237	2001 2001
GN51 MBF	99709 919052-9	4wDM	R/R	_(DAF (SRS	231236	2001 2001
GN51 MBV	99709 919054-5	4wDM	R/R	_(DAF (SRS	231201	2001 2001
X188 MKN	99709 919069-3	4wDM	R/R	_(DAF (SRS	221955	2001 2001
X164 MKN	99709 919074-3	4wDM	R/R	_(DAF (SRS	219897	2000 2000

DUBLIN LUAS, DUBLIN TRAMWAY SYSTEM
(Operated by Transdev Dublin Light Rail Ltd)
Gauge 4ft 8½in   www.luas.ie

97 D 67550	4wDM	R/R	Landrover		1997
97 D 66701	4wDM	R/R	Nissan		1997
10 D 962	4wDM	R/R	_(Mercedes (SRT		2010 2010
09 LS 853	4wDM	R/R	_(Mitsubishi (Sanline		2009 2018

Broombridge Depot, Dublin
Gauge 4ft 8½in

–	4wBE	R/R	Zephir	2734	2017
–	4wBH		BBM	268	2009

Red Cow Depot, Dublin
Gauge 4ft 8½in

–	4wBH		BBM	155	2002
–	4wDH		Unilok	4019	2009
162 D 16509	4wDM	R/R	_(Unimog (CMAR	241358	2016 2016

Sandyford Depot, Dublin
Gauge 4ft 8½in

–	4wBH		BBM	156	2002	
–	4wBE		SET	1007/1	2003	OOU
–	4wDH		Unilok	4013	2004	

S. DUFFY PLANT HIRE LTD,
STEPHENSTOWN INDUSTRIAL ESTATE, BALBRIGGAN
Gauge 5ft 3in   www.sduffyplanthire.ie

	T216 UGE	99609 976014-1	4wDM	R/R	_(Landrover		1999	
					(Perm		1999	
RRJ 02	P580 JVC	99609 976006-7	4wDM	R/R	_(Landrover	999690	1997	
					(Perm		1997	
RRJ 03	T586 RGE	99609 976007-5	4wDM	R/R	_(Landrover	165224	2000	
					(Perm		2000	
RRD 02	04 LH 3097	99609 943055-4	2w-2DH	R/R	Dieci		2004	a
	77 LH 521	(RO M 186)	4wDMR	R/R	IFA		1977	OOU
TPR 01	99609 919003-4		4wDMR	R/R	Manitou	739502	2010	b
	NK54 SDY		2-2wDH	R/R	_(SchörlingB	05-1949	2005	
	99609 943084-4				(Bergman Baufix		2005	
ACL 126	NK54 SDZ		2-2wDH	R/R	_(SchörlingB	04-1761	2005	
	99609 943083-6				(Bergman Baufix		2005	
	V493 EBC		4wDM	R/R	_(Thwaites	3-95209	1999	
					(Rexquote 1229			
UM 01	20		4wDM	R/R	Unimog	107598	c1983	
UM 02	Plant No.LO 51		4wDM	R/R	Unimog			

a   converted dumper
b   on hire to Iarnród Éireann, North Wall Yard, Dublin

Gauge 4ft 8½in

	R913 MAU		4wDM	R/R	_(Landrover		1998	
					(Perm		1998	OOU
	DK02 ORX		4wDMR	R/R	_(Landrover	609746	2002	
					(Contracked		2013	
016	BD03 LVH	99709 976031-3	4wDM	R/R	_(Landrover	656671	2003	
					(Harsco		2003	
1228	V494 EBC	99709 943047-9	4wDM	R/R	_(Thwaites	7-95400	1998	
					(Rexquote 1228			OOU

GUNNING JOINERY LTD,
RED COW INDUSTRIAL ESTATE, NAAS ROAD, DUBLIN
Gauge 3ft 0in

LM 91		4wDM	RH	371962	1954

IARNRÓD ÉIREANN
www.irishrail.ie
Departmental Stock
Railcars are based at :–   C - Connelly Station, Dublin
                               G - Greystones Station, Co. Wicklow
                               H - Heuston Carriage Sidings, Dublin
                               I - Inchicore Works, Dublin
                               K - Kildare Plant Depot, Co.Kildare
                               LJ - Limerick Junction, Co.Tipperary
                               NW - North Wall, Dublin

Gauge 5ft 3in

700		4w-4wDMR	Plasser	26	1974	K
720	0531 005217	4w-4wDHR	Geismar	1181	2019	I
721	JESS	2w-2DMR	HPET	1068	1992	LJ
722	99609 429722-2	4wDMR	MatisaSPA	0214	1994	NW
723		4wDMR	MatisaSPA	0215	1994	K
728		2w-2DH	Plasser	1140	1974	
	rebuilt as	2w-2DHR	Inchicore		1999	LJ
734		2w-2DM	Plasser	1772	1980	
	rebuilt as	2w-2DMR	Inchicore		1998	K
790		4w-4wDHR	Geismar	999	2009	K

DART Overhead Maintenance Depot, Connolly Station, Dublin
Gauge 5ft 3in

25	02 D 13723	99609 919002-6	2w-2DM	R/R	_(Volvo		2002
					(SRS		2002
(FR 220)	02 D 3124	99609 919001-8	2w-2DM	R/R	_(Volvo		2002
					(SRS		2002

DART System Fairview Depot, Fairview Park, Dublin
Gauge 5ft 3in

–		2w-2PM		_(Zagro	80138	2010
				(Zweihoff		2010

Inchicore Works, St. Patrick's Terrace, Inchicore, Dublin
Gauge 5ft 3in

134		Bo-BoDE		GM	26284	1960	a
–		2w-2PM	R/R	_(Zagro		2010	
				(Zwiehoff		2010	OOU
–		4wDH	R/R	Zephir	2793	2018	

a   owned by RPSI

Infrastructure Department, North Wall Yard, Dublin
Gauge 5ft 3in

TPR 01	99609 919003-4	4wDMR	R/R	Manitou	739502	2010	a

a   on hire from S. Duffy Plant Hire Ltd, Jenkinstown, near Dundalk, Louth

SANLINE SYSTEMS LTD,
12th LOCK, GRAND CANAL, NEWCASTLE ROAD, LUCAN, DUBLIN
Gauge 5ft 3in

–		4wDMR	R/R	Bedford		1981	
92 MH 4823		4wDMR	R/R	Landrover		1992	Dsm
00 LH 9282		4wDM	R/R	Landrover		2000	
02 D 76075		4wDM	R/R	Landrover		2002	
06 D 84465		4wDMR		_(Landrover		2006	
				(Geismar		2006	
–		2w-2PMR		Sanline		1999	OOU

PRESERVATION SITES

DEPARTMENT OF MECHANICAL ENGINEERING, PARSONS BUILDING, TRINITY COLLEGE, DUBLIN
Gauge 1ft 9in   www.tcd.ie/mecheng

–	0-6-0	IC	Kennan	1855	a

a   working model; used for instruction only

GUINNESS IRELAND GROUP LTD, GUINNESS STOREHOUSE MUSEUM, ST JAMES'S GATE BREWERY, CRANE STREET, DUBLIN
Gauge 1ft 10in   www.guinness-storehouse.com

17	0-4-0T	IC	Spence	1902	Pvd
47	4wDM	FH	3444	1950	Pvd

PRIVATE OWNER, DUBLIN
Gauge 1ft 0in

–	4-4-0T	OC	u/c

RAILWAY PRESERVATION SOCIETY OF IRELAND, CONNOLLY DEPOT, DUBLIN
Gauge 5ft 3in   www.steamtrainsireland.com   No Public Access

No.85 MERLIN	4-4-0	3C	BP	6733	1932
131	4-4-0	IC	NR	5757	1901
B141	Bo-BoDE		GM	27467	1962
175	Bo-BoDE		GM	27501	1962

Other visiting preserved locomotives may occasionally be present

GALWAY

INDUSTRIAL SITES

BORD NA MÓNA, Derryfadda R.T.C.   M 800432
Some locomotives are out based at the following locations – Daleysgrove, Gowla, Boughill Bog, Castlegar, Area 16 and Area 17 Killaderry Tea Centres.

For locomotive details see the Bord na Móna fleet list at the end of this Section.

CARRA PLANT SERVICES, NEW INN
Gauge 5ft 3in

> Road/rail plant usually present

DECLAN MAHER PLANT LTD, t/a CARRA PLANT SERVICES, CARRA, BULLAUN, LOUGHREA
Gauge 5ft 3in

> Road/rail plant usually present

UNILOKOMOTIVE LTD, IDA INDUSTRIAL ESTATE, DUNMORE ROAD, TUAM
New locomotives under construction are usually present.   www.unilok.ie

PRESERVATION SITES

BALLYGLUNIN STATION ASSOCIATION, STATION ROAD, COOLFOWERBEG, BALLYGLUNIN
Gauge 5ft 3in   www.ballyglunin.com

-		2w-2PMR	Bryce	2017	a

a   stored off site

CONNEMARA RAILWAY PROJECT, MAEM CROSS
Gauge 3ft 0in   www.connemararailway.ie

LM 194	0-4-0DM	Dtz	57136	1960

GALWAY MINIATURE RAILWAY (LEISURELAND EXPRESS), SALTHILL, GALWAY
Operated by Galway City Council Leisure Services Department. Operates June to September
Gauge 2ft 0in   www.leisureland.ie   M 278237

LEISURELAND EXPRESS	4-4-0+4-4wDH	s/o	SL	2155	2003

M. MITCHELL, DUNSANDLE STATION, DUNSANDLE
Gauge 5ft 3in   M 597242

E428	6wDH	Inchicore	1962

GARRY SKELTON, GALWAY
Private site. No casual visitors.
Gauge 1ft 10in   **Private Site**

21	0-4-0T	OC	Spence	1905	Dsm

KERRY

PRESERVATION SITES

KERRY COUNTY COUNCIL, TRALEE & DINGLE STEAM RAILWAY CO LTD, BLENNERVILLE, near TRALEE
Gauge 3ft 0in   **(Closed)**

No.5T	2-6-2T	OC	HE	555	1892	OOU
LM 92 L	4wDM		RH	371967	1954	OOU

LARTIGUE MONORAILWAY, JOHN B KEANE ROAD, LISTOWEL
Gauge monorail   www.lartiguemonorail.com

L.B.R. 4	0-2-0+2wDH	s/o	AK	62	2002	a

a   carries worksplate dated 2000

KILDARE

INDUSTRIAL SITES

BORD NA MÓNA

Almhain North R.T.C.	**N 801211**
Ballydermot	**N 659216, 656215**
Locomotives stored at Ballydermot Works pending disposal.	
Gilltown Landsale R.T.C.	**N 796338**
Kilberry (Closed - on care and maintenance basis)	**S664998**
Prosperous R.T.C.	**N 831292**
Ummeras (Closed)	**N 623147**

For all locomotive details see the Bord na Móna fleet list at the end of this Section.

DEPARTMENT OF DEFENCE, IRISH ARMY, ENGINEER MECHANICAL WORKSHOPS, No.1 MAINTENANCE YARD, CURRAGH CAMP, near KILDARE
Gauge 2ft 0in   R.T.C.   www.military.ie

–	4wDM	MR	8970	1945	OOU

GPX RAIL LTD, OSBERSTOWN BUSINESS PARK, NAAS
Gauge 4ft 8½in   www.gpxrail.com

ENG 5873 13 006	4wDM	Richard		1960	Pvd	+

+   worksplate carries ES12M

former LULLYMORE BOG, near CULLENTRA, LULLYMORE
Gauge 3ft 0in

(18)	4wDM	RH	211687	1941	OOU

LYONS & BURTON NEW HOLLAND AGRICULTURAL, THE MOUNT, CLANE ROAD, BALLYBRACK, near KILCOCK
Gauge 5ft 3in   www.lnb.ie   **N 864374**

99 KK 723 99609 945012-3	2w-2DM	R/R	Zetor 112450302	1992

McCABE TRANSPORT & HAULAGE SERVICES (IRELAND) LTD, TIMAHOE
Gauge 3ft 0in   www.mccabetransport.com

LM 134	4wDM	RH	382811	1955	OOU

PRESERVATION SITES

LULLYMORE HERITAGE & DISCOVERY PARK, RATHANGAN
(Note : Rathangan is located in Co. Offaly, but the park is located in Co. Kildare)
Gauge 3ft 0in   www.lullymoreheritagepark.com

LM 75		4wDM	RH	326047	1952	Dsm
LM 85		4wDM	RH	329693	1952	
LM 97	T	4wDM	RH	375332	1954	Dsm
LM 139		4wDM	RH	392137	1955	
LM 309	LM 3	0-4-0DM	HE	8545	1977	Dsm
C 72	LM 9	4wPMR	BnM		1972	OOU

LAOIS

INDUSTRIAL SITES

BORD NA MÓNA, Coolnamóna (Closed - on care and maintenance basis)   **S 456947**

Supplies locomotives to the surrounding area bogs, with some locomotives outbased at the following locations – Togher Bog Tea Centre, Cashel Bog Tea Centre, Coolcarton Bog and Dogs Head Bog.

For all locomotive details see the Bord na Móna fleet list at the end of this Section.

IARNRÓD ÉIREANN, CHIEF MECHANICAL ENGINEERS DEPARTMENT, LAOIS TRAINCARE FACILITY, near PORTLAOISE

Gauge 5ft 3in   www.irishrail.ie

621 RBL-020-400	DRIVER TOM LYNAM	4wBE	Scul	2008

IARNRÓD ÉIREANN, ENGINEERING SERVICES DIVISION, PORTLAOISE RAIL SUPPLY & INFRASTRUCTURE FACILITY, PORTLAOISE

Gauge 5ft 3in   www.irishrail.ie

02 D 78046	4wDM	R/R	Unimog	201419	2002	OOU

LINMAG RAIL GmbH, c/o IARNRÓD ÉIREANN, ENGINEERING SERVICES DIVISION, PORTLAOISE RAIL SUPPLY & INFRASTRUCTURE FACILITY, PORTLAOISE

Gauge 5ft 3in / 4ft 8½in (Dual Gauge)   www.linmag.com

2	162 D 25725	4w+4w-6wDM	R/R	_(MAN	2016	
OAJ 001	99809 900005-6 ERIN			(Linsinger	2016	a

a   MAN tractor unit couples to Milling Machine (Dual Gauge for IR/LUAS)

ROD PEARCE TRANSPORT LTD, near BALLYCOOLAN

Rail vehicles occasionally present in this yard.

PRESERVATION SITES

PAUL CUSHEN ENGINEERS LTD, STRADBALLY

Gauge 3ft 0in

(C 42)	2w-2PMR	Wkm	7129	1955
	reb	Cushen		2020

JIM DEEGAN, CROOKED VALLEY RAILWAY, LUGGACURREN, near STRADBALLY

Private Site. No casual visitors.

Gauge 3ft 0in   **Private Site**

LM 194	0-4-0DM	Dtz	57136	1960	a

a   currently at the Coonemara Railway Project, Maem Cross, Galway

IRISH STEAM PRESERVATION SOCIETY LTD,
STRADBALLY WOODLAND EXPRESS, STRADBALLY HALL, STRADBALLY

Gauge 3ft 0in   www.irishsteam.net   **S 568966**

No.2 LM 44 RÓISIN	0-4-0WT	OC	AB	2264	1949	
NIPPY	4wDM		FH	2014	1936	
(4)	4wDM		RH	300518	1950	
1	4wDM		RH	326051	1952	OOU
4	4wDM		RH	326052	1952	
LM 55	4wDM		RH	259205	1948	Dsm
(LM 167) Q	4wDM		RH	402978	1956	
(LM 191)	0-4-0DM		Dtz	57134	1960	Dsm
(LM 317)	4wDM		SMH	60SL742	1980	

STRADBALLY STEAM MUSEUM, MAIN STREET, STRADBALLY

Gauge 1ft 10in   **S 575961**

15	0-4-0T	IC	Spence		1912	a

a   carries incorrect plate dated 1895

LEITRIM

PRESERVATION SITES

THE CAVAN & LEITRIM RAILWAY CO LTD,
DROMOD STATION, STATION ROAD, DROMOD

Gauge 5ft 3in   www.cavanandleitrimrailway.com

720	2w-2DMR		P&T	322	1971	
		reb	BRE		1984	a
JZA 979	2-2wDM		_(Scammell		1959	
			(Cold Chon		1963	b

a   stored in IE Dromod yard and not on public display
b   conversion of a Scammell road lorry by Cold Chon Ltd, Oranmore, Co. Galway

Gauge 3ft 0in

NANCY	0-6-0T	OC	AE	1547	1908	
LM 180	0-4-0DM		Dtz	57122	1960	OOU

LM 186		0-4-0DM	Dtz	57132	1960	OOU
LM 253		0-4-0DM	Dtz	57834	1965	OOU
LM 258		0-4-0DM	Dtz	57839	1965	OOU
LM 260		0-4-0DM	Dtz	57841	1965	OOU
LM 262		0-4-0DM	Dtz	57843	1965	OOU
–		4wDM	HE	2280	1941	Dsm
–		4wDM	HE	6075	1961	Dsm
F511	DINMOR	4wDM	JF	3900011	1947	OOU
9		4wDH	MR	115U093	1970	
1		4wDM	RH	249526	1947	Dsm
LM 49		4wDM	RH	259189	1948	
(213)	"MARIE"	4wDM	RH	283896	1949	
2		4wDM	RH	314223	1951	Dsm
LM 87		4wDM	RH	329696	1952	Dsm
LM 101		4wDM	RH	379059	1954	
LM 106		4wDM	RH	375345	1954	OOU
LM 114	JOE ST LEGER	4wDM	RH	379070	1954	
LM 116	M 15	4wDM	RH	379077	1954	Dsm
(LM 131)		4wDM	RH	379086	1955	Dsm
LM 138		4wDM	RH	382819	1955	Dsm
(212)		4wDM	RH	394022	1956	OOU
LM 161		4wDM	RH	402175	1956	
LM 174		4wDM	RH	402986	1957	Dsm
LM 175		0-4-0DM	RH	420042	1958	OOU
LM 11		0-4-0DM	Ruhrthaler	1348	1934	OOU
LM 349		4wDM	SMH	60SL743	1980	OOU
(LM 351)		4wDM	SMH	60SL745	1980	Dsm
LM 361		4wDH	DunEW	LM361	1984	OOU
LM 362		4wDH	DunEW	G4464	1984	OOU
LM 369		4wDH	DunEW	S4484	1984	OOU
C 11		2w-2PHR	SolHütte		1934	
	rebuilt as	2w-2PMR	BnM(Cs)		c1963	Dsm
C 47		4wPMR	_(BnM	3	1958	
			(SM		1958	OOU
(C 51)		4wPMR	_(BnM	7	1958	
			(SM		1958	Dsm
(C 66)		4wDMR	BnM		1972	OOU
C 74		4wPMR	BnM		1972	OOU
5		2w-2PMR	DC	1495	1927	
	rebuilt as	2w-2DHR	AK		1995	
W6/11-4		2w-2PMR	Wkm	9673	1964	

Gauge 2ft 0in

(No.1)	4wDM		HE	2239	1940
		reb	HE	7340	1972
(No.3)	4wDM		HE	2304	1940
		reb	HE	7341	1972
–	4wDM		HE	2659	1942
(No.2)	4wDM		HE	2763	1943
		reb	HE	7342	1972

LM 16	B	4wDM		RH	200075	1940	
LM 42	C	4wDM		RH	252849	1947	Dsm
LM 198		4wDM		RH	398076	1956	OOU

Gauge 1ft 10in

22		0-4-0T	IC	Spence		1912	Dsm
–		4wDM		FH	3068	1947	OOU
31		4wDM		FH	3446	1950	OOU
–		4wDM		FH	3447	1950	Dsm

ANDREW MARSHALL, STRACARNE, MOHILL
Gauge 3ft 0in

LM 178	0-4-0DM	Dtz	57120	1960	
LM 187	0-4-0DM	Dtz	57133	1960	Dsm

LIMERICK

INDUSTRIAL SITES

IARNRÓD ÉIREANN,
LIMERICK WAGON WORKS, ROXBOROUGH ROAD, LIMERICK
Gauge 5ft 3in   www.irishrail.ie

42-048-087	4wBH	R/R	AVB	AVB5735/1	2010	
42-048-088	2w-2wBH		Mechan	6063/3	2008	OOU

LONGFORD

INDUSTRIAL SITES

BORD NA MÓNA, Mountdillon   N 048688
Locomotives are out based at the following locations : Derraghan, Mountdillon p-way and track repair depot, Old wagon repair facility (known as Mountdillon Old Works), Derrycolumb Bog, Corlea Bog, Derryaroge Bog, Begnagh Branch, Erenagh Tea Centre, Clontusket Tea Centre, Derryad Bog No 1, Derryad Bog No 2, Derryad Bog No 3, Killashee, Derryshangoe Bog Tea Centre, Coolumber Tea Centre, Edera, Loughbannow, Cloonbonny (various spellings), Knappoge, Cloneeny, Clondara, Moher, Curraghroe, Derycashel, Grannaghan, Cloonshannagh and Derrymoylin Bogs.

For all locomotive details see the Bord na Móna fleet list at the end of this Section.

McCORMICK BROS. RAIL & MACHINERY HIRE LTD,
LAUREL LODGE, TERLICKEN, BALLYMAHON
Gauge 5ft 3in

98 LD 1484		4wDM	R/R	Case		1998
98 LD 1594	99609 945008-1	4wDM	R/R	Case	JAJ2000	1998

PRESERVATION SITES

LONGFORD VINTAGE CLUB LTD, LONGFORD BARRACKS, LONGFORD
Gauge 3ft 0in

LM 66	5	4wDM	RH	259750	1948

LONGFORD COUNTY COUNCIL, LANESBOROUGH GREENWAY, LANESBOROUGH
Gauge 3ft 0in   www.longfordcoco.ie

(C 71)	4wPMR	BnM		1972	DsmT a

a   rebuilt as an ambulance car

LOUTH

PRESERVATION SITES

CARLINGFORD BREWING CO,
THE OLD MILL, RIVERSTOWN, CARLINGFORD, near DUNDALK
Gauge 4ft 8½in   www.carlingfordbrewing.ie   **J 168063**

No.4	ROBERT NELSON No.4	0-6-0ST	IC	HE	1800	1936	OOU
	(BREL 75)	4wDMR		_(BRE		c1984	
				(Leyland RB002/001	c1984		OOU

MAYO

PRESERVATION SITES

BORD NA MÓNA plc,
BELLACORICK WINDFARM VISITOR FACILITY, near BANGOR ERRIS
Gauge 3ft 0in

LM 81	H	4wDM	RH	329686	1952	Pvd

MAYO NORTH OLD ENGINE AND TRACTOR CLUB, ENNISCOE HERITAGE CENTRE, ENNISCOE HOUSE AND GARDENS, near CROSSMOLINA
Gauge 3ft 0in   www.northmayo.ie

LM 129		4wDM	RH	383264	1955

WESTPORT HOUSE & COUNTRY ESTATE, QUAY ROAD, WESTPORT
Gauge 1ft 3in   www.westporthouse.ie

WESTPORT HOUSE EXPRESS	2-6-0DH	s/o	SL	80.10.89	1989

MEATH

INDUSTRIAL SITES

BORD NA MÓNA, Kinnegad (Closed, R.T.C.)   N 600441
This site is known locally by BNM employees as the Rosen Bog Works.
For all locomotive details see the Bord na Móna fleet list at the end of this Section.

BREFFNI HIRE PLANT LTD, RAYSTOWN BUSINESS PARK, ASHBOURNE
Gauge 5ft 3in   www.breffnigroup.ie

Road/Rail plant may be present at this location

JOHN CONATY CONSTRUCTION MACHINERY AND TRACTOR PARTS, R163, BALNAGON LOWER, KELLS
Gauge 3ft 0in   www.johnconaty.com

LM 380		4wDH	HE	9252	1985
LM 213		0-4-0DM	HE	6246	1963

DIXON BROS. (AGRICULTURE & PLANT) LTD, CULLENTRA ROAD, RATHMORE, ENFIELD
Gauge 5ft 3in   www.dixonbros.ie

XLR 8103	97 MH 6832 [R648 JFE]	4wDMR	R/R	_(Landrover 136483	1998
				(Raynesway	
XLR 8106	97 MH 6833 [R904 WBC]	4wDMR	R/R	_(Landrover 136517	1998
				(Raynesway	
RJ 02	99609 976003-4	4wDM	R/R	Landrover	

RRJ03	T584 RGE	99609 976001-8	4wDMR	R/R	_(Landrover 163895	2000		
					(Perm	2000		
	99609 976024-0		4wDM	R/R	_(Nissan 033148	2016		
					(Fitz FPSNRRV001	2021	a	
T302 VMA	99609 945017-2		4wDM	R/R	New Holland	1999		
01 D 38049	99609 945003-2		4wDM	R/R	New Holland	2002	b	
RRV 1	99609 979006-4		4wDM	R/R	_(Unimog 199936	2001		
					(Zweihoff	2020		
01 MH 15254			4wDM	R/R	Mercedes	2001	c	
–			4wDM	R/R	Unimog			(U1200)
–			4wDM	R/R	Unimog			(U1200)
RRV02			4wDM	R/R	Unimog			(U1600)

a demonstration vehicle
b also carries T302 VMA 99609 945017-2 in error
c rail vactor unit

OLIVER DIXON (HEDGECUTTING & PLANT HIRE) LTD, ST. OLIVER'S ROAD, LONGWOOD
Gauge 5ft 3in

84 MH 585	99609 945004-0		4wDM	R/R	Ford		1984	
89 MN 2449	99609 945002-4		4wDM	R/R	Ford	BC86514	1989	
89 MH 3147	99609 945001-6		4wDM	R/R	Ford	BC63664	1989	
90 MH 5687	99609 945005-7		4wDM	R/R	Ford		1990	
90 WD 2041	99609 945013-1		4wDM	R/R	Fiat	265311	1990	
01 WX 2649	99609 945011-5		4wDM	R/R	New Holland 160182B	2001		
98 MH 8010	99609 976010-9	RRJ01	4wDM	R/R	_(Landrover154626	1999		
					(Raynesway	1999		

P.F. DIXON PLANT HIRE, RATHCORE, ENFIELD
Gauge 5ft 3in   www.pfdixonplanthire.com

Plant Yard with Road/Rail vehicles usually present

NEW BOLIDEN GROUP, TARA MINES, KNOCKUMBER ROAD, NAVAN
Gauge 5ft 3in   www.boliden.com   **(Closed)**

021		4wDM	R/R	Unilok	3028	1987	
			reb	Unilok		2014	
			reb	Unilok		2017	

Closed - on care and maintenance basis

PRESERVATION SITES

BRIAN DARLINGTON
Gauge 2ft 0in

–	4wDM		LB	54183	1964

EMERALD PARK, KILBREW, ASHBOURNE
Gauge 2ft 0in   www.emeraldpark.ie

1954	4-4-0+4w-4wDH	s/o	SL	15982	2015

MONAGHAN

PRESERVATION SITES

BALLINASCARRA BRIDGE COMMITTEE,
former GNR(I) COOTHILL RAILWAY EMBANKMENT, near LISNALONG, COOTHILL
Gauge 3ft 0in   **H 648163**

SLIABH gCUILLEAN	0-4-0DM	s/o	Dtz	57131	1960

UNKNOWN OWNER
Gauge 3ft 0in

(LM 99)	4wDM		RH	375336	1954
rebuilt as	4wDH		Bnm(Be)		1988

OFFALY

INDUSTRIAL SITES

BORD NA MÓNA
Bellair   **N 190329**
Site serves Bellair Bog Tippler, Bellair North and Bellair South and Lemanaghan. An allocation of locomotives for both haulage and service use is based here.

Blackwater System   **N 002251**
In addition to the main works, locomotives are normally found out based at the following locations : , Area 4 Ballyhurt, Area 1 & 2 Bloomhill and circuit, Derryharney, Bunnahinley Bogs, Kylemore Lock, Cornaveagh Tea Centre, Blackwater P-Way Depot,, Derry Bratt Tea Centre, Clongowney More Bog, , Clonfert Bridge and Clonfert Wagon Works, Cloonburren Bogs, Cloonburren Depot,. The following areas also use locomotives on a regular basis : Drumlosh, Cornafulla, Conniff, Cloonbeggan, Creggan, Lismanny and Kilgarven.

Boora System
N 181196

Locomotives can be also found out based at Derrinlough (former Briquette Plant)

Croghan Works & Tip Head (Closed)
N 515276

Locomotives are stored here, additional locomotives are allocated to here for use on maintenance works at Rathdrum, Derries and Ballanakill.

Derrygreenagh System
N 495382, 495384

This is the largest single system of BNM, thus the whole area is split into four sub divisions. See also Ballydermot Works under Co Kildare, the remaining three sites are located within Co Offaly, see also Edenderry EPL and Croghan Works and Tip Head. Derrygreenagh Works supplies locomotives for Derryhinch, Toar Bog, Ballybegg, under PICAS operations only.

Edenderry EPL & Power Station
N 613273

Site supplies locomotives for use at the station of the same name. Ash is delivered to Cloncreen Ash Tip and Cells where locomotives are out based for service duties as required.

Lemanaghan
N 147259

See under Bellair Operations for locomotive allocations.

For all locomotive details see the Bord na Móna fleet list at the end of this Section.

PRESERVATION SITES

BORD NA MÓNA plc,
LOUGH BOORA PARKLANDS, LOUGH BOORA DISCOVERY PARK, BOORA
Gauge 3ft 0in   www.loughboora.com

LM 23		4wDM		RH	244788	1946	Pvd
LM 51	1	4wDM	s/o	RH	259191	1948	Pvd
LM 127		4wDM		RH	379928	1954	Pvd
LM 160	6	4wDM		RH	402176	1956	
	rebuilt as	4wDH	s/o	BnM		1995	Pvd
LM 171	Q	4wDM		RH	402983	1956	Pvd
LM 181		0-4-0DM		Dtz	57123	1960	Pvd

Lough Boora Parklands Restoration Workshops,
c/o Bord na Möna, Boora Works, Boora
Gauge 3ft 0in
N 181196

No.5 LM 38E	4wDM		RH	252246	1947	
LM 74 F 6	4wDM		RH	259760	1948	OOU
LM 105 U	4wDM		RH	375344	1954	
LM 111	4wDM		RH	379079	1954	
(LM 126)	4wDM		RH	379927	1954	OOU
LM 184	0-4-0DM		Dtz	57130	1960	
LM 278	0-4-0DM		HE	7243	1972	
LM 343	4wDM		SMH	60SL746	1980	
LM 372	4wDH		DunEW			
–	4wDM		RH	422567	1958	
F 308	4wDM		BnM			OOU

BORD NA MÓNA plc,
MOUNT LUCAS WINDFARM VISITOR FACILITY, BALLYCON
Gauge 3ft 0in   www.mountlucaswindfarm.ie

LM 37	E	4wDM	RH	252245	1947	Pvd

BORD NA MÓNA VINTAGE MACHINERY MUSEUM,
BLACKWATER WORKS, SHANNONBRIDGE
Gauge 3ft 0in   (Closed)   N 002252

LM 254		0-4-0DM	Dtz		57835	1965
LM 354		4wDH	DunEW			
LM 367		4wDH	DunEW	LM367	1984	
LM 368		4wDH	DunEW	LM368	c1984	
C 80		4wPMR	BnM(BI)		1972	
F 630	341 NRI	4wDM	BnM			

DAVID BRADY, WALSH ISLAND, GEASHILL
Gauge 3ft 0in

LM 197	0-4-0DM	Dtz	57139	1960
LM 78	4wDM	RH	329682	1952
(LM 94)	4wDM	RH	373377	1954
LM 136	Q	4wDM	RH	382815	1955
C45	2w-2PMR	Wkm	7132	1955

MARTIN DUFFY, BEEHIVE LODGE, WALSH ISLAND
Gauge 3ft 0in

LM 14	SIR 275	4wDM	RH	198290	1940

TULLAMORE COLLEGE - COLÁISTE THULACH MHÓR,
RIVERSIDE, TULLAMORE
Gauge 3ft 0in   www.tullamorecollege.ie

LM 305	0-4-0DM	HE	8544	1977

WALSH ISLAND HERITAGE COMMITTEE, BORD Na MÓNA GREEN, WALSH ISLAND
Gauge 3ft 0in   N 516197

LM 28	G 12	4wDM	RH	249543	1947

ROSCOMMON

PRESERVATION SITES

ARIGNA MINING EXPERIENCE, DERREENAVOGGY, ARIGNA
Gauge 2ft 0in   www.arignaminingexperience.ie

No.8023	ARIGNA MINES ST. ELBA	4wDM	L	8023	1936	Pvd
–		4wDM	MR	5861	1934	a
41		4wPM	Montania/OK 2563	1927	a

   a   not on public display

HELL'S KITCHEN PUBLIC HOUSE,
CASTLEREA RAILWAY MUSEUM, MAIN STREET, CASTLEREA
Closed - open by appointment only; intending visitors should ring 00353 87 230 8152 for Mr Sean Browne
Gauge 5ft 3in   www.hellskitchenmuseum.com

A55	Co-CoDE	MV	941	1956

SLIGO

PRESERVATION SITES

QUIRKY NIGHTS GLAMPING VILLAGE, ENNISCRONE
Gauge 4ft 8½in

62411 (1498)	4w-4wRER	York(BRE)	1971

SLIGO VEHICHLE MUSEUM, near BALLYMOATE
Gauge 3ft 0in

LM 215	0-4-0DM	HE	6248	1963
LM 324	0-4-0DM	HE	8931	1980

TIPPERARY

INDUSTRIAL SITES

BORD NA MÓNA
Littleton System
Urlingford / Lisheen (Temporary track lifting work with one locomotive allocated)
For locomotive details see the Bord na Móna fleet list at the end of this Section.

MIKE LYNCH EXCAVATIONS LTD,
Junction of LIMERICK / TIPPERARY ROAD, MONARD
Gauge 5ft 3in

98 TS 4363	99609 976008-3	4wDMR	R/R	_(Landrover	1998	
				(Harsco	1998	
	99609 943033-1	4wDHR	R/R	_(Thwaites1-96791		
				(Keltec		

MOLONEY AGRICULTURAL SERVICES,
KILLBALLYBOY, CLOGHEEN, near CAHIR
Gauge 5ft 3in

05 TS 2115	99609 945010-7	4wDM	R/R	New Holland	196348B	2005
08 TS 3156	99609 945009-9	4wDM	R/R	New Holland	ZOBD053922	2008
09 TS 1499	99609 945015-6	4wDM	R/R	New Holland	29BK03298	2009

PRESERVATION SITES

IRISH TRACTION GROUP, CARRICK-ON-SUIR STATION, CARRICK-ON-SUIR
Gauge 5ft 3in www.irishtractiongroup.com

B103	A1A-A1ADE	BRCW	DEL22	1956	
G601	4wDH	Dtz	56118	1956	
G616	4wDH	Dtz	57227	1961	
226	Bo-BoDE	MV	972	1956	

WATERFORD

PRESERVATION SITES

TRAMORE AMUSEMENT & LEISURE PARK,
'THE TRAIN', TRAMORE MINIATURE RAILWAY, STRAND ROAD, TRAMORE
Gauge 1ft 3in www.tramoreamusements.com

–		2-8-0PH	s/o	SL	22	1973

WATERFORD & SUIR VALLEY HERITAGE RAILWAY,
KILMEADON STATION, KILMEADON
Gauge 3ft 0in www.wsvrailway.ie

LM 179		0-4-0DM		Dtz	57121	1960	
LM 256		0-4-0DM		Dtz	57837	1960	OOU
LM 259		0-4-0DM		Dtz	57840	1965	Pvd
No.3	ENTERPRISE	4wDM		MR	60S382	1969	
	rebuilt			AK		2000	
	rebuilt as	4wDH		AK		2004	
(LM 96)		4wDM		RH	375314	1954	
	rebuilt as	4wDH		HE	9904	2011	
LM 348		4wDM		SMH	60SL744	1980	
	99709 901051-1	2w-2PMR	R/R	Geismar	M44/001	2005	DsmT
–		2w-2PMR		Geismar	101038	2010	

WESTMEATH

INDUSTRIAL SITES

BORD NA MÓNA
Ballivor (R.T.C.) **N 644541**
Site is on Care and Maintenance basis

Coolnagun (R.T.C.) **N 384702**
Locomotives are kept at the main works and at the tip head on the Coole Road. Some locomotives may be outbased on the Coole Bog, Kiltareher bog and on the South Bog from time to time.

For locomotive details see the Bord na Móna fleet list at the end of this Section.

JOHN DIXON PLANT HIRE LTD, SARSFIELDTOWN, KILLUCAN
Gauge 5ft 3in www.johndixonplanthire.ie

88 G 6156	99609 945016-4	4wDM	R/R	Ford	BB86513	1988
90 WH 3006	99609 945020-6	4wDM	R/R	Ford	BC47870	1990

90 WH 4006		4wDM	R/R	Ford	1990	
06 WH 4510	99609 976004-2	4wDMR	R/R	_(Landrover 713641	2006	
				(Aquarius	2006	
99 WH 4717	99609 976012-5	4wDMR	R/R	_(Landrover 162647	1999	
				(Hy-Rail 0307-B1	1999	
SJ53 CTX	99709 975050-4 RRF01	4wDH	R/R	_(JCB 642766	2002	
				(Chieftain	2002	
Q847 HFR	99609 979001-5	4wDM	R/R	_(Unimog 122401	1985	
				(Zweiweg 1106	1985	

WEXFORD

INDUSTRIAL SITES

DOYLE AGRICULTURAL & RAILPLANT SERVICES LTD, BALLYCOURSEY, ENNISCOURTHY
Gauge 5ft 3in

DA 082 90 WX 5496	99609 945006-5	4wDM	R/R	Ford	1990	Dsm
(X527 OVV)		2w-2DM	R/R	Benford EY05HD203	2001	

Dealer of new or used locomotives and rail plant, with vehicles for hire, re-sale or rail conversion occasionally present.

PRESERVATION SITES

OFFICE OF PUBLIC WORKS, J.F.KENNEDY ARBORETUM, NEW ROSS
Gauge 2ft 0in

–	4wDM	RH	371538	1954	Pvd

PHILIP PARKER, BALLYCANEW
Gauge 3ft 0in

LM 141	4wDM	RH	392142	1955

UNKNOWN OWNER
Gauge 3ft 0in

LM 130 P	4wDM	RH	382807	1955
rebuilt as	4wDH	BnM(K)		1993

BORD NA MÓNA

BORD NA MÓNA plc — IRISH TURF BOARD

The Bord operates peat bogs throughout the Irish Midlands, with locomotives kept at the locations listed below. (See also County Headings for additional sub headings within each area group or site).

Various of the larger sites (e.g. Bl, Bo, Dg & M) have constructed their own locomotives & railcars www.bordnamona.ie

Al	Almhain North, Co.Kildare	R.T.C.	N 801211
Bd	Ballydermot, Co.Kildare		N 659216, 656215
Be	Bellair, Co.Offaly		N 190329
Bi	Ballivor, Co.Westmeath	R.T.C.	N 644541
Bl	Blackwater, Co.Offaly	R.T.C.	N 002251
Bo	Boora, Co.Offaly		N 181196
Cg	Coolnagun, Co.West Meath	R.T.C.	N 384702
Cr	Croghan, Co.Offaly	(Closed)	N 515276
Cm	Coolnamona, Co.Laois	(Closed)	S 456947
De	Derryfadda, Co.Galway	R.T.C.	M 800432
Dg	Derrygreenagh, Co.Offaly		N 495382, 495384
Ed	Edenderry, Co.Offaly		N 613273
Gi	Gilltown Landsale, Co.Kildare	R.T.C.	N 796338
K	Kilberry, Co.Kildare	R.T.C.	S 664998
Kd	Kinnegad, Co.Meath	(Closed)	N 600441
Le	Lemanaghan, Co.Offaly		N 147259
Li	Littleton, Co. Tipperary		S 207535
M	Mountdillon, Co.Longford		N 048688
Pr	Prosperous, Co.Kildare	R.T.C.	N 831292
U	Ummeras, Co.Kildare	(Closed)	N 623147

LOCO FLEET LIST

Gauge 3ft 0in

LM 30	G		4wDM	RH	249545	1947	OOU	Gi
(LM 34)	No.4		4wDM	RH	252239	1947	OOU	Dg
LM 36	E	2	4wDM	RH	252241	1947		
		rebuilt as	4wDH	BnM			OOU	M
LM 39	E		4wDM	RH	252247	1947	OOU	M
LM 40	E		4wDM	RH	252251	1947		
		rebuilt as	4wDH	BnM(M)		1991		M
LM 41	E	(No.1)	4wDM	RH	252252	1947	OOU	Dg
LM 46			4wDM	RH	259184	1948		
		rebuilt as	4wDH	BnM			OOU	Dg
LM 48	F		4wDM	RH	259186	1948	OOU	Dg
LM 50			4wDM	RH	259190	1948		
		rebuilt as	4wDH	BnM(K)			OOU	Cg
LM 57			4wDM	RH	259203	1948		
		rebuilt as	4wDH	BnM(Dg)			OOU	Gi
LM 62			4wDM	RH	259743	1948		
		rebuilt as	4wDH	BnM(U)				Kd

ID	Type	Builder	Works No	Year	Status		Loc
(LM 65)	4wDM	RH	259749	1948	Dsm		Bl
LM 67	4wDM	RH	259751	1948	Dsm		Cg
LM 69 F	4wDM	RH	259755	1948	OOU		Gi
LM 70	4wDM	RH	259756	1948	Dsm		Dg
LM 71 F	4wDM	RH	259757	1948	OOU		M
(LM 72 F)	4wDM	RH	259758	1948	Dsm		M
LM 73 8	4wDM	RH	259759	1948	OOU		M
LM 86 J	4wDM	RH	329695	1952	OOU		M
LM 100	4wDM	RH	375341	1954	OOU		Bd
LM 109	4wDM	RH	379064	1954	OOU		Bo
(LM 113) U	4wDM	RH	379068	1954	OOU		Dg
LM 121	4wDM	RH	379922	1954			
rebuilt as	4wDH	BnM			OOU		Bd
LM 123	4wDM	RH	379925	1954			
rebuilt as	4wDH	BnM		1996	OOU		Pr
LM 128 X	4wDM	RH	383260	1955	OOU		Dg
LM 132 P	4wDM	RH	382809	1955	OOU		Kd
LM 135 Q	4wDM	RH	382814	1955	OOU	a	M
LM 144 Q	4wDM	RH	392149	1956	Dsm		Dg
(LM 145)	4wDM	RH	394023	1956	Dsm		Dg
LM 148 X	4wDM	RH	394026	1956			
rebuilt as	4wDH	BnM			OOU		Al
LM 150 X	4wDM	RH	394027	1956			
rebuilt as	4wDH	BnM(Dg)			OOU		Bd
LM 153 Q	4wDM	RH	394029	1956			
rebuilt as	4wDH	BnM			OOU		Bd
LM 154 X	4wDM	RH	394030	1956			
rebuilt as	4wDH	BnM(Bl)		1991	OOU		Bl
LM 155 X	4wDM	RH	394031	1956	OOU		M
LM 156 X KNIGHTRIDER	4wDM	RH	394032	1956			
rebuilt as	4wDH	BnM(Dg)			OOU		Cr
LM 157 X	4wDM	RH	394033	1956			
rebuilt as	4wDH	BnM(Dg)			OOU		Cr
LM 158	4wDM	RH	394034	1956	Dsm		Dg
LM 159 X	4wDM	RH	402174	1956			
rebuilt as	4wDH	BnM		2001	OOU		Pr
LM 163 X	4wDM	RH	402178	1956			Be
LM 164 Q 2	4wDM	RH	392152	1956	OOU		Dg
LM 165	4wDM	RH	402179	1956			
rebuilt as	4wDH	BnM(Dg)			OOU		Cr
LM 166	4wDM	RH	402977	1956			
rebuilt as	4wDH	BnM(Bi)		2014	OOU		Kd
LM 169 Q	4wDM	RH	402981	1956	Dsm		Dg
LM 173	4wDM	RH	402985	1957	OOU		Dg
LM 176	0-4-0DM	BnM		1961			
rebuilt as	4wDH	BnM(Dg)		1999			
rebuilt		BnM(Dg)		2016			Dg
LM 199	0-4-0DM	HE	6232	1962			
rebuilt as	4wDH	BnM(Dg)		1995	OOU		Ed
LM 200	0-4-0DM	HE	6233	1962	OOU		Bo
LM 201	0-4-0DM	HE	6234	1962	Dsm		Dg
LM 202	0-4-0DM	HE	6235	1962	OOU		Dg

LM 203			0-4-0DM	HE	6236	1962		
		rebuilt as	0-4-0DH	BnM(Bo)		2008	OOU	Bl
LM 204			0-4-0DM	HE	6237	1963		
		rebuilt as	4wDH	BnM(Dg)		1993		
		rebuilt		BnM(Dg)		2018		Cg
(LM 206)	LM 335		0-4-0DM	HE	6239	1963		
		rebuilt as	0-4-0DH	BnM(Bo)		2008	OOU b	Al
LM 208			0-4-0DM	HE	6241	1963		
		rebuilt as	4wDH	BnM(Dg)		1995	OOU	Cr
LM 209			0-4-0DM	HE	6242	1963	OOU	M
LM 210			0-4-0DM	HE	6243	1963		
		rebuilt as	4wDH	BnM(Dg)		1992		Dg
LM 211			0-4-0DM	HE	6244	1963		
		rebuilt as	0-4-0DH	BnM(Bo)		2010	OOU	Dg
LM 212			0-4-0DM	HE	6245	1963		
		rebuilt as	0-4-0DH	BnM(Bo)		2012		Bl
(LM 214)			0-4-0DM	HE	6247	1963	OOU	Be
LM 216			0-4-0DM	HE	6249	1963	Dsm	Gi
LM 217			0-4-0DM	HE	6250	1963	OOU	Bl
LM 218			0-4-0DM	HE	6251	1963		
		rebuilt as	4wDH	BnM(Dg)		1994	OOU	Bl
LM 219			0-4-0DM	HE	6252	1963	OOU	Bo
LM 221	2111-KE-SAM	KILDARE	0-4-0DM	HE	6254	1963	OOU	Bd
LM 222			0-4-0DM	HE	6255	1963	OOU	Dg
LM 223			0-4-0DM	HE	6256	1963		
		rebuilt as	4wDH	BnM(Dg)		1993		Ed
LM 225			0-4-0DM	HE	6304	1964		
		rebuilt as	4wDH	BnM(Dg)		1994		
		rebuilt		BnM(Dg)		2018	OOU	Dg
LM 226			0-4-0DM	HE	6305	1964		
		rebuilt as	4wDH	BnM(M)		1994		
		rebuilt		BnM(Dg)		2018		Cg
LM 227			0-4-0DM	HE	6306	1964	OOU	Bl
LM 228			0-4-0DM	HE	6307	1964	OOU	Bl
LM 229	(LM 294)		0-4-0DM	HE	8531	1977	OOU	Bo
LM 230			0-4-0DM	HE	6309	1964	OOU	Bl
LM 231			0-4-0DM	HE	6310	1964	OOU	Bl
LM 232			0-4-0DM	HE	6311	1964	OOU	Dg
LM 233			0-4-0DM	HE	6312	1965	OOU	Bl
LM 234			0-4-0DM	HE	6313	1965	OOU	Bl
LM 235			0-4-0DM	HE	6314	1965	OOU	Bl
LM 236			0-4-0DM	HE	6315	1965		
		rebuilt as	0-4-0DH	BnM(Bo)		2010	OOU	Bi
LM 237			0-4-0DM	HE	6316	1965	OOU	Cm
LM 238			0-4-0DM	HE	6318	1965		
		rebuilt as	4wDH	BnM(Dg)		2000		
		rebuilt		BnM(Dg)		2017		Ed
LM 239			0-4-0DM	HE	6317	1965	OOU	Al
LM 240			0-4-0DM	HE	6319	1965		
		rebuilt as	0-4-0DH	BnM(Bo)		2008	OOU	Gi
LM 242			0-4-0DM	HE	6321	1965	OOU c	Dg

LM 243	A 3		0-4-0DM	HE	6322	1965	Dsm	Dg
LM 244			0-4-0DM	HE	6323	1965	OOU	Bo
LM 245			0-4-0DM	HE	6324	1965		
		rebuilt as	0-4-0DH	BnM(Bo)		2012	OOU	Bl
LM 246			0-4-0DM	HE	6325	1965	OOU	Bl
LM 247	A 3		0-4-0DM	HE	6326	1965	OOU	Dg
LM 248			0-4-0DM	HE	6328	1965		
		rebuilt as	0-4-0DH	BnM(Bo)		2015		M
LM 249			0-4-0DM	HE	6327	1965	OOU	M
LM 250			0-4-0DM	HE	6329	1965	OOU	Bl
LM 252			0-4-0DM	HE	6331	1965		
		rebuilt as	4wDH	BnM		2000		
		rebuilt		BnM(Dg)		2017		M
(LM 255)			0-4-0DM	Dtz	57838	1965	OOU	Cg
LM 261			0-4-0DM	Dtz	57842	1965	Dsm	Bl
LM 266			0-4-0DM	HE	7232	1971	OOU	Bl
LM 267			0-4-0DM	HE	7233	1971	OOU	M
LM 269			0-4-0DM	HE	7235	1971	OOU	Bo
LM 270			0-4-0DM	HE	7237	1971		
		rebuilt as	4wDH	BnM		2000		
		rebuilt		BnM(Dg)		2017		Le
LM 272			0-4-0DM	HE	7239	1971	OOU	M
LM 273			0-4-0DM	HE	7246	1972	OOU	Bl
LM 274			0-4-0DM	HE	7238	1971		
		rebuilt as	0-4-0DH	BnM(Bo)		2012	OOU	Kn
LM 275	(LM 328)		0-4-0DM	HE	7240	1972	Dsm	Bo
LM 277			0-4-0DM	HE	7242	1972	OOU	Be
LM 280			0-4-0DM	HE	7245	1972	OOU	Bl
LM 281			0-4-0DM	HE	7247	1972	OOU	Bl
LM 282			0-4-0DM	HE	7248	1972	OOU	Bl
LM 283			0-4-0DM	HE	7250	1972	OOU	Bo
LM 284			0-4-0DM	HE	7249	1972		
		rebuilt as	0-4-0DH	BnM(Bo)		2009	OOU	Al
LM 286			0-4-0DM	HE	7254	1972		
		rebuilt as	0-4-0DH	BnM(Bl)		2004	OOU	Bl
LM 288			0-4-0DM	HE	7256	1972		Bl
LM 289			0-4-0DM	HE	7252	1972	OOU	U
LM 290			0-4-0DM	HE	7251	1972	OOU	Dg
LM 292			0-4-0DM	HE	8529	1977	Dsm	Bl
LM 293			0-4-0DM	HE	8530	1977	OOU	Bl
(LM 294	LM 229) 158		0-4-0DM	HE	6308	1964	OOU	Bl
LM 295			0-4-0DM	HE	8532	1977		
		rebuilt as	4wDH	BnM(Bl)		1993	OOU	Dg
LM 296			0-4-0DM	HE	8534	1977	OOU	Bl
LM 297			0-4-0DM	HE	8533	1977	OOU	Bl
LM 298			0-4-0DM	HE	8538	1977	OOU	Bo
LM 299			0-4-0DM	HE	8537	1977	OOU	Dg
LM 300			0-4-0DM	HE	8535	1977	OOU	M
LM 301			0-4-0DM	HE	8536	1977		
		rebuilt as	0-4-0DH	BnM(Bo)		2009	OOU	Kd

LM 302			0-4-0DM	HE	8539	1977			
		rebuilt as	0-4-0DH	BnM(Bo)		2015	OOU		Dg
LM 303			0-4-0DM	HE	8540	1977			
		rebuilt as	4wDH	BnM(Dg)		2000			
		rebuilt		BnM(Dg)		2017			Ed
LM 304			0-4-0DM	HE	8543	1977	OOU		Bl
LM 307			0-4-0DM	HE	8542	1977			
		rebuilt as	0-4-0DH	BnM(Bo)		2010			Bi
LM 308			0-4-0DM	HE	8546	1977	OOU		Ed
LM 310			0-4-0DM	HE	8547	1977	Dsm		Dg
LM 311	(LM 332)		0-4-0DM	HE	8551	1977	OOU		Bl
LM 311			0-4-0DM	HE	8930	1980	OOU		Bl
LM 312			0-4-0DM	HE	8550	1977			
		rebuilt as	0-4-0DH	BnM(Bo)		2014			Bl
LM 313			0-4-0DM	HE	8549	1977	OOU		Bl
LM 315			0-4-0DM	HE	8922	1979	OOU		Bl
LM 318			0-4-0DM	HE	8925	1979	OOU		Cg
LM 319			0-4-0DM	HE	8926	1979			Be
LM 320			0-4-0DM	HE	8939	1980	OOU		Bl
LM 321	TURBO		0-4-0DM	HE	8927	1977			
		rebuilt as	0-4-0DH	BnM(Bo)		2014	OOU		Bl
LM 322	(PHOENIX)		0-4-0DM	HE	8923	1979	OOU		Bl
LM 323			0-4-0DM	HE	8924	1979	OOU		Bl
LM 326			0-4-0DM	HE	8932	1980	OOU		Bl
LM 327			0-4-0DM	HE	8933	1980			
		rebuilt as	0-4-0DH	BnM(Bo)		2010	OOU		Dg
LM 328	(LM 275)		0-4-0DM	HE	8935	1980			
		rebuilt as	0-4-0DH	BnM(Bo)		2014			K
LM 330			0-4-0DM	HE	8937	1980	OOU		M
LM 331			0-4-0DM	HE	8934	1980	OOU		Bl
LM 334			0-4-0DM	HE	8941	1981	OOU		Bl
LM 335			0-4-0DM	HE	8938	1980	OOU		Cm
LM 336			0-4-0DM	HE	8943	1981			Bo
LM 337			0-4-0DM	HE	8944	1981	OOU		Dg
LM 339			0-4-0DM	HE	8946	1981			
		rebuilt as	0-4-0DH	BnM(Bo)		2010			Bd
LM 340			0-4-0DM	HE	8928	1980	OOU		K
LM 342			0-4-0DM	HE	8929	1980			
		rebuilt as	0-4-0DH	BnM(Bo)		2014	OOU		Bl
LM 355			4wDH	DunEW			OOU		M
LM 357			4wDH	DunEW			Dsm		Bl
LM 358			4wDH	DunEW			OOU		Kn
LM 359			4wDH	DunEW			OOU		M
LM 364			4wDH	DunEW		1984	OOU	d	Bl
LM 365			4wDH	DunEW	T4494	1984	OOU		Bl
LM 374			4wDH	HE	9239	1984			M
LM 375			4wDH	HE	9240	1984	OOU		M
LM 376			4wDH	HE	9241	1984	OOU		M
LM 377			4wDH	HE	9243	1984	OOU		M
LM 378			4wDH	HE	9242	1984	OOU		Bl
LM 379			4wDH	HE	9251	1985	OOU		Ed

LM 381			4wDH	HE	9253	1985	OOU	Bl
LM 382			4wDH	HE	9254	1986	OOU	Dg
LM 383			4wDH	HE	9255	1986	OOU	Dg
LM 384			4wDH	HE	9256	1986	OOU	Cg
LM 385			4wDH	HE	9257	1986		M
LM 386			4wDH	HE	9258	1986	OOU	Bl
LM 387			4wDH	HE	9259	1986		Cm
LM 388	"LOT 80"		4wDH	HE	9272	1986	OOU	Bl
LM 389			4wDH	BnM(Bl)		1994		Li
LM 390			4wDH	BnM(Bl)		1994		De
LM 391			4wDH	BnM(Bo)		1994	OOU	Bl
LM 392			4wDH	BnM(Bo)		1995		Bl
LM 393			4wDH	BnM(Bo)		1995	OOU	Bl
LM 394			4wDH	BnM(Bo)		1996		Bo
LM 395			4wDH	BnM(Bl)		1995		
		rebuilt		BnM(Dg)		2018		Bo
LM 396			4wDH	BnM(Bl)		1995	OOU	Bl
LM 397			4wDH	BnM(Bo)		2005		Bo
LM 398			4wDH	BnM(Bo)		2005		
		rebuilt		BnM(Dg)		2018	OOU	Bl
LM 399			4wDH	BnM(Bo)		2004	OOU	Bl
LM 400			4wDH	BnM(Bo)		2004		
		rebuilt		BnM(Dg)		2018		Bo
LM 401			4wDH	BnM(M)		1996		M
LM 402			4wDH	BnM(M)		1996		M
LM 403			4wDH	BnM(De)		2000		
		rebuilt		BnM(Dg)		2017	OOU	Bl
LM 404			4wDH	BnM(De)		2000		
		rebuilt		BnM(Dg)		2018		M
LM 405			4wDH	BnM(Bl)		2000		
		rebuilt		BnM(Dg)		2016		Bl
LM 406			4wDH	BnM(Bl)		2000		
		rebuilt		BnM(Dg)		2016		Le
LM 407			4wDH	BnM(Li)		2002		
		rebuilt		BnM(Dg)		2019		Be
LM 408			4wDH	BnM(Bl)		2001		
		rebuilt		BnM(Dg)		2017		Be
LM 409			4wDH	BnM(Bl)		2001		
		rebuilt		BnM(Dg)		2017		Be
LM 410			4wDH	BnM(Bo)		1998		
		rebuilt		BnM(Dg)		2017	OOU	Bl
LM 411			4wDH	BnM(Bo)		1999		
		rebuilt		BnM(Dg)		2016		Be
LM 412			4wDH	BnM(Dg)		2000		
		rebuilt		BnM(Dg)		2018		Ed
LM 413			4wDH	BnM(Dg)		2000		
		rebuilt		BnM(Dg)		2015		Bd
LM 414			4wDH	BnM(Dg)		2000		Ed
LM 415			4wDH	BnM(Dg)		2000		
		rebuilt		BnM(Dg)		2017		Ed
LM 416			4wDH	BnM(Bo)		2000		
		rebuilt		BnM(Dg)		2017		Ed

LM 417		4wDH	BnM(Bo)		2000			
	rebuilt		BnM(Dg)		2017			Ed
(LM 418)		4wDH	BnM(Bo)		2000			
	rebuilt		BnM(Dg)		2019			Cr
LM 419		4wDH	BnM(Bo)		2000			
	rebuilt		BnM(Dg)		2019			Ed
LM 420		4wDH	BnM(Bo)		2000			
	rebuilt		BnM(Dg)		2017			Ed
LM 421		4wDH	BnM(Bo)		2000			
	rebuilt		BnM(Dg)		2017	Dsm		Ed
LM 422		4wDH	BnM(De)		2000			
	rebuilt		BnM(Dg)		2017			Ed
LM 423		4wDH	BnM(De)		2000			
	rebuilt		BnM(Dg)		2018			Dg
LM 424		4wDH	BnM(De)		2000			
	rebuilt		BnM(Dg)		2017	OOU		Ed
LM 425		4wDH	BnM(Bl)		2002			
	rebuilt		BnM(Dg)		2016	OOU		Bl
LM 426		4wDH	BnM(Bl)		2002			
	rebuilt		BnM(Dg)		2018			M
LM 427		4wDH	BnM(M)		2002			
	rebuilt		BnM(Dg)		2017			M
LM 428		4wDH	BnM(M)		2001			
	rebuilt		BnM(Dg)		2017			M
LM 429		4wDH	BnM(Bo)		2004			
	rebuilt		BnM(Dg)		2018			Be
LM 430		4wDH	BnM(Bo)		2004			
	rebuilt		BnM(Dg)		2018			Be
LM 431		4wDH	BnM(Dg)		2004			
	rebuilt		BnM(Dg)		2017	OOU		Bo
LM 432		4wDH	BnM(Dg)		2004			
	rebuilt		BnM(Dg)		2018	OOU		Bl
LM 433		4wDH	BnM(Dg)		2004			
	rebuilt		BnM(Dg)		2018	Dsm		Bl
LM 434		4wDH	BnM(Dg)		2004			Bl
LM 435		4wDH	BnM(Bo)		2005			
	rebuilt		BnM(Dg)		2017	OOU		Bl
LM 436		4wDH	BnM(Bo)		2005			
	rebuilt		BnM(Dg)		2018			Be
LM 437		4wDH	BnM(Bo)		2006			Cm
LM 438		4wDH	BnM(Bo)		2006			
	rebuilt		BnM(Dg)		2017			M
LM 439		4wDH	BnM(Bo)		2006			
	rebuilt		BnM(Dg)		2017	OOU		Bl
LM 440		4wDH	BnM(Bo)		2006			
	rebuilt		BnM(Dg)		2017	OOU		Bl
LM 441		4wDH	BnM(Bo)		2007			
	rebuilt		BnM(Dg)		2017			M
LM 442		4wDH	BnM(Bo)		2011			M
C 49		4wPMR	BnM	5	1958	OOU	e	M
C 65		2w-2PMR	BnM		1972	DsmT		M
C 77		4wPMR	BnM(Bl)		1972	OOU		Cg

(F 230)		4wDM	BnM		DsmT	f	Bl
F 353		4wDM	BnM	1983	OOU		Bl
F 842		4wDM	BnM		OOU		Bl
F 866		4wDM	BnM	c1975	OOU		Bl
RM 1		4wDH	BnM(Bl)	2000	OOU		Bl
RL 1		4wDH	BnM(M)	1992	OOU		M
(RL 2)	RM 7	4wDH	BnM(TAE/M)	2003	OOU		Bl
RM 3		4wDH	BnM(M)	2004			M
RM 4		4wDH	BnM(M)	2004	OOU	g	Dg
RM 5		4wDH	BnM(M)	2004			M
RM 6		4wDH	BnM(M)	2004			Le

a   carries plate 382841
b   incorrectly numbered following rebuilding
c   carries correct w/n HE 6321 on one side, and HE 6331 on the other
d   loco stored at BnM Vintage Machinery Museum Siding, but not part of the collection
e   dumped at Loading Point on Derryaroge Bog (Mountdillon System) N 028718
f   converted to a wagon for use with RM 1
g   loco stored off track at Ballycon Depot (stationary use only)

SECTION 5 – OFFSHORE ISLANDS

CHANNEL ISLANDS

PRESERVATION SITES

ALDERNEY RAILWAY SOCIETY, MANNEZ QUARRY, ALDERNEY

Gauge 4ft 8½in   www.alderneyrailway.gg   **601087**

(6 ALD 40)		4wVBT	VCG	S	6909	1927	Dsm
ALD 7 MOLLY 2		4wDM		RH	425481	1958	
D100 ELIZABETH		0-4-0DM		_(VF	D100	1949	
				(DC	2271	1949	
1044		2-2w-2w-2RER		MetCam		1960	
1045		2-2w-2w-2RER		MetCam		1960	
(PWM 3776)		2w-2PMR		Wkm	6655	1953	DsmT
PWM 3954 MARY LOU		2w-2PMR		Wkm	6939	1955	a
1 GEORGE	RLC/009025	2w-2PMR		Wkm	7091	1955	
(8 9028) SHIRLEY		2w-2PMR		Wkm	7094	1955	
7 (9022)		2w-2PMR		Wkm	8086	1958	
(9)		2w-2PMR		Wkm	9359	1963	

a   currently in store c/o Nick Best, National Westminster Bank Ltd, Victoria Street, St.Annes

THE L.C. PALLOT TRUST, THE PALLOT STEAM, MOTOR & GENERAL MUSEUM, RUE DE BECHET, TRINITY, JERSEY
JE3 5BE

Gauge 4ft 8½in   www.pallotmuseum.co.uk   **652532**

LA MEUSE	0-6-0T	OC	LaMeuse	3442	1933	
–	0-4-0ST	OC	P	2085	1948	
–	0-4-0ST	OC	P	2129	1952	
J.T.DALY	0-4-0ST	OC	WB	2450	1931	
(D1)	0-4-0DH		NBQ	27734	1958	

Gauge 2ft 0in

–	4wDM	s/o	MR	11143	1960	
–	4wDM		MR	60S383	1969	Dsm a

a   in use as a brake van

ISLE OF MAN

INDUSTRIAL SITES

AULDYN CONSTRUCTION LTD, (part of the Colas Group),
PLANT YARD, off PEEL ROAD, BRADDAN, ST. JOHNS   **(IM4 4LH)**
Gauge 3ft 0in   www.colas.co.uk

LM 350		4wDM		SMH	60SL748	1980	OOU
	SAINT BERNARD	4wDM	R/R	Thwaites			
–		2w-2BER		Consillia		2018	

Railway contractors to Isle of Man Transport railways

NATIONAL AIR TRAFFIC SERVICES LTD, LAXEY   **IM4 7AZ**
Gauge 3ft 6in   www.nats.aero   **SC 432847**

–	4wDHR		Wkm R	11730	1991
		reb	CE	B4630	2018

PRESERVATION SITES

GROUDLE GLEN RAILWAY CO LTD,
LHEN COAN, GROUDLE GLEN, KING EDWARD ROAD, ONCHAN
Gauge 2ft 0in   www.ggr.org.uk   **SC 418786**

	ANNIE	0-4-2T	OC	Booth R		1997	a
	BROWN BEAR	2-4-0T	OC	_(GGR		2019	
				(JF(B)	102	2019	
	OTTER	0-4-0ST	OC	NBRES		2019	b
	SEA LION	2-4-0T	OC	WB	1484	1896	
			reb	‡		1987	
	MALTBY	0-4-0DM	s/o	Bg	3232	1947	
			reb	NBRES		2017	
1	DOLPHIN	4wDM		HE	4394	1952	OOU
2	WALRUS	4wDM		HE	4395	1952	
No.313	POLAR BEAR	4wBE		WR	556801	1988	
			reb	AK	72R	2004	c

‡ rebuilt by British Nuclear Fuels Ltd, Windscale Factory, Sellafield, Cumbria
a currently at John Fowler & Co (Leeds) Ltd, Bouth Cumbria
b frames constructed in 2017
c carries worksplate BEV 313

ISLE OF MAN MOTOR MUSEUM, JURBY INDUSTRIAL ESTATE, JURBY   **IM7 3BZ**
Gauge 3ft 0in   www.isleofmanmotormuseum.com   **SC 362988**

16	MANNIN	2-4-0T	OC	BP	6296	1926	

ISLE OF MAN TRANSPORT, DEPARTMENT OF TOURISM & TRANSPORT
Isle of Man Steam Railway

Locomotives are kept at :

Douglas   IM1 1BR SC 374754, 375755
Port Erin IM8 2WE SC 198689

Gauge 3ft 0in   www.visitisleofman.com

No.4	LOCH	2-4-0T	OC	BP	1416	1874		
No.5	MONA	2-4-0T	OC	BP	1417	1874	Pvd	a
No.8	FENELLA	2-4-0T	OC	BP	3610	1894	OOU	
No.9	DOUGLAS	2-4-0T	OC	BP	3815	1896	OOU	b
No.10	G.H. WOOD	2-4-0T	OC	BP	4662	1905		c
No.11	MAITLAND	2-4-0T	OC	BP	4663	1905		
No.12	HUTCHINSON	2-4-0T	OC	BP	5126	1908	OOU	
No.13	KISSACK	2-4-0T	OC	BP	5382	1910		
16	MANNIN	2-4-0T	OC	BP	6296	1926		d
	CALEDONIA No.15	0-6-0T	OC	D	2178	1885		
No.17	VIKING	4wDH		Schöma	2086	1958	OOU	
18	AILSA	4wDH		_(HAB	770	1990		
				(HE	9446	1990		
		reb		HE	9342	1995		
19		0-4-0+4DMR		WkB/Dundalk		1950	OOU	
20		0-4-0+4DMR		WkB/Dundalk		1951	OOU	
21		Bo-BoDE		MPES	550/1	2013		e h
24	BETSY	4wDM		MR	22021	1959		f
No.25	SPROUT	4wDM		MR	40S280	1966		g
LM 344		4wDM		SMH	60SL751	1980		
		reb		BoothWKelly		2008		
–		2w-2PMR		Wkm	8849	1961	OOU	
	943032-1	4wDM	R/R	_(Thwaites1-96542		1999		
				(Rexquote 1323		2000		

a	stored at Port Erin Carriage Shed, Port Erin
b	in store for Isle of Man Railway & Tramway Preservation Society
c	currently at Alan Keef Ltd, Ross-on-Wye, Herefordshire
d	currently at Isle of Man Motor Museum workshops, Jurby
e	uses parts from an unidentified General Electric locomotive;
f	based at Port Erin
g	based at Castletown Goods Shed
h	stored at Bus Vannin, Banks Circus Bus Depot, Douglas

Manx Electric Railway, Laxey Car Sheds, Laxey   IM4 7AZ

Gauge 3ft 0in   www.manxelectricrailway.co.uk   SC 432846

No.23	DR.R.PRESTON HENDRY	4w-4wWE	MER		1900	a
34		4w-4wDE	MER		2004	b
No.22		2w-2PMR	Wkm	7442	1956	
	rebuilt as	4wDHR	M-TEK	7224	2014	

a	stored on un-motorised bogies, property of Isle of Man Railway & Tramway Preservation Society
b	normally based at Derby Caste Depot and Works, Douglas

Port Erin Railway Museum, Railway Station, Port Erin

IM8 2WE
SC 198689

Gauge 3ft 0in   www.visitisleofman.com

No.1	SUTHERLAND	2-4-0T	OC	BP	1253	1873	
No.6	PEVERIL	2-4-0T	OC	BP	1524	1875	

Snaefell Mountain Railway, Laxey

IM4 7AZ
SC 432844

Gauge 3ft 6in   www.visitisleofman.com

–		4wDMR		Wkm	10956	1976	a
	SHERPA	4wDM	R/R	_(Thwaites	18-93130	1997	
				(Rexquote	1149	1997	b

a   visits National Air Traffic Services, Laxey for maintenance
b   normally stabled at Bungalow Station, SC 395867

LAXEY & LONAN HERITAGE TRUST,
GREAT LAXEY MINE RAILWAY, off MINES ROAD, LAXEY

IM4 7NL
SC 433847

Gauge 1ft7in   www.laxeyminerailway.im

ANT	0-4-0T	IC	GNS	20	2004	a
BEE	0-4-0T	IC	GNS	21	2004	b
WASP	4wBE		CE	B0152	1973	

a   carries worksplate Lewin 684
b   carries worksplate Lewin 685

MANX TRANSPORT TRUST LTD, JURBY TRANSPORT MUSEUM,
HANGAR 230, JURBY INDUSTRIAL ESTATE, JURBY

IM7 3BD
SC 360988

Gauge 3ft 0in   www.jtmiom.im

–	4wPM	s/o	FH	2027	1937	a

a   locomotive may also be found at Queens Pier Tramway, Ramsey

QUEENS PIER RESTORATION TRUST,
QUEENS PIER RAILWAY, STANLEY MOUNT EAST, RAMSEY

IM8 1EW
SC 456940

Gauge 3ft 0in   www.ramseypier.im

–	4wPM	s/o	FH	2027	1937	a

a   locomotive may also be found at Jurby Transport Museum, Jurby

SECTION 6
MINISTRY OF DEFENCE

DEPOT TYPES :

DM	Defence Munitions
DSDC	Defence Storage & Distribution Centre

LOCATIONS :

ASH	DSDC Ashchurch, Gloucestershire (R.T.C.)	SO 932338
BIS	DSDC Bicester, Oxfordshire	SP 581203
CD	Copehill Down Battle Training Area, Salisbury Plain, Wiltshire	SU 014455
GD	DM Glen Douglas, near Arrochar, Argyll & Bute	NS 279998
KIN	DM Kineton, Warwickshire	SP 373523, 374524
LON	DM Longtown, Cumbria	NY 363682
LUD	Ludgershall Railhead, Wiltshire	SU 261507
LUL	Lulworth Ranges, East Lulworth, Dorset	SY 863822

LOCOMOTIVES
Gauge 4ft 8½in

	PERCY	0-4-0DM	JF	22503	1938	Pvd		KIN
01510 (272)		4wDH	TH	V320	1987			LON
01511 (275)		4wDH	TH	V323	1988			LON
01512 (301)	CONDUCTOR	4wDH	TH	V319	1988			
			rep	LH Group	76699	2002		BIS
01513 (302)	GREENSLEEVES	4wDH	TH	V318	1987			
			rep	LH Group	76638	2002		KIN
01514 (303)		4wDH	TH	V332	1988			
			rep	LH Group	76634	2002		LON
01521 (278)	FLACK	4wDH	TH	V333	1988			BIS
01522 (254)		4wDH	TH	272V	1977			LUD
01524 (261)		4wDH	TH	301V	1982			BIS
01525 (264)	DRAPER	4wDH	TH	306V	1983			BIS
01527 (256)		4wDH	TH	274V	1977			BIS
01528 (267)		4wDH	TH	309V	1984			GD
01548 (257)		4wDH	TH	275V	1978			GD
01549 (258)		4wDH	TH	298V	1981			KIN
01550 (271)	STOREMAN	4wDH	TH	V324	1987			BIS
430		0-6-0DH	RH	466621	1961	OOU		CD
	ANNA	4wRE	Wkm	11547	1987			LUL
	FIONA	4wRE	Wkm	11548	1987			LUL
	BELLA	4wRE	Wkm	11549	1987			LUL
	DEBBIE	4wRE	Wkm	11550	1987			LUL
	ENID	4wRE	Wkm	11551	1987			LUL
	CLAIRE	4wRE	Wkm	11552	1987			LUL
–		4wDMR	Wkm	11621	1986			LUL
9121		4wDHR	BD	3710	1975			KIN
(9123) 1		4wDHR	BD	3712	1975	DsmT		KIN
9128	THE HORNET	4wDHR	BD	3744	1976			LON
9129		4wDHR	BD	3745	1976			BIS

761	(TNS 102)		2w-2DMR		Robel			
					56.27-3.AF33	1983		LON
764	(TNS 101)		2w-2DMR		Robel			
					56.27-10-AF32	1983		ASH
08 CP 05			4wDM	R/R	_(Ford		1986	
					(Wkm	11618	1986	KIN
50 KM 19	(DRC 730J KAR 536V)		4wDM	R/R	Unimog	004971	1970	OOU	KIN

ELECTRIC MULTIPLE UNITS
Gauge 4ft 8½in

| 62910 | (319220) | | 4w-4wRER/WER | York(BRE) | | 1988 | |

SECTION 7
UNKNOWN LOCATIONS

The following locomotives may be found at locations which are not recorded in IRS records. Some were previously recorded as 'Unknown Owner, Unknown Location' within the main text of the EL Handbook. Some locomotives listed below are at locations which the owner does not wish to be publicised.

Where known, the last known county or country of the locomotive is listed at the end of each entry.

Gauge 1524mm

794		0-6-0T	OC	TK	350	1925	Greater London

Gauge 4ft 8½in

47406		0-6-0T	IC	VF	3977	1926	ex Leicestershire
	"HARRY"	0-4-0ST	OC	AB	1823	1924	Gloucestershire
–		0-4-0F	OC	AB	1944	1927	ex Shropshire
1928	ALAN GLADDEN	2-6-4T	OC	Nohab	2229	1953	East Sussex
	ANNIE	0-4-0ST	OC	P	1159	1908	ex Bucks
–		0-4-0ST	OC	WB	2702	1943	North Yorks
567		4-4-0	IC	567LG		2018	u/c Warks
–		4wDM		Bg/DC	2136	1938	East Sussex
4		0-4-0DM		JF	22889	1939	Hampshire
18242	ROF CHORLEY 4	0-4-0DH		JF	4220022	1962	Warwickshire
97701		4wDE		Matisa	2655	1975	
			reb	Kilmarnock		1986	East Sussex
–		4wDM		RH	294268	1951	East Sussex
7	"SWANSEA JACK"	4wDM		RH	393302	1955	Mid-Wales
LOT 26	YARD No.766	0-4-0DM		RH	414300	1957	Gtr. Manchester
	CHARLES	4wDM		RH	417889	1958	Lincolnshire
11348/C	W.L. No.1	4wDH		RR	10268	1967	Shropshire
D1995	5 HEATHER	0-4-0DM		_(VF	D293	1955	
				(DC	2566	1955	ex Norfolk
11	PPM 30	2w-2F/BER		PPM	11	1995	ex Shropshire
(54256	14256)	2w-2-2-2wRER		BRCW		1939	Essex
3007		2-2w-2w-2RER		MetCam		1967	Beds
3022		2-2w-2w-2RER		MetCam		1968	Beds
3107		2-2w-2w-2RER		MetCam		1967	Beds
3122		2-2w-2w-2RER		MetCam		1968	Beds
	"EROS"	4wFER		SUSTRACO		2005	West Midlands
5	N5 WGM MARSHY	4wDM	R/R	_(Landrover 989185		1996	
				(Hy-rail 29932		1996	Cheshire
A456 NWX	L84	4wDM	R/R	_(Unimog 101335		1983	
				(Zweiweg 1035		1983	North Yorkshire
OXS 433		4wDM		Ford (Jeep)		1942	a
OXS 434		4wDM		Ford (Jeep)	81265	1942	a Surrey
RTU 6647A	99709 901033-9	2w-2PMR		Geismar ST/01/05		2001	Suffolk
	99709 901010-7	2w-2PMR		Lesmac LMS020		2006	North Wales
(DE 320467	DB 965049)	2w-2PMR		Wkm	7564	1956	West Yorkshire
68/020		2w-2PMR		Wkm	7600	1957	DsmT N.Yorks
DB965949	DX 68005	2w-2PMR		Wkm	10645	1972	Derbyshire

a   road vehicle with exchangable bolt-on rail wheels

Gauge 3ft 0in

E6	4wBEF		CE	B1850A	1979		Derbyshire
–	4wDM		JF	3930044	1950	a	Gloucs

a   has occasionally appeared at Welland Steam Rally, Worcestershire

Gauge 900mm

–	4wDH	Schöma	6257	2008	London
–	4wDH	Schöma	6258	2008	London
–	4wDH	Schöma	6259	2008	London
–	4wDH	Schöma	6260	2008	London
–	4wDH	Schöma	6261	2008	London

Gauge 2ft 6in

695		0-6-4T	OC	KS	4408	1928
666		4-6-2	OC	NB	17111	1906
AK 16		2-6-2T	OC	WB	2029	1916
No.3	CONQUEROR	0-6-2T	OC	WB	2192	1922
10		4wDM		HE	2260	1940	Lincolnshire
8		4wDH		HE	8830	1979	Cheshire
11		4wDH		HE	8968	1980	Cheshire
L4134		2w-2PM		Wkm	3174	1942

Gauge 750mm

–	0-6-0WT	OC	Chrz	3326	1954	North Wales

Gauge 700mm

–	0-4-4-0T	4C	OK	3770	1909	

Gauge 2ft 2in

–	4wDH	HE	8972	1979	Derbyshire
–	4wDH	HE	8970	1979	Derbyshire

Gauge 2ft 0in

121		2-8-2	OC	AFB	2668	1951	
(779)		4-6-0T	OC	BLW	44657	1916	Derbyshire
	SSE 1912	4-4-0	OC	FE	265	1897	
	LISBOA	4-4-0	OC	FE	266	1897	
–		0-4-0ST	OC	Hayling			a North Wales
1		0-8-0T	OC	Hen	15540	1917	North Wales
146		2-8-2	OC	Hen	29587	1951	South Wales
–		0-4-2T	OC	JF	16341	1924	Derbyshire
–		0-6-0T	OC	Porter	6465	1920	Australia
–		0-4-0WT	OC	SLM		1944	b Northants
NG 50		4wDH		AB	719	1987	
			reb	HAB		1996	Lincolnshire
NG 52		4wDH		AB	721	1987	
			reb	HAB		1996	
(T2)	SALLY	4wDH		AK	No.8	1982	Cumbria
–		4wDH	s/o	AK	14	1984	Northants
–		4wDM		AK	26	1988	Cumbria

51	No.646	0-4-0PM	BgC	646	1918	Scotland
	CLARA	4wPM	Bonnymount		c1986	
(1835)		4wDM	HE	1835	1937	
9		4wDM	HE	2259	1940	Lincolnshire
	YARD No.1076	4wDM	HE	7448	1976	Shropshire
14		4wDH	HE	9081	1984	Lincolnshire
–		4wDM	L	38296	1952	
–		4wDM	LB	55070	1966	
		reb	Gartell	1001	1987	ex Hampshire
12	1 ARCHER	4wDM	MR	4709	1936	Bucks
–		4wDM	MR	5853	1934	Leicestershire
–		4wDM	MR	7128	1936	Dsm Bucks
	LIDDEL	4wDM	MR	7188	1937	Cumbria
	GELT	4wDM	MR	8696	1941	Cumbria
	TITCH	4wDM	MR	8729	1941	Devon
	MOLE	4wDM	MR	22031	1959	Lincolnshire
MPU 9	149	4wDM	MR	22119	1961	Dsm Norfolk
87025	L201N	4wDM	MR	22238	1965	Cumbria
87032	L202N	4wDM	MR	40S412	1973	Dsm
(6)		4wFER	PPM	6	1993	West Midlands
(9)		2w-2FER	PPM	9	1995	West Midlands
–		4wDM	RH	174542	1935	North Wales
(AD40	LOD 758366)	4wDM	RH	202000	1940	Lincolnshire
X 025		4wDH	Schöma	5574	1998	Staffordshire
026		4wDH	Schöma	5575	1998	Staffordshire
PLM 35		4wDH	Schöma	5702	2001	
–		4-2wPMR	StanhopeT		2005	West Midlands
No.9		0-6-0DM	WB	3124	1957	
RTT 767165	(RTT/767149)	2w-2PM	Wkm	3238	1943	
–		4wDM	#			Essex
–		4wBE	CE	5590	1969	
		reb	CE	B4174B	1996	
–		4wBE	CE	5590/6	1969	
		reb	CE	B4174A	1996	
4		4wBE	CE	5949A	1972	
	263 021	4wBE	CE	B0465	1974	
03		4wBE	CE	B1524/2	1977	ex Gtr. Manc
	263 076	4wBE	CE	B3766C	1991	
PTL 02		4wBE	CE	B4246A	1998	
		reb	CE	B4381B	2002	Lancashire
PTL 01		4wBE	CE	B4246B	1998	
		reb	CE	B4381A	2002	Lancashire
–		4wBE	CE			ex Gtr. Manc
37	CAPITAL LETTERS	2w-2-2-2wRE	EEDK	760	1930	ex Oxfordshire
44	[221 222]	2w-2-2-2wRE	EEDK	812	1930	ex Oxfordshire
S26	25	0-4-0BE	WR	D6886	1964	Cornwall
–		0-4-0BE	WR	8079	1980	
	rebuilt as	0-4-0DM	BrownGM		1993	Durham
–		0-4-0BE	WR			Cornwall

#	replica Lister locomotive
a	frames only - loco under construction
b	either SLM 3854 or SLM 3855

Gauge 610mm / 760mm

–	2w-2FER		PPM	8	1994	

Gauge 600mm

SMT T907	0-10-0T	OC	OK	10956	1925	Derbyshire
6 LA HERRERA	0-6-0T	OC	Sabero		c1937	
(NG) 35	4wDH		HE	7010	1971	
	reb		HAB	6941	1988	Lincolnshire

Gauge 1ft 11½in

120 NG15 BEDDGELERT	2-8-2	OC	AFB	2667	1951	Surrey
(GELLI)	0-4-0VBT	VC	DeW		1893	Dsm Hants

Gauge 1ft 6in

GNR No.1	2-2-2	IC	?		c1863	Wiltshire

Gauge 1ft 3in

(PRINCE OF WALES)	4-4-2	OC	BL	11	1908	South Yorkshire
111 (YVETTE)	4-4-0	OC	CravenEA		1946	Co. Durham
–	0-4-0T	OC	FMB	004		Hampshire
42869	2-6-0	OC	GibbonsCL		1993	Cheshire
4 BLUE PACIFIC	4-6-2VB	OC	GuinnessNL		c1935	Cheshire
3205 EARL OF DEVON	4-4-0	IC	Prestige		2001	Devon
–	4-6-0	OC	SmithEL			a Wiltshire
–	4-4-0	OC	ThurstonTS			a Berkshire
4 LENKA	4-4wDHR		Longleat		1984	

a   incomplete loco

Gauge 1ft 0¼in

2 EXMOOR	2-4-2T	OC	ESR	293?	1993	ex Surrey
PRINCE EDWARD	4-4-2	OC	FlooksG		1935	
	reb		Clarkson		c1967	ex Surrey
–	2-8-0	OC	Iron Horse		1990	
	completed by		SfP		1998	
SOUTHERN QUEEN	4-6-2	OC	Thurston		1953	Dsm Hants
T.J. THURSTON	4-6-2	OC	Thurston		1948	South East
8100 JULIET	0-4-0T+T	OC	?		?	
1000 SPRINGBOK	4-6-0		?		?	
LADY BARBARA	2w-2-2-2wDH		EvansC		1997	
a rebuild of [SOUTHERN BELLE]	4-2-2PH	s/o	DugginCG		1988	
–	4w-4wPE		HuntTG		c1960	
3015	2-4-0DE	s/o	Maxfield?			
4020	2-4-2BE	s/o	Maxfield		1969	Staffordshire
4030 ZION	4w-4wPM		Walker&Etherington	1971		
–	4wPM		?		?	Co.Laois, Ireland

Gauge monorail

–	2a-2DH		AK	M001	1988	Wiltshire

SECTION 8
LOCOMOTIVE INDEX

386

394

POSTCODE INDEX

THE INDUSTRIAL RAILWAY SOCIETY

ABOUT US

The **Industrial Railway Society** was founded in 1949 to specialise in the study of privately owned railways, their motive power and rolling stock. Initially this meant those railways serving private industry but over the years the heritage railway movement has developed so we now devote much effort to recording that, especially as so many heritage lines depend on former industrial locomotives for their motive power. We produce

- A bi-monthly **UK News Bulletin**, both as an A5 summary booklet and in a larger full colour A4 digital edition, covering current developments on industrial and heritage railways in the UK and Ireland;
- A quarterly **Overseas News Bulletin** covering industrial and preserved railways throughout the rest of the world
- A well-researched and illustrated quarterly magazine, the **Industrial Railway Record**, featuring articles of historical interest on subjects seldom covered by the mainstream railway press but which represent an integral part of railway history. Twelve issues per volume, with index. Our members often have their copies bound.

We are active **publishers**, producing not just this EL series but books covering specific items of industrial railway interest. All these publications are available to members at discounted prices (we also sell books of industrial railway interest from other publishers). We have a substantial archive and encourage members to participate in research. We also organise **visits** to industrial and heritage railways, especially areas normally off-limits.

THE INDUSTRIAL RAILWAY SOCIETY

Researching the Past, Recording the Present - For the Benefit of the Future
Why not join us and come along for the ride!

There are several different levels of membership to suit different interests. You can join at our E-Shop by scanning the QR code or go to https://irsshop.co.uk/membership or, for more information, please email subs@irsociety.co.uk or write to the Sales Officer at 24 Dulverton Road, Melton Mowbray LE13 0SF

However, if you not interested in the wider benefits of joining the Society but do want to keep this EL edition up to date you can, for a modest one-off fee, subscribe to the bi-monthly amendment lists, sent as PDF downloads. This option is only available online. If you wish to subscribe please scan this QR code or go to https://irsshop.co.uk/ELAmendmentlists

[1] ARMY 220 (AB 359 of 1941) at the Rotherwas Industrial Estate in Hereford on 21st September 2023.
(Adrian Booth)

[2] 2-8-0 DH Locomotive SL 73.35 of 1973 is ready for the first train of the day at the Thorpe Light Railway, Whorlton, Co.Durham on 16 July 2023. (Alan Kemp)

[3] 01529 (TH 310V of 1984) owned by Harry Needle Railroad Co Ltd on display at Long Marston, Warwickshire on 21st June 2023.   (Anthony Coulls)

[4]   Ex-BR class 73 locomotive No.73001 (Elh of 1962) at the Ecclesbourne Valley Railway, Wirksworth, Derbyshire on 21st January 2023.   (Bernard Caddy)

[5] 4wDH r/r Zephir 2136 of 2008 at Tarmac plc's Mountsorrel site in Leicestershire on 26th June 2021. (Chris Weeks)

[6] ARCHER (MR 4709 of 1936) formerly owned by the late John Butler whilst being transferred by Vic Haynes Transport to its new owner on 21st December 2021.
(Claire Conway-Crapp)

[7] VICTOR (WB 2996 of 1951) was a visitor to the North Tyneside Steam Railway, Northumberland on 30th April 2022. (Michael Denholm)

[8] CATHRYN (HC 1884 of 1955) in use at Peak Rail, Rowsley, Derbyshire on 1st October 2023. (Dennis Graham)

[9] The crew of AB 1016 of 1904 take a break during a visit to the Foxfield Railway by this Tanfield based locomotive on 15th September 2023.   (Derek Horton)

[10] 4wDH locomotive (AK 19 of 1985) glistens in the sunshine at the Silverleaf Poplar Railway, Old Leake, Lincolnshire on 26th August 2023.   (Ed Copcutt)

[11] TM-LT-02 (Niteq B334 of 2012) hauling a Eurostar unit at Temple Mills Depot, London on 2nd March 2019. (Edward Barnes)

[12] SMH 60SL751 of 1980 owned by Auldyn Construction Ltd at Douglas, Isle of Man on 20th April 2023. (Jim Ballantyne)

[13] RH 441424 of 1960 on the narrow gauge system at the Chasewater Railway, Brownhills, Staffordshire on 18th June 2023.   (John Browning)

[14] PAM (AK 52 of 1996) poses in front of the station sign at the Wotton Light Railway, near Aylesbury, Buckinghamshire on 6th August 2023.   (John Maskell)

[15] TWIZELL (RS 2720 of 1891) provides a glimpse of its everyday work at Handon Hold Colliery, Co. Durham on 21st May 1965. (John Scholes)

[16] Ex-BR class 08 diesel locomotive (D4038, Dar of 1960) at Eastern Rail Services, Great Yarmouth, Norfolk on 21st June 2021. (Ken Scanes)

{17} Following the recent completion of its overhaul, RENISHAW IRONWORKS No.6 (Hudswell Clarke 1366/1919) is moving wagons in the Tanfield Railway's Marley Hill yard during 14th October 2023 to release Austerity 49. (Cliff Shepherd)

[18] Abandoned WR/BEV/Pikrose loco at Honister Slate Mine, Borrowdale, Cumberland in August 2023. Type WR7, weight 2000KG. (Kevin Straddon)

[19] A view inside the locomotive shed at the Tanfield Railway, Co. Durham on 16th June 2023. The AB 040ST is visiting engine No.1219 of 1910. (Malcolm Braim)

[20] Stabled alongside the station at Tywyn on 26th October 2022 was St CADFAN (BD 3779 of 1963) in an elegant maroon livery. (Mark Hambly)

[21] Rebuilt ex-BR class 08 locomotive 03 (D3378, Der of 1957) on the turntable at Barrow Hill Roundhouse on 25th November 2023 whilst examples of somewhat older electric technology look on. (Martin Shill)

[22] No.3 (HL 3581 of 1924) from the Foxfield Railway visiting Marley Hill, Co. Durham on 17th June 2023. (Michael Denholm)

[23] Ex-BR 4wDM ZM32 HORWICH (RH 416124 of 1957) on a train at the Steeple Grange Light Railway, Wirksworth, Derbyshire on 29th August 2023.　　(Mike Shaw)

[24] Restored FH 1830/1933, owned by Peter Nicholson until 2017, given to Jason Keswick and restored to operational use on Pete's 75th birthday on 1st May 2023, now based at the Westonzoyland Light Railway, near Bridgwater, Somerset.　(Josh Brinsford)

[25] DUNLOP No.6 (WB 2648 of 1941) on a rake of MGR wagons at the Chasewater Railway, Brownhills, Staffordshire on 15th October 2023. (Pete Stamper)

[26] Ex-BR departmental (S&H 7505 of 1967) owned by R. & N. Cessford, Brechin, Angus, but in store off site. Photographed on 4th June 2023. (Simon Guppy)

[27] INDIAN RUNNER (RH 200744 of 1940) being controlled by the driver walking alongside. Seen during the IRS visit on 22nd April 2023 to North Ings Farm, near Ruskington, Lincolnshire.
(Sydney Leleux)

[28] S 6515 of 1926 at the Cholsey & Wallingford Railway, Wallingford, Oxfordshire on 7th July 2021.
(Tom Dovey)

[29] No.29 (K 4263/1904) shunting wagons at Levisham on the North Yorkshire Moors Railway on 25th June 2023 prior to hauling some to Pickering.   (Cliff Shepherd)

[30] 27 (MR 5863 of 1934) freshly repainted in green and black at Amberley Museum, Arundel, West Sussex on 10th April 2022.   (Warren Hardcastle).

[31] SUE (MR 40SD502 of 1975), owned by Alan Keef, with an engineering train at the Leighton Buzzard RPS's deviation contract, Leighton Buzzard, Bedfordshire on 30th January 2022. (William Shelford)

[32] No.10 (AK 91 of 2022) running round its train at the Corris Railway, Maespoeth, Gwynedd on 14th October 2023. (Robert Pritchard)

[33] ENTERPRISE (RR 10282 of 1968) stabled with a rake of oil wagons at the Ribble Steam Railway, Preston, Lancashire during the IRS AGM on 22nd April 2022.

(Andrew Smith)

[34] An impressive lineup of locomotives at Liberty Steels, Aldwarke, Yorkshire on 18th March 2019.
(Chris Weeks).

[35] HE 1026 of 1910 after being unloaded at the Statfold Barn Railway, Statfold, Staffordshire on 13th January 2022. (Roy Etherington)

[36] Plinthed HL 3827 of 1934 at the East Carlton Country Park, near Corby, Northamptonshire on 20th February 2022. (Tom Dovey)

[37] LADY PATRICIA 10916 (TK946 of 1955) at David Buck, Fifield on 6th August 2023.
(John Maskell)

[38] CARROLL (HC D631 of 1846) inside the shed at the Middleton Railway Trust, Hunslet, Leeds, West Yorkshire on 11th June 2023.
(Adrian Booth)

[39] EEV D1230 of 1969 at the Cambrian Heritage Trust railway at Llynclys, Shropshire, on 25th September 2023.   (Ed Copcutt)

[40] Ex-BR class 14 diesel hydraulic locomotive D9520 (Sdn of 1964) stabled at Haworth on the Keighley & Worth Valley Railway, West Yorkshire on 28th October 2023.

(Warren Hardcastle)

[41] Unilok/Jung A114 of 1965 at the RPSI depot at Whitehead, Co.Antrim, Northern Ireland on 6th July 2012. (Danny Sheehan)

[42] LM 263 (RH 7002/0600-1 of 1968) inside the closed St Connells Museum, Glenties, Co. Donegal on 28th May 2023. (Simon Guppy)

[43] LM 322 and 323 (HE 8923 & 8924 of 1979 on a Railtour on 10th March 1996 at the Esker Tea Centre on the Bord na Móna's Blackwater System, Co.Offaly.   (Paul Rafferty)

[44] LM 379 (HE 9251 of 1985) hauling the final passenger train at the Grand Canal Lift Bridge on the Bord na Móna's Derrygreenagh System, Co.Offaly on 20st September 2022.   (Ted McAvoy)

[45] 4wDH LM 396 (BnM(Bl) of 1995) ex-works after overhaul at the Bord na Móna's Blackwater Works, Co.Offaly on 23rd April 2023. (Andrew Waldron)

[46] Spence 23L of 1921 on display at Amberley Museum. Amberley, West Sussex on 12th September 2010. (Paul Carpenter)

[47] *Above*: LM46 (RH 259184 of 1946, rebuilt as 4wDH by BnM) at the Bord na Móna's, Croghan Works, Co.Offaly on 24th May 2016. (Geoff Warcup)

[48] *Left:* Zephir 2793 of 2018 at Iarnród Éireann's Inchicore Works, Dublin on 22nd October 2022.
(Sean Cain)